3rd International Conference on Design, Mechanical and Material Engineering (D2ME 2018)

MATEC Web of Conferences Volume 237 (2018)

Phuket, Thailand
27-29 September 2018

Editor:

Raj Das

ISBN: 978-1-5108-7562-3

Printed from e-media with permission by:

Curran Associates, Inc.
57 Morehouse Lane
Red Hook, NY 12571

Some format issues inherent in the e-media version may also appear in this print version.

No changes have been made to the content of these proceedings. There may be changes to pagination and minor adjustments for aesthetics.

Printed by Curran Associates, Inc. (2019)

For additional information, please contact EDP Sciences – Web of Conferences
at the address below.

EDP Sciences – Web of Conferences
17, Avenue du Hoggar
Parc d'Activité de Courtabœuf
BP 112
F-91944 Les Ulis Cedex A
France

Phone: +33 (0) 1 69 18 75 75
Fax: +33 (0) 1 69 28 84 91

contact-edps@webofconferences.org

Additional copies of this publication are available from:

Curran Associates, Inc.
57 Morehouse Lane
Red Hook, NY 12571 USA
Phone: 845-758-0400
Fax: 845-758-2633
Email: curran@proceedings.com
Web: www.proceedings.com

TABLE OF CONTENTS

PREFACE

D2ME 2018 was held in Phuket during September 27-29, 2018. D2ME 2018 is co-organized by RMIT University, Swinburne University of Technology, The Australian National University and Hong Kong Society of Mechanical Engineers. The conference provides a useful and wide platform both for display the latest research and for exchange of research results and thoughts in broad Mechanical, Material and Design Engineering. The participants of the conference were from almost every part of the world, with background of either academia or industry, even well-known enterprise. The success and prosperity of the conference is reflected high level of the papers received.

The proceedings are a compilation of the accepted papers and represent an interesting outcome of the conference. This book covers 3 chapters: Mechanical Engineering; Materials Engineering; Design Engineering.

We would like to acknowledge all of those who supported D2ME 2018. Each individual and institutional help were very important for the success of this conference. Especially we would like to thank the organizing committee for their valuable advices in the organization and helpful peer review of the papers.

We sincerely hope that D2ME 2018 will be a forum for excellent discussions that will put forward new ideas and promote collaborative researches. We are sure that the proceedings will serve as an important research source of references and the knowledge, which will lead to not only scientific and engineering progress but also other new products and processes.

Dr. Raj Das, RMIT University, Australia

COMMITTEES

General Chair

Dr. Raj Das, RMIT University, Australia

Conference Co-Chairs

Prof. Zhan Chen, AUT University, Auckland, New Zealand

Prof. LAU Kin Tak Alan, Swinburne University of Technology, Australia

Prof. Qinghua Qin, The Australian National University, Australia

Advisory Co-Chair

Prof. Guy Littlefair, Deakin University, Australia

Program Committee Chair

Prof. Gong Hao, National University of Singapore, Singapore

International Technical Committee

Prof. Wong Shaw Voon, Universiti Putra Malaysia, Malaysia

Prof. Li Chang, The University of Sydney, Australia

Prof. Alan C. Lin, National Taiwan University of Science and Technology, Taiwan

Dr. Abhishek Kundu, Cardiff University, UK

Dr. Khalid Arif, Massey University, New Zealand

Dr. Pierre Kerfriden, Cardiff University, UK

Dr. Chi Kit Au, The University of Waikato, New Zealand

Dr. Kheng-Lim Goh, Newcastle University, UK

Dr. Hao Wang, Northestern University, China

Dr. Prodip Das, Newcastle University, UK

Dr. Yat Wong, Swinburne University of Technology, Australia

Dr. Jie Zhang, University of Bristol, UK

Dr. Zhangming Wu, Cardiff University, UK

Dr. Alokesh Pramanik , Curtin university, Australia

Dr. Pooya Sareh, Imperial College London,UK

Dr. Ching L. Ko, University of Oklahoma, USA

Dr. Mohsen Motahari-Nezhad, Zabol University, Zabol, Iran

edp sciences

MATEC Web of Conferences

Statement of Peer review

In submitting conference proceedings to MATEC Web of Conferences, I certify to the Publisher that I adhere to the **Policy on Publishing Integrity** of the journal in order to safeguard good scientific practice in publishing.

1. All articles have been subjected to peer review administered by the proceedings editors.
2. Reviews have been conducted by expert referees, who have been requested to provide unbiased and constructive comments aimed, whenever possible, at improving the work.
3. Proceedings editors have taken all reasonable steps to ensure the quality of the materials they publish and their decision to accept or reject a paper for publication has been based only on the merits of the work and the relevance to the journal.

Title, date and place of the conference

2018 3rd International Conference on Design, Mechanical and Material Engineering (D2ME 2018)

September 27-29, 2018 Phuket, Thailand

Proceedings editor(s):

Dr. Raj Das, RMIT University, Australia

Date and editor's signature

1-Nov.-2018

Research and Application of Optimal Combination of Quick Drop Valve Flow Regenerative Performance Based on Orthogonal Experimental Design

Xiaoxiang Li[1,a], Anlin Wang[1] and Jiadong Lin[2]

[1]*School of Mechanical Engineering, Tongji University, 4800 Caoan Highway, Shanghai, China*
[2]*Research & Development Center, Shantui Construction Machinery Co.,Ltd, No. 58, Highway G327, Jining, Shandong, China*

Abstract. Based on the simulation model of quick drop valve flow regeneration measurement data, the influence factors of quick drop valve are analysed through two orthogonal experimental designs to obtain the optimal combination of the efficiency of oil filled β for flow regeneration. Simulation and experimental verification results show that: In the structure of the original quick drop valve, through the range R, the influence factor valve opening C has the greatest influence on the efficiency of oil filled β.The efficiency of oil filled β of the optimal combination $A_1A_2A_3$ is 78.9% lower than the standard requirement of 85%. Based on the method of controlling the oil return variable according to the load feedback and orthogonal experimental design, the improved structure makes the efficiency of oil filled β for flow regeneration increase from 57% of the measured data to more than 85% of the measured verification. Through 1000 hours performance verification, the performance of the quick drop valve is stable, no failure rate, and the efficiency of oil filled β is still higher than 85%. At the same time, the efficiency of the earth-moving work after the quick drop of the blade is increased by more than 28%.

1 Introduction

In some construction machinery such as bulldozer blades lowering process, as one of the important hydraulic components of the hydraulic system, the quick drop valve can connect the rod-less cavity and the rod cavity of the hydraulic cylinder, reduce the flow of the oil return tank in the hydraulic cylinder, thereby shortening the time lag before the blade starts to dig. The flow regeneration performance of the quick drop valve has an important influence on the operation efficiency, and at the same time, its function can make the blade lower quickly to complete the freezing operation of some frozen soils in extreme cold conditions. At present, there is very little information on the theoretical mechanism of the flow regeneration of the quick drop valve. Most of the information is the introduction to the principle of its function.

The principle of paralleling the hydraulic pipe in the quick drop valve pipe to increase the blade's lowering speed was elaborated[1]. However, the flow regeneration performance of the quick drop valve had not been researched from the theoretical mechanism. Kong only tested the blade lowering

[a] Corresponding author : lixiaoxiang@tongji.edu.cn

time with or without a quick drop valve in bulldozer's working hydraulic system[2]. Wang and Tang introduced the function and principle of quick drop valve for loader, and did not analyze the performance theoretically[3]. Li et al introduced the failure of the quick drop valve during the lifting of the bulldozer blade[4-8]. Simple analysis of the selection of the return spring in the quick drop valve[9].The author had theoretically analyzed the influence degree of multi-valve flow characteristic coefficient on flow regeneration performance in working hydraulic system and did not consider the performance of the quick drop valve[10].

Based on the analysis of the influence of two orthogonal test designs on the influence factors of the quick drop valve, the improvement of the structure is achieved by using the method of load feedback to control the return oil variable. The efficiency of oil filled β of the quick drop valve is increased from 57% of the measured data to 85% of the measured verification, and the efficiency of the earth-moving work after lowering quickly is increased by more than 28%. The quick drop valve is verified by a failure-free performance test of more than 1000 hours.

2 Modeling and calibration

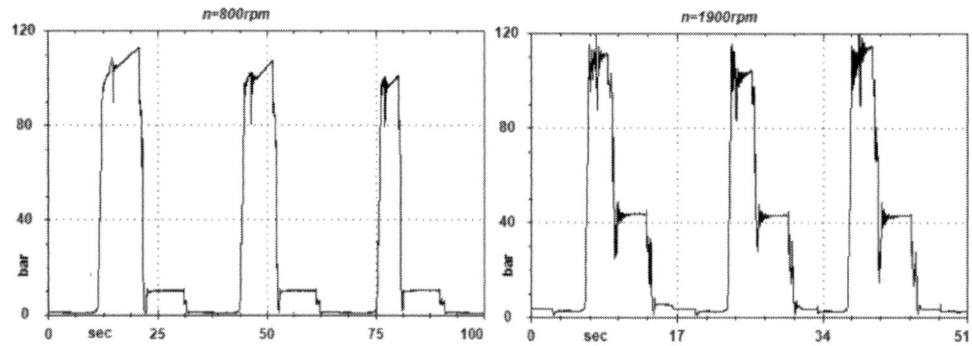

Figure 1. Pressure curve of rod-less cavity of cylinder

The performance test of the quick drop valve is based on the engine speed n=800rpm and n=1900rpm of a certain type of bulldozer. The pressure curve of rod-less cavity of cylinder is shown in Figure 1. According to the Figure 1, It can be seen that during the process of the blade lowering quickly: t=7.8s for n=800rpm, and t=3.7s for n=1900rpm.Calculated by the cylinder stroke and the outlet flow rate of the quick drop valve, the efficiency of oil filled β are 56.6% and 56.9% respectively. It is known that the engine speed has little effect on the flow regeneration performance of the quick drop valve.

The system simulation software Simulation X is used to model the system and the system model is calibrated by the measured value. The pressure curve of rod-less cavity of cylinder of the hydraulic system simulation model is shown in Figure 2. Based on the simulation curves of flow regeneration and pressure build-up time, the efficiency of oil filled β of the flow regeneration is calculated. The efficiency of oil filled β at the engine speed is 56.7% and 57% respectively, and the error between simulation and measurement is less than 0.5%. It can be considered that the simulation results of the system model have a high consistency with the measured data.

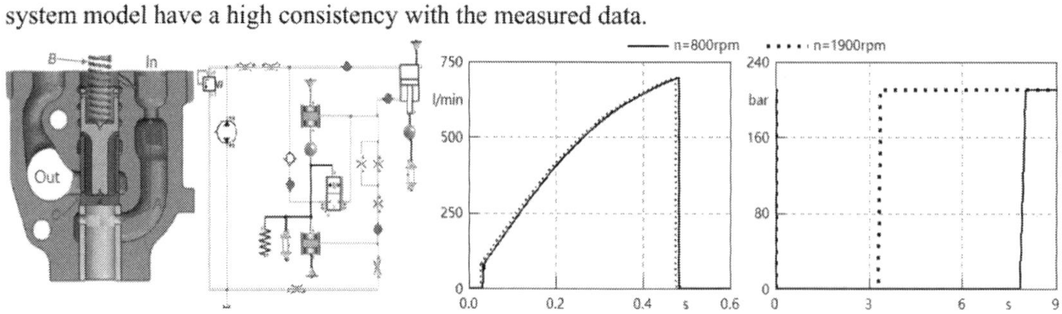

Figure 2. System simulation model, curves of flow regeneration and pressure build-up time

3 Orthogonal experimental design

It can be seen from the structure diagram of the quick drop valve in Figure 2.The flow regeneration performance of the quick drop valve is related to the structural parameters such as the orifice A, the return spring stiffness B and the valve opening C. The Orthogonal experimental design is used to analyze the influence of the above three parameters on flow regeneration and the optimal level combination. Considering factors such as pressure loss, spool leakage requirement and space structure limitation, the selected parameters value are shown in Table 1.

Table 1. Parameter value table

Parameter	A_1		A_2		A_3	
Value	2-Φ8 *mm*+Φ13 *mm*		2-Φ10 *mm*+Φ13 *mm*		2-Φ12 *mm*+Φ13 *mm*	
Parameter	B_1	B_2	B_3	C_1	C_2	C_3
Value	3 *N/mm*	5 *N/mm*	7 *N/mm*	7 *mm*	8 *mm*	9 *mm*

According to the L9-orthogonal table, 9 orthogonal experimental designs are performed. The data is shown in Table 2. According to the K_i (i=1,2,3) value, the optimal level combination is, and the corresponding the efficiency of oil filled β for flow regeneration is 78.9%. It can be seen from the range R that the structural parameter of the valve opening C has the greatest influence on the efficiency of oil filled β. There is still a certain gap between the efficiency of oil filled β and the standard requirement of 85% of the quick drop valve under the optimal level combination condition.

Table 2. Efficiency of oil filled β experimental data and calculation

Experiment number	Factors				
	Orifice A	Return spring stiffness B	Valve opening C		Experiment result
	Level				
	1	2	3	4	Efficiency of oil filled β(%)
1	1	1	1	1	78.9
2	1	2	2	2	71.4
3	1	3	3	3	57
4	2	1	2	3	66.2
5	2	2	3	1	52.2
6	2	3	1	2	74.3
7	3	1	3	2	49.1
8	3	2	1	3	71.9
9	3	3	2	1	63
K_1	69.1	64.8	75.1		
K_2	64.3	62.2	66.9		T=584
K_3	61.3	64.7	52.7		
R	7.8	2.6	22.4		

It is difficult to increase the efficiency of oil filled β on the basis of the current limit of the valve opening and the space structure. Taking the first two items in the influence factor that have a large influence on the efficiency of oil filled β as the research object, the disadvantages of the quantitative feedback structure of the orifice A and the separated structure of the valve spool are analyzed. The method of variable control based on feedback from the load and the integrated structure of the valve spool structure are used to improve the performance of the quick drop valve. The following section will model and simulate the flow regeneration performance of the quick drop valve with improved structure and analyze orthogonal experimental data.

4 Improve performance

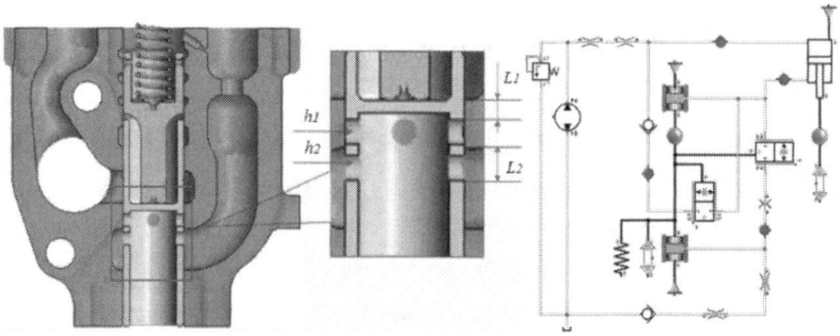

Figure 3. Improved quick drop valve structure diagram and system simulation model

Based on the analysis in the previous section, the structure of the quick drop valve is improved and a simulation model is established. As shown in Figure 3, the parameters of the variables are h_1, h_2, L_1, and L_2. Under the conditions of pressure loss, spool leakage requirement and space structure limitation, the Orthogonal experimental design is used to analyze the influence of the above four parameters on flow regeneration and the optimal level combination. The specific parameters are shown in Table 3.

Table 3. Parameter value table

Parameter	h_{11}	h_{12}	h_{13}	h_{21}	h_{22}	h_{23}	L_{11}	L_{12}	L_{13}	L_{21}	L_{22}	L_{23}
Value(*mm*)	4-Φ8	5-Φ8	6-Φ8	2-Φ8	2-Φ9	2-Φ10	3	4	5	12	15	18

According to the L9-orthogonal table, 9 orthogonal experimental designs are performed. The data is shown in Table 4. According to the K_i (i=1,2,3)value, the optimal level combination is $h_{13}h_{21}L_{11}L_{21}$, and corresponding the efficiency of oil filled β for flow regeneration is 94.3% which meets the standard requirement of 85%. It can be seen from the range R that the influence factor under-lap L_2 has the greatest influence on the efficiency of oil filled β. Considering the orifice h_1 from the layout of the valve spool structure, h_{11} is the best. Considering the orifice h_2 from the pressure loss, h_{23} is the best. Considering the amount of valve spool leakage from over-lap L_1, L_{13} is the best. Since the influence factor under-lap L_2 has the greatest influence on the efficiency of oil filled β, the parameter L_2 is adjusted based on $h_{11}h_{23}L_{13}$.

Table 4. Efficiency of oil filled β experimental data and calculation

Experiment number	Factors				
	Orifice h_1	Orifice h_2	Over-lap L_1	Under-lap L_2	Experiment result
	Level				
	1	2	3	4	Efficiency of oil filled β(%)
1	1	1	1	1	92.3
2	1	2	2	2	74
3	1	3	3	3	61.1
4	2	1	2	3	73
5	2	2	3	1	91.3
6	2	3	1	2	78.9
7	3	1	3	2	78.9
8	3	2	1	3	76.7
9	3	3	2	1	93
K_1	75.8	81.4	82.6	92.2	
K_2	81.1	80.7	80	77.3	T=719.2
K_3	82.9	77.7	76.4	75.7	
R	7.1	3.7	6.2	16.5	

Simulation analysis is performed in the range of 12mm≤ L_2 ≤24mm, and the efficiency of oil filled β is shown in Table 5. With the increase size of L_2, the efficiency of oil filled β gradually decreases and tends to be stable, and the decreasing amplitude Δβ gradually decreases. Take L_2 as 12mm (L_2 = 12.5mm, β = 84.9%) and use the parameter combination $h_{11}h_{23}L_{13}L_{21}$ to improve the structure. The following section will verify the improved performance of the quick drop valve.

Table 5. Efficiency of oil filled β at 12mm≤ L_2 ≤24mm

Value(*mm*)	12	13	14	15	17	18	19	20	21	22	23	24
β(%)	89	81	74.3	69.3	65.6	63	61.2	60.1	59.4	59.3	59.3	59.3
Δβ(%)		8	6.7	5	3.7	2.6	1.8	1.1	0.7	0.1	0	0

5 Verification

Based on the analysis in the previous section, the structure of the quick drop valve is improved with $h_{11}h_{23}L_{13}L_{21}$. The performance of the quick drop valve is also tested on the basis of the same bulldozer engine speed n=800rpm and n=1900rpm.The improved quick drop valve and the pressure curve of the tested rod-less cavity of cylinder are shown in Figure 4. It can be seen that during the process of the blade lowering quickly, t=5.55s for n=800rpm, and t=2.36s for n=1900rpm. Calculated by the relevant data, the efficiency of oil filled β are 87.9% and 88.1% respectively. The error between the test data and the simulation theory data is about 1%. Considering the errors in parts manufacturing and testing, the results meet the requirements.

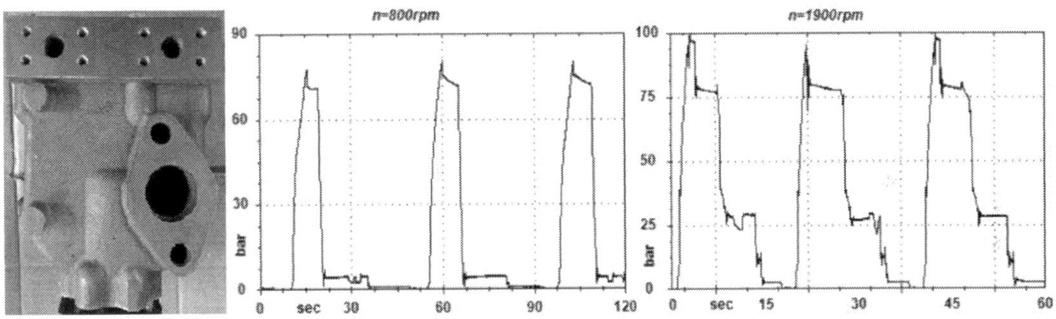

Figure 4. The improved quick drop valve and the pressure curve of the tested rod-less cavity of cylinder

So far, the quick drop valve has passed the verification of more than 1000 hours, and the performance is stable and no failure. The efficiency of oil filled β is still higher than 85%. The improvement of quick drop valve performance can effectively solve the problem that the efficiency of oil filled β is lower than the standard requirement of 85%.

6 Conclusions and future work

Through the theoretical analysis and verification of the flow regeneration performance of the quick drop valve, the following conclusions can be drawn:

(1) In the structure of the original quick drop valve, the influence factor valve opening C has the greatest influence on the efficiency of oil filled β, the optimal level combination is $A_1B_1C_1$, and the efficiency of oil filled β=78.9% is lower than the standard requirement of 85%.

(2) In the improved structure, the influence factor under-lap L_2 has the greatest influence on the efficiency of oil filled β, the optimal level combination is $h_{11}h_{23}L_{13}L_{21}$, and the corresponding efficiency of oil filled β=94.3% is higher than the standard requirement of 85%.

(3) Considering the layout of the spool structure, pressure loss and the spool leakage requirement, we finally use the parameter combination $h_{11}h_{23}L_{13}L_{21}$ to improve the structure.

(4) The structural improvement based on the method of controlling the oil return variable according to the load feedback makes the quick drop valve efficiency of oil filled β for flow regeneration increase from 57% of the measured data to more than 85% of the measured verification, and the efficiency of the earth-moving work after lowering quickly is increased by more than 28%.

We propose, and are currently involved in, the following extensions of the current work. First, consider the influence of the flow regeneration function of the quick drop valve on the performance of the working hydraulic system from the aspect of matching the entire machine power performance. Second, the feedback control of the flow regeneration function of the quick drop valve should be changed from mechanical to electrical. Real-time detection and tracking of pressure changes when the blade is lowering quickly, forming an effective closed-loop control system.

References

1. Y.S. Jiang, L.L. Song, K. Jin, J.J. Zhao, J. Wang, A.L. Lv, C. Mechanization, **2**, 57(2013)
2. L.P. Kong, Y.X. Gao, C. Mechanization, **10**, 36(2017)
3. D.P. Wang, R.X. Tang, H. Lin, Y.W. Feng, Mach Tool Hydrau, **44**,89 (2016)
4. Q.Y. Li, J.B. Song, J.M. Ding, X.L. Guan, W.Z. Zhang, C.MACHINERY, **9**,104, (2013)
5. L.X. Geng, J.M. Gao, CM&M **9**,146, (2014)
6. C.P. Chen, CM&E, **33**,50, (2002)
7. N. Yang, CM&M, **5**,160, (2007)
8. Y.P. Tian, CM&M, **5**,159, (2005)
9. C.L. Li, X.X. Song, CM&E, **2**,14, (1987)
10. X.X. Li, G.P. Li, S.B. Wang, P.Hu, Chin Hydrau Pneumatics, 9, 49(2009)

Research on Energy Efficiency of Typical Construction Equipment for Transmission Lines in Tibet

XIA Yongjun[1] and ZHAO Fei[1, a]

[1] *China Electric Power Research Institute, Haidian District, Beijing 100192, China*

Abstract. The construction of transmission line in Tibet high altitude area has the characteristics of low pressure, low oxygen and low temperature. In the special environmental condition, the transmission line construction equipment with engine as the power source is highly effective, which affects the efficiency and safety of construction[1].The experimental study on engineering mechanical industry plateau transmission line is still in the blank stage,thus relying on the Tibet networking project selects 5 different high altitude test points to plateau 500kV transmission line construction commonly used types of pullers, tensioners and powered winch as an object of study to carry out the typical construction equipment energy efficiency test of field measurements effective output power equipment in different altitude environment from the perspective of practical application and analysis of law and effect mechanism of falling equipment.

1 Introduction

The Tibet area has a special environment with high altitude, high cold, anoxia and so on. The bad construction conditions cause the engineering construction personnel and the machinery to reduce the efficiency seriously. With the increase of EHV power transmission project construction projects in Tibet area, 500kV transmission line voltage will gradually become the backbone network in Tibet power grid, the special construction environment, increase construction quantity and construction workers in the Tibet area that will be more dependent on machinery[2]. Therefore, in the Tibet area to carry out research on typical construction equipment adaptability, analysis of its height decreases and regularity of effect mechanism in different altitude, and puts forward some concrete measures for typical construction equipment and reduce efficiency reduction principle, equipment capacity, construction personnel can reduce the labor intensity and improve the construction efficiency for future power grid construction in high altitude area to give full play to.

2 Test situation

2.1 Test object and method

[a] Corresponding author : 316970757@qq.com

The energy efficiency test of the typical construction equipment is aimed at the environmental characteristics of the high altitude environment in Tibet, and carries out the research on the pullers, tensioners and powered winch application with the engine as the power source in the transmission line construction equipment.The experimental research samples are commonly used in the 500kV transmission line projects in Tibet area, such as pullers, tensioners and powered winch as the research sample [3], the details are shown in Table 1.

Table 1. Test sample table

Name	Model	Rated load	Rated speed
The pullers	SA-QY-250	250kN	2.4km/h
The tensioners	SA-ZY-2×40	2×40kN	2.5km/h
The powered winch	CJM-5	50kN	0.066m/s

The typical construction equipment efficiency test method is to use the test sample equipment as the traction terminal, the auxiliary equipment as the braking terminal to pull through the wire rope, the test layout is shown in Figure 1 to 3, and the test condition is shown in Table 2.The test is carried out by using the sample machine to drive the auxiliary machinery at the maximum speed at the same load and engine speed. At the same time, the dynamic wireless sensor and the wire rope speed measuring device are used to record the tension and speed data during the whole test.

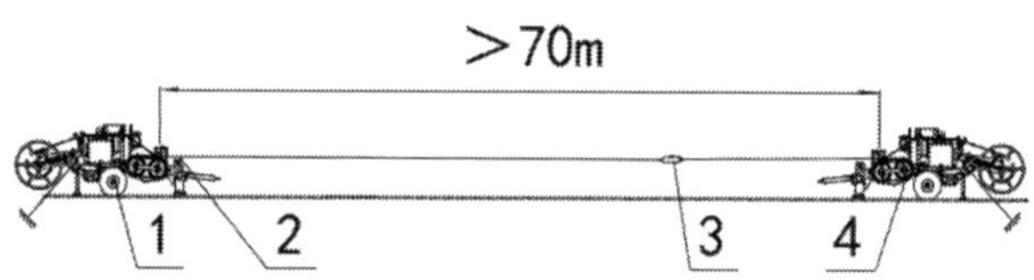

1-sample pullers; 2-speed measuring device; 3-dynamic tension sensor; 4-auxiliary puller

Figure 1. Schematic diagram of energy efficiency test for pullers

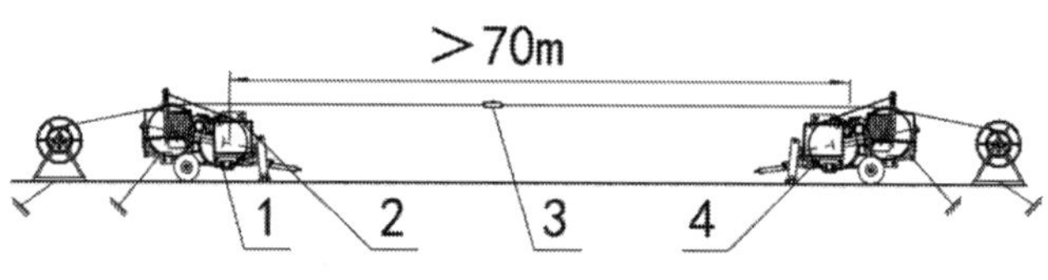

1-sample tensioners; 2-speed measuring device; 3-dynamic tension sensor; 4-auxiliary tensioners

Figure 2. Schematic diagram of energy efficiency test for tensioners

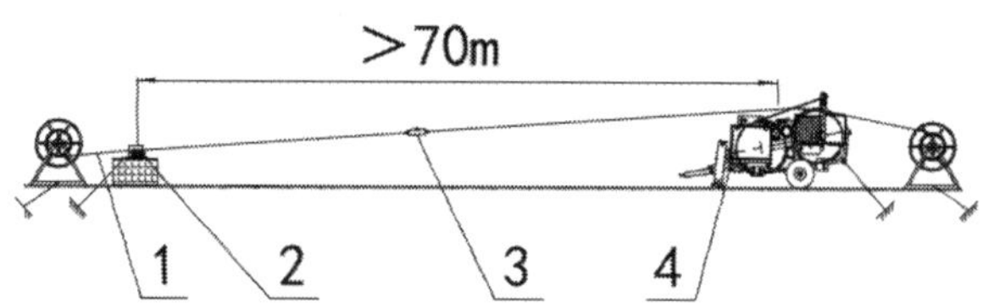

1-speed measuring device; 2-sample powered winch; 3-dynamic tension sensor; 4-auxiliary tensioners

Figure 3. Schematic diagram of energy efficiency test for powered winch

Table 2. Test condition table

Serial number	Test sample	Test condition	
		Engine speed	Test load
1	The puller of SA-QY-250	1500rpm	150kN
		1500rpm	175kN
2	The tensioner of SA-ZY-2×40	1500rpm	24kN(reverse)
		1500rpm	28kN(reverse)
3	The powered winch of CJM·5	I gear	30kN
		I gear	35kN

2.2 Test implementation

Select 5 different high altitudes test points as the data acquisition point in the area of Tibet [4]; select 1 plain test points as a reference point in Zhengzhou. The typical construction equipment efficiency test were carried out a total of 6 test points, the scene is shown in Figure 4, the specific implementation place as shown in table 3.

Figure 4. Field test photo of typical construction equipment

Table 3. Test implement site table

Serial number	Position	Actual altitude
1	Reference point	162m
2	Test point near 3300m	3313m
4	Test point near 3700m	3740m
5	Test point near 4100m	4147m
6	Test point near 4500m	4553m
7	Test point near 4900m	4990m

3 Test result

3.1 Data processing method

The data collected by the dynamic wireless tension sensor and the wire rope speed measuring device used during the test are all the database with frequency of 5Hz.The tension time curve and the speed time curve are drawn into two databases, and the two curves are compared by the time zero point reclosed.From the whole process of the test selection of tension relatively stable and relatively stable period of time speed as the effective time period, as shown in Figure 5.

The data in the effective time segment is calculated according to the power point formula, and the average value of all data points in the effective time period is taken as the last effective output power [5], as shown in Figure 6.

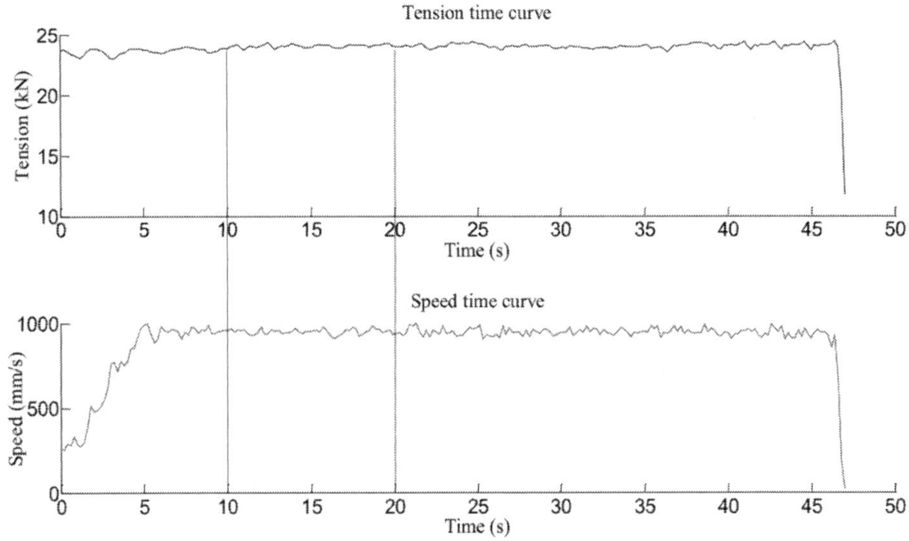

Figure 5. 24kN working condition data curve of tensioners at altitude 162m

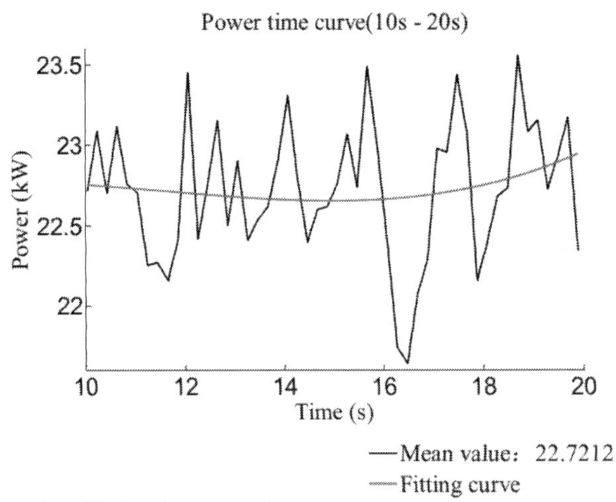

Figure 6. Power-time curve in effective time period

3.2 Data results

A summary of the effective output power collation of 6 test points is summarized as shown in Table 4.

Table 4. Summary table of effective output power of typical construction equipment

Working condition		Effective output power at the altitude test point(kW)					
		162m	3313m	3740m	4147m	4553m	4990m
The pullers	175kN	166.578	161.724	153.939	151.775	146.092	139.937
	150kN	146.954	129.567	128.725	125.232	122.821	114.250
The tensioners	28kn	26.036	22.146	19.679	19.470	18.769	18.617
	24kn	22.721	19.454	18.583	17.860	17.052	17.021
The powered winch	35kN	2.781	2.652	2.507	2.262	2.107	1.787
	30kN	2.552	2.296	2.229	2.087	1.838	1.672

From table 4, it can be seen that the effective output power of the typical construction equipment is gradually reduced with the elevation of the altitude test points.

4 Result analysis

The test results of high altitude test are compared with the results of the benchmark datum in the plain. The average reduction range of different load cases is used to reflect the descending effect of the typical construction equipment at the altitude test point. It is summarized and summarized as shown in Table 5, and is drawn up and made into a chart as shown in Figure 7.

Table 5. Energy efficiency scale table of typical construction equipment

Altitude	The pullers	The tensioners	The powered winch
3313m	7.37%	14.66%	7.32%
3740m	10.00%	21.31%	11.24%
4147m	11.83%	23.31%	18.43%
4553m	14.36%	26.43%	26.09%
4990m	19.12%	26.79%	35.11%

In typical construction equipment samples, the engine of traction and tension machine is a diesel engine with turbocharger, and the motor form of engine is a turbocharger diesel engine.

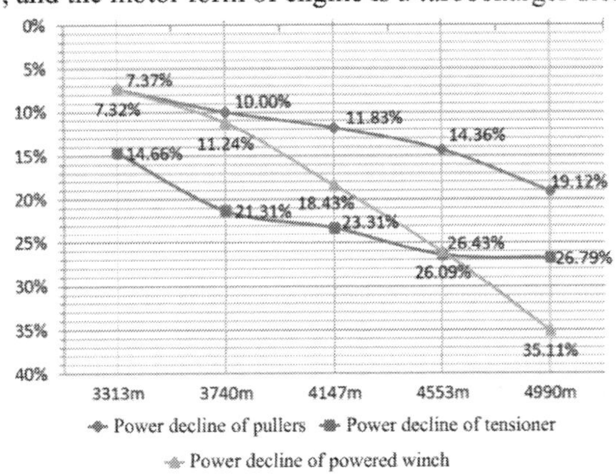

Figure 7. Reduced efficiency curve diagram of energy test of typical construction equipment

5 Conclusion

The adaptability of typical construction equipment is studied based on the Tibet Internet project. The eff effective output power of the typical equipment is taken as the research object, and the effective output power of the equipment is measured:

(1) The typical construction equipment energy efficiency test results representation in the special environment of Tibet low, hypoxia and low temperature, choose to adapt to the harsh construction environment to maintain a certain amount of reserve power construction equipment can be better with a turbocharged engine and use conditions.

(2) Under the restriction of construction hardware conditions, we can also take the efficiency reduction test curve of typical construction equipment in Figure 7 as reference for selection, and consider the factors such as old equipment, engine calibration power and other factors.

(3) In the research and design of special plateau type construction equipment, it is recommended to consider comprehensively from two aspects of engine power margin and practical measures, and enhance the applicability of equipment through overall design accounting.

Acknowledgements

This research was support by Science and Technology Research Project of State Grid(Research on technical standardization of ultra high altitude EHV and super high difference power grid construction)

References

1. YIN Dengfeng, ZHOU Hongbing, HAO Peng. Computer Simulation. Forecast and analysis for diesel engine plateau performance of engineering machinery. 2014 (2014. 09): 422-426.
2. SHEN Weichun,TIAN Jinbu,ZHANG Tianwen, et al. Power Technology.Study on the efficienty reduction of artificial labor in Tibet power grid. (1674-4586), 2010 (9).
3. YU Qihao,LIU Houxiong, QIAN Jin, et al. Journal of Engineering Geophysics. Analysis on the engineering problem of the ±500kV transmission line of the Qinghai Tibet direct currentnetwork. 2009, 6(6): 806-812.
4. DONG Hongyu. Industrial Technology in Shandong. Analysis of construction machinery reduction efficiency in power grid engineering in alpine area. 2015 (19): 92-92.
5. SUN Miaomiao, ZHAO Jinliang, LI Xiuzhi, et al. Electronic Test. Countermeasures for the use of low temperature environment of lead acid battery. 2016 (7X): 165.

Experimental Investigation and Performance Analysis of Ceramic Inserts in Laser Assisted Turning of Waspaloy

Martyna Wiciak[1,a], Tadeusz Chwalczuk[1] and Agata Felusiak[1]

[1]*Poznan University of Technology, Piotrowo 3 St., 60-965 Poznan, Poland*

Abstract. In this paper, the influence of laser assisted turning of hard-to-cut nickel-based superalloy on tool cutting ability was presented. The conducted research involved the machining performance along with tool life of ceramic inserts during turning of heat-resistant alloy under a trade name Waspaloy. The ceramic insert with geometry in accordance with the ISO – RPGX 120700 T01020 were applied during longitudinal turning with laser beam. The investigations has been completed with various cutting conditions such as laser power *P*, cutting speed *vc*, feed *f* and depth of cut a_p. In order to determine the relations between the tool wear and cutting time, the tool life *T* has been selected. The increment of tool wear was correlated with the change of vibration signals and the critic points of tool wear was presented. In addition, the shape and form of chip was evaluated based on macroscopic observation and SEM analyses. The conducted research was primarily focused on effective application of ceramic inserts during turning Waspaloy with laser beam and comparison this technology with conventional machining.

1 Introduction

Mechanical processing of hard-to-cut materials, ex. superalloys and aluminum-ceramic composites is a challenge for cutting tools [1, 2]. An important aspect is a replacement of the tool at an appropriate moment, before the catastrophic tool wear. In parallel, the process is aimed to maximal utilization of cutting tools and thereby obtaining longer tool life. The right selection of tool material and insert's geometry allows to achieve required dimensional and shape precision. Moreover, it may lead to improve machining conditions and maximize the tool life [3, 4, 5].

Cutting inserts' wear is related with different mechanisms which accompany machining processes, ex. abrasion and adhesion. In case of machining hard-to-cut materials such as heat-resistant alloys, most of research suggest that abrasive wear is the most evident and it results from hard phases occurring in microstructure of superalloys [6]. In paper [1] the influence of two hard-to-cut materials microstructures: Waspaloy and Inconel 718, on the wear of cutting inserts [7] was investigated. It was presented that high temperatures occurring in cutting zone have the main impact on precipitated wear. On the other hand, in paper [5] wear of polycrystalline diamond (PCD) tools during machining of chromium carbide semi-finished product was investigated. Researches focus as well on carbides milling [8] and examination of surface machined using CBN cutting edges [9]. Basing on the

[a] Corresponding author : martyna.r.wiciak@doctorate.put.poznan.pl

researches it was concluded that the material is hard-to-cut and application of methods which improve efficiency of machining is recommended. To improve machining conditions and to decrease tool wear new methods of cooling (ex. cryogenic cooling) or special methods which assist machining are applied [10]. One of the popular technologies is laser assisted machining which enables effective forming of superalloys [11, 12, 13]. Due to located heating and softening of material which resulted in reduction of its durability in cutting zone, lower cutting forces and wear of cutting tools with simultaneously improve of machined surface quality were obtained. [2, 14] Researches in which laser assist was applied to improve machining conditions present that proper selection of parameters, like laser beam power, cutting speed and feed rate, have influence on temperature in the cutting zone [14].

In paper [14] the comparison of cutting inserts' durability during conventional turning and laser assisted turning was presented. 50% increase in inserts' durability during machining of Waspaloy was found due to applying the laser assist. Moreover, decrease of cutting forces by 20% was spotted. Increasing amount of researches focus on analysis of vibrations which accompany the machining process [15]. Generating of too heavy vibrations may lead to a loss of process stability. Various technologies are applied to reduce values of vibrations' accelerations. In paper [16] the decrease in amplitude of vibrations during laser assisted turning of cemented carbides was spotted. For examination of hard-to-cut materials machinability the analysis of chip morphology is being carried out [17, 18]. It is commonly known that the shape of chip is influenced by machining conditions, tool's geometry, machined material and the type of coolant. Appropriate forming and flow of chip is significant for machining process, thus selection of proper cutting parameters eliminates generating short and thick chips and decrease cutting forces and mechanical vibrations. In paper [19] conventional machining was compared with LAM during turning of Inconel 718. It was found that besides reduction of cutting forces, tool wear and improving machined surface quality, during laser assisted machining reduction of chip thickness occur and the tendency to creation segmented chip is significantly lower.

Presented research focus on different effects of laser assisted turning of hard-to-cut material (Waspaloy). Selection of proper process parameters which allow to obtain the best tool life and process efficiency is a key element for this paper. On the other hand, analysis of mechanical vibrations during the machining process and characteristics of chip morphology led to define improvement of machining conditions using laser assist.

2 Experimental

2.1 The research stand

In this paper the wear of round ceramic (6220) cutting inserts of type RPGX 120700 T01020, produced by Sandvik, was investigated. This grade of ceramics is adjusted to machining of superalloys, ex. Waspaloy, thus it was selected for the research. Examination of tool life was carried out on the station for laser assisted machining. This station includes turning centre CTX 310 ecoline coupled with diode laser TRUMP TruDiode 3004 which is integrated with robot Kuka KR 16-2. During the research, material was cooled using carbon dioxide under a pressure of 4 bar.

Next step was the unit for measuring acceleration of mechanical vibrations. The test stand is presented in Figure 1. Measurement of mechanical vibrations' acceleration was carried out using three-axis vibration sensor Brüel & Kjær of type 4321 which was placed on the lathe's turret. Vibrations were measured in three dimensions, in accordance with Cartesian system of the machine. Measurement of accelerations was executed in three directions: main (compliant with v_c vector), feed (compliant with v_f vector) and thrust direction. Measurement of wear was carried out with workshop microscope after each machining pass. For evaluation of wear VB_c criterion was chosen. This criterion indicates the value of flank face wiping. Moreover, macroscopic and microscopic pictures of chips were presented.

Figure 1. The test stand used for a test.

In order to select proper cutting parameters which enable to obtain the best tool life and process efficiency, machining was carried out with different parameters. Performed tests and various parameters are given in Table 1. Before this test, these cutting parameters were generally applied to conventional machining of Waspaloy.

Table 1. Cutting conditions.

Test number	Cutting speed vc [m/min]	Cutting depth a_p [mm]	Feed rate f [mm/rev]	Laser beam power P [W]
I	120	0,2	0,2	2000
II	120	0,12	0,2	1500
III	150	0,12	0,2	1500
IV	150	0,2	0,2	2000
V	150	0,2	0,2	1500

For measurement of mechanical vibrations equipment shown in simplified scheme of circuit in Figure 2 was used. For amplification and processing signal amplifier Brüel & Kjær NEXUS of type 2692-C was carried out.

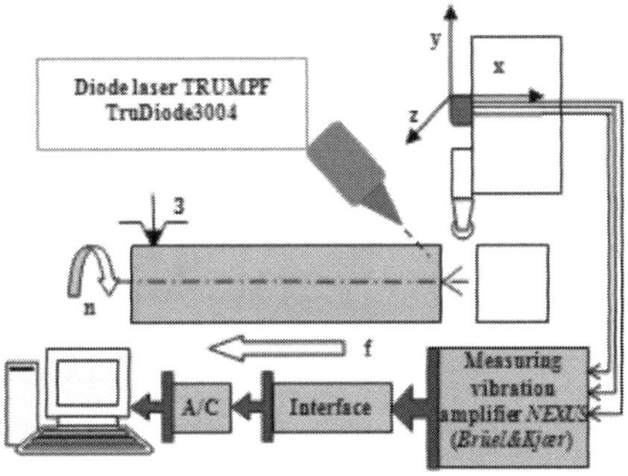

Figure 2. The set-up of experimental apparatus.

3 Results and discussion

3.1 Analysis of tool wear

Measurement of tool wear was carried out after each machining pass for all performed tests. By evaluating the wear increase it was possible to define tool life T for particular test and to correlate values of cutting edge wear with microscopic pictures of chips.

Firstly, exemplary progress of VB_c wear curve as a function of machining time t_s was presented with determined wear criterion VB_w=0,6 mm. Parameters with which the test was performed and the wear of round cutting insert are shown in Figure 3. Flank face wear was also presented in this figure.

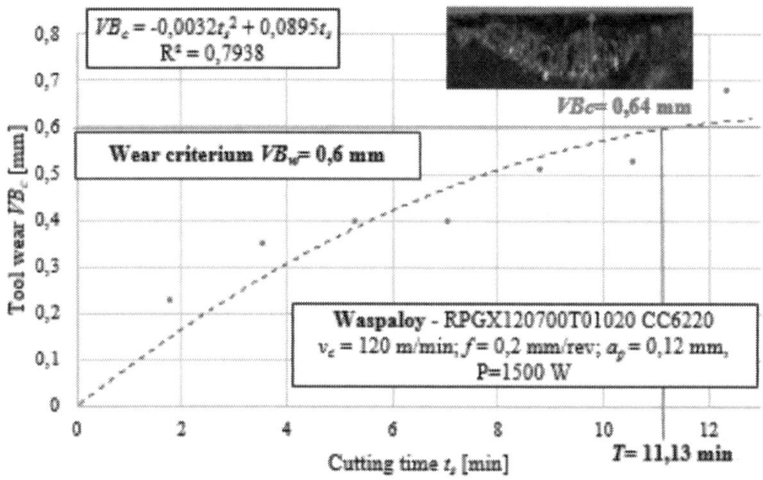

Figure 3. Tool wear in function of time for the investigated ceramic inserts under laser assisted turning.

For parameters vc= 120 m/min, f=0,2 mm/rev, a_p=0,12 mm, P=1500 W the best durability of cutting inserts was obtained. On the other hand, by analyzing all the passes, it was spotted that the best efficiency is obtained for lower laser beam power (1500 W) and lower cutting speed (120 m/min).

On the basis of macroscopic observation of chips, the correlation between the value of tool wear and the chip character was spotted. In Figure 4 the character of chip in dependence upon modification of cutting insert wear is presented. At initial stage of wear, ribbon chip occurs and its brown color signifies the low increase in machining temperature. With increasing tool wear, character and color of chip is changing.

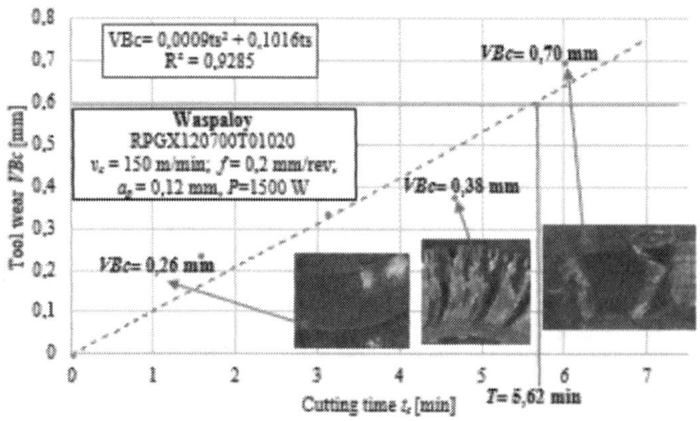

Figure 4. Changes in shape of chip in dependence upon increase of tool wear.

The occurrence dark blue and purple color signifies the significant increase in machining temperature which is result of higher tool wear. With a wear VB_c=0,38 mm the significant change in shape of chip is spotted. Segmented chip occurs and due to side flow, at the final stage of wear, it contains anchored particles of machined material.

On the basis of comparative analysis of durability of cutting inserts T and machining process performance Q, two combinations of cutting parameters were selected . As the result of applying these parameters the best process performance and tool life were obtained. The evaluation of both criteria is incredibly important to the development of process with the best efficiency. For parameters vc=150 m/min, f=0,2 mm/rev, a_p=0,5 mm, P=1500 W the best machining efficiency was obtained and using parameters vc=120 m/min, f=0,2 mm/rev, a_p=0,12 mm, P=1500 W the highest ceramic insert durability was obtained. Figure 5 shows comparison of tool life and performance for different cutting parameters.

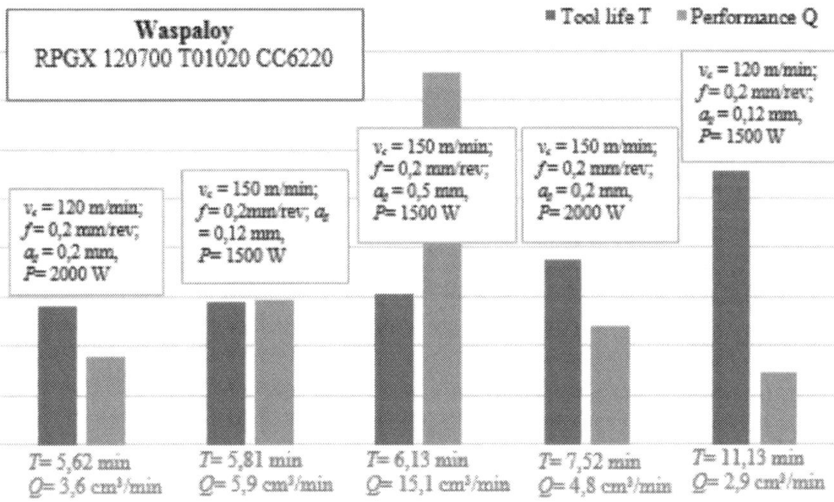

Figure 5. The comparison of process efficiency and durability of cutting inserts for different cutting parameters.

The last step of research was the analysis of vibrations' accelerations which were measured during the process of machining with laser assisted. By analyzing root mean square values A_{p_RMS} of diagnostic measures in three dimensions in accordance with axes of coordinate system, increase in mechanical vibrations with incremental value of flank face wear was spotted. Figure 6 depicts the exemplary chart containing the relation between increment of tool wear and value of root mean square A_{p_RMS}.

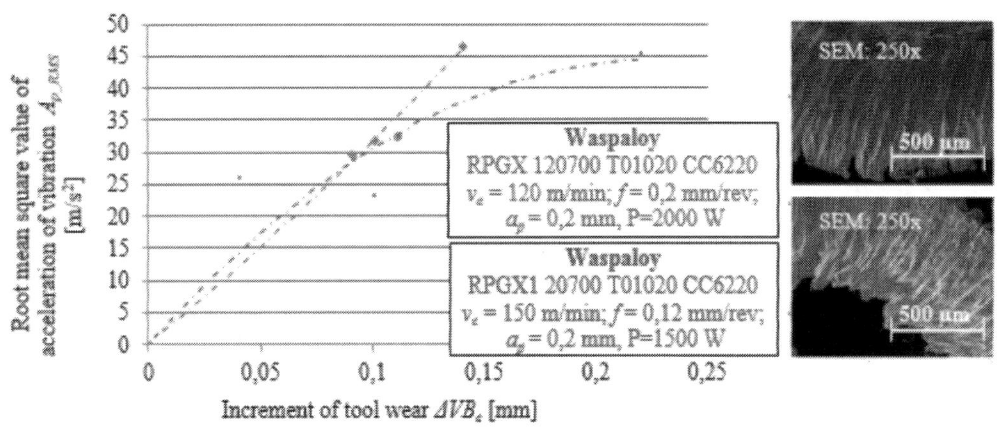

Figure 6. Relations between increment of flank tool wear ΔVB_c and value of acceleration A_{p_RMS}.

The higher increment in tool wear, the larger mechanical vibrations, whereby during the test single drops in vibrations were spotted and they occurred as the result of insignificant increase in tool wear after subsequent machining passes. Built-up edge and reinforcement of Waspaloy could have caused a loss of vibration signals. With increase of vibrations also change in shape of chip occurs and its segmentation is more visible (figure 6).

4 Conclusions

On the basis of research following conclusions were formulated:

- Laser assist of turning Waspaloy which is hard-to-cut material, with round ceramics inserts improve material forming and increase durability of cutting inserts, especially if proper cutting parameters are selected (vc=150 m/min, f=0,2 mm/rev, a_p=0,5 mm, P=1500 W).
- On the basis of chip morphology it can be spotted that shape of chip changes with increasing tool wear and temperature in cutting zone. Chip changes its color and its segmentation occurs. Moreover, anchoring of machined material on cutting edge occurs due to changes in temperature. Built-up edge leads to decrease of vibrations during machining, on the other hand the higher incremental value of flank wear the higher vibration signals.
- Volumetric machining process efficiency and the tool life of ceramics inserts significantly improves if lower laser beam power (1500 W) is applied. The increase of laser beam power it is not necessary to use, the reinforcement of Waspaloy is more visible and its machinability is not notables.

Acknowledgments

The presented research results, executed under the domestic project LIDER of No. 164/L-6/14/NCBR/2015 with grants for education allocated by the National Centre for Research and Development.

References

1. P. Hoier, A. Malakizadi, P. Stuppa, S. Cedergren, U. Klement, Wear, Vol.**400-401**(2018)
2. T. Chwalczuk, S. Legutko, R. Talar, Machine Manufacturing Technology, Vol.**29**, p.19(2015)
3. Y. Fan, Z. Hao, J. Lin, Z. Yu, Materials and Manufacturing Processes, 01 January 2014
4. S. Olovsjö, L. Nyborg, Wear, Vol.**282-283**, pp.12-21(2012)
5. D. Przestacki, P. Kieruj, Advances in Manufacturing Science and Technology, Vol.**38**(4)(2014)
6. D. Przestacki, T. Chwalczuk, MATEC, Vol.**136**, 02006(2017)
7. R. Polvorosa, A. Suarez, L.N. Lopez de Lacalle, I. Cerrillo, A. Wretland, F. Veiga, Journal of Manufacturing Processes, Vol.**26**, pp.44-55(2017)
8. S. Wojciechowski, D. Przestacki, T. Chwalczuk, MATEC, Vol.**136**, 01006(2017)
9. S. Wojciechowski, P. Twardowski, T. Chwalczuk, J.of Phys.: Conference, Vol.**483**, pp.1-8(2014)
10. Isik, Yahya, Advances in Mechanical Engineering, Vol.**8**(5)(2016)
11. S. Xavierarockiaraj, P. Kuppan, Procedia Engineering, Vol.**97**(2014)
12. D. Przestacki, M. Kuklinski, A. Bartkowska, The Inter. J. of Adv. Manuf. Tech., Vol. **93**(2017)
13. D. Przestacki, T. Chwalczuk, S. Wojciechowski, The Inter. J. of Adv. Manuf. Tech, Vol.**91**(9)(2017)
14. Kim, Jong-Do;Lee, Su-Jin ; Suh, Jeong, J. of Mech. Sc. and Tech., Vol.**25**(4), pp.995-1001(2011)
15. S. Wojciechowski, P. Twardowski, T. Chwalczuk, J. of Phys.: Conf., Vol.**483**, pp.1-8(2014)
16. P. Kieruj, D. Przestacki, T. Chwalczuk, Mechanik , Vol.**8-0**/2016
17. M. Rybicki, P. Szablewski, MATEC Web of Conferences, Vol.**121**, 03020(2017)
18. Y. Xiong, W. Wang, R. Jiang, K. Lin, M. Shao, Materials, Vol.**11**(4)(2018)
19. K. Venkatesan, R. Ramanujam, P. Kuppan, Int J Adv ManufTechnol, Vol.**91**:3807-3821(2017)

Surface Roughness Characterization of Inconel 718 after Laser Assisted Turning

Agata Felusiak[1,a], Tadeusz Chwalczuk[1] and Martyna Wiciak[1]

[1]*Poznan University of Technology, Piotrowo 3 St., 60-965 Poznan, Poland*

Abstract. This paper describes the surface roughness parameters evaluation of Inconel 718 (43 HRC) after turning under laser assisted conditions. The experiment was conducted for cemented carbides inserts under various cutting and laser heating parameters. Tesets based on central composite research program allows to describe an interaction between input (cutting speed v_c, laser power P, feed f) and output (roughness parameters according to PN-ISO 4288: *Ra, Rz, Rt, Rsk, RSm, Rdq*) parameters. The results showed that laser power density decrease values of amplitude type of roughness parameters. Such occurrence is not observed for horizonal and hybrid parameters of surface. The function is not described by monotonic projection. Therefore optimization by usability function was performed to observe complex interaction between input and output values. The findings of this work define which one of values from cutting speed, laser power and feed values have crucial impact on surface roughness constitution.

1 Introduction

Inconel 718 is one of the most frequently applied nickel–based superalloy. The reason of this are, among others, strength and anti–corrosive properties at high temperatures [1]. Another important feature is higher amount of iron which significantly decreases the cost of material. However, high amount of iron lowers the work temperature of the 718 alloy. Mass fractions of elements in Inconel 718 are given in Table 1.

Table 1. Chemical composition of Inconel 718 alloy (weight %).

Ni	Cr	Nb	Mo	Ti	Al	Co	Si	Mn	Cu	C	P	S	B	Fe
50,00-55,00	17,00-21,00	4,75-5,50	2,80-3,30	0,65-1,15	0,20-0,80	max 1,00	max 0,35	max 0,35	max 0,30	max 0,08	max 0,015	max 0,015	max 0,006	balance

Nickel–based superalloys are mainly applied in aviation and marine industry as parts of gas turbines, space shuttles and nuclear reactors. They are characterized by a great durability at high temperatures, wear and corrosion resistance, and a superior creep resistance below 700°C [2, 3].

[a] Corresponding author : agata.z.felusiak@doctorate.put.poznan.pl

Alloys of Inconel group, including 718 alloy, are considered to be hard–to–cut materials. It is related with work-hardening and high temperature occurring in the cutting zone which eventually results in accelerated tool wear. Inconel exhibits strong affinity for formation of adhesive bonds with the cutting edge. This results in creation built–ups on the flank face and the rake face. Moreover, this is the reason of the side material flow. According to the research [4], roughness of machined surface is mainly influenced by the initiation of plastic chip side flow, equal progress of tool wear and formation of built–up on the cutting edge.

In relation to difficult machining of 718 alloy, researchers are searching for new methods to make this process easier. Another trouble is maintaining low roughness of obtained surfaces which is a common requirement for parts made of nickel–based superalloys [5]. One of the methods is application of high speed machining (HSM). It was proved in researches [6] that usage of cutting speed equal to approximately 190 m/min leads to a lowering of roughness with insignificant increase of tool wear, in comparison with the usage of lower cutting speeds. However, application of higher cutting speed shorten the tool life significantly and worsen the condition of machined surface.

Another method which leads to improve machinability of nickel–based superalloys, including Inconel 718, is application of laser assisted machining (LAM) [7, 8]. The usage of LAM is based on heating the workpiece with laser beam of high power in order to soften the material which results in decreasing cutting forces and durability of workpiece during machining [9, 10]. As the researches show [11, 12], the usage of laser assist improve the quality of machined surface– approximately 3–times decrease in roughness was noted. Furthermore, in comparison with conventional machining, longer tool life was spotted.

2 Experimental details

2.1 Applied process technology

In this research shaft made of Inconel 718 was machined. For the machining process, cemented carbide WNMG 080408EN inserts produced by CERATIZIT were used and they were attached with MNLNL2020 K08 tool–holder. The research was carried out on a numerically controlled CTX310 Ecoline lathe with self–centric hydraulic three–jaw chuck. For laser assisted machining, diode laser *TRUMPF* TruDiode 3004 was applied. Machining was carried out in accordance with the scheme shown in Figure 1. The treatment was carried out in accordance to the research plan given in Table 2. Depth of cut and laser spot diameter were constant and equal to respectively a_p= 0,15 mm and d_f= 1 mm. Variable laser beam power and its constant geometry result in variable surface laser beam power density. In this research, machining with simultaneously heating using laser beam was applied.

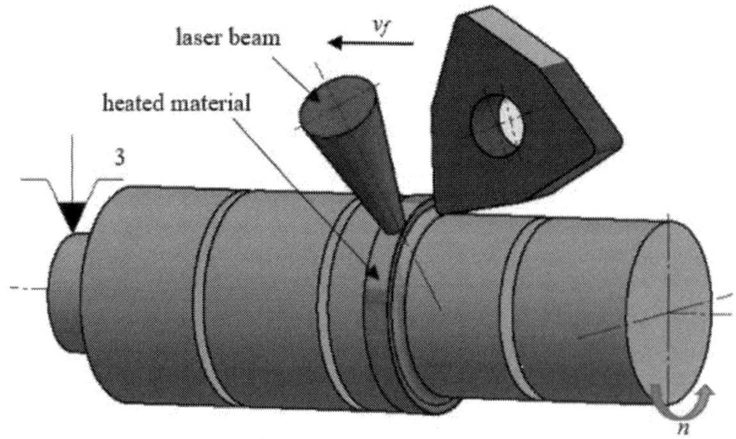

Figure 1. The scheme of machining with laser assistant.

Table 2. Applied plan of research.

nr	CODED VALUE			REAL VALUE		
	v_c	f	P	v_c	f	P
	[-]	[-]	[-]	[m/min]	[mm/rev]	[W]
1	-1,682	0,000	0,000	20	0,100	1400
2	-1,000	-1,000	-1,000	32	0,073	1162
3	-1,000	-1,000	1,000	32	0,073	1638
4	-1,000	1,000	-1,000	32	0,127	1162
5	-1,000	1,000	1,000	32	0,127	1638
6	0,000	-1,682	0,000	50	0,055	1400
7	0,000	0,000	-1,682	50	0,100	1000
8	0,000	0,000	0,000	50	0,100	1400
9	0,000	0,000	0,000	50	0,100	1400
10	0.000	0,000	1,682	50	0,100	1800
11	0,000	1,682	0,000	50	0,145	1400
12	1,000	-1,000	-1,000	68	0,073	1162
13	1,000	-1,000	1,000	68	0,073	1638
14	1,000	1,000	-1,000	68	0,127	1162
15	1,000	1,000	1,000	68	0,127	1638
16	1,682	0,000	0,000	80	0,100	1400

2.2 Carried out measurements

For each parameters set from Table 2 five roughness measurements were taken. For the surface topography assessment, stationary profilographometer T8000 was carried out. Measuring length used was equal to ln = 4 mm and elementary section was lr= 0,08 mm according to the PN-ISO 4288:1998 standard. Basing on obtained surfaces' profiles following roughness parameters were calculated: *Ra, Rz, Rt, Rsk, Rq, RSm, Rdq*.

3 Results and discussion

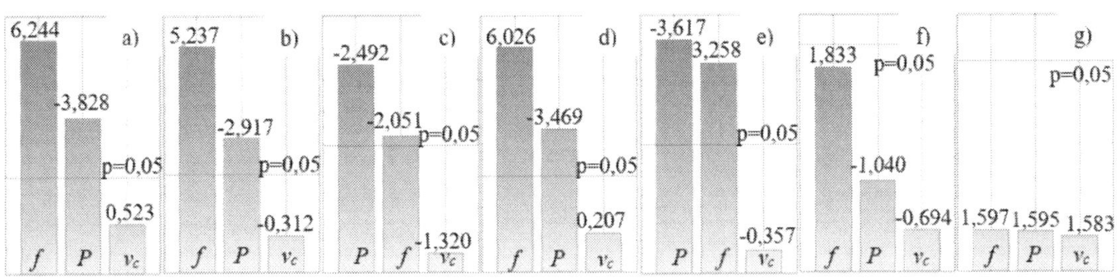

Figure 2. Pareto chart of effects of input parameters influence on roughness parameters: a) *Ra*, b) *Rz*, c) *Rsk*, d) *Rq*, e) *Rt*, f) *RSm*, g) *Rdq*.

In Figure 2 the significance of impact of different variables on obtained roughness parameters. As it results from carried out Pareto analysis of effects for confidence interval equal to p= 0,05, roughness is most significantly influenced by feed rate and laser beam power, whereby the feed rate mainly influences two most commonly considered parameters: *Ra* and *Rz*. It is seen that the cutting speed has no influence on surface roughness in the preset range of cutting parameters. Moreover, no influence of feed rate on the average offset between profile's peaks (*RSm*) which theoretically should be approximately equal to the feed rate value. In this case the occurrence of material side flow is possible and this phenomenon may have the prevailing impact on the frequency of the profile's peaks formation (*RSm*). Also the initial state of cutting edge, wear and cutting edge nickness which is reflected on the machined surface, have significance for this roughness parameter. None of the examined parameters have no impact on roughness parameters which describe the tilt of the profile. Feed rate and laser beam power are mostly responsible for surface roughness, however, in further research additional parameters, ex. tool geometry, should be taken into consideration.

The relation of influence of laser beam power and feed rate on values of roughness parameters: average square profile deviation (*Rq*) and the tilt of the profile (*Rsk*) are shown in Figures 3a and 3b respectively.

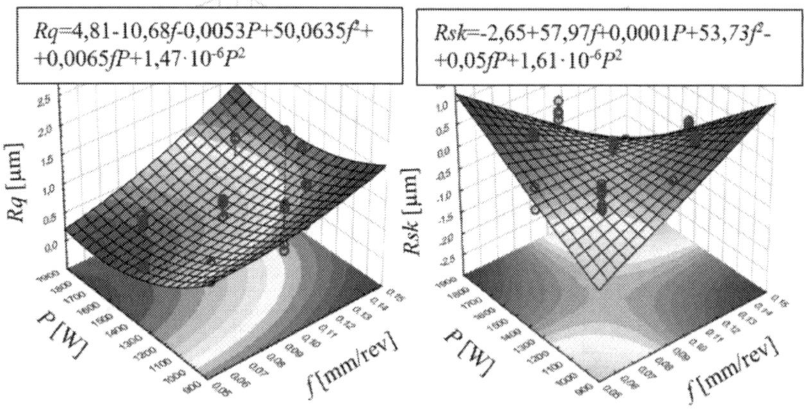

(a) (b)

Figure 3. Laser power and feed influence on value: a) *Rq*, b) *Rsk*.

During the analysis of the impact of laser beam power and feed rate on roughness parameters it was spotted that the roughness increases with increasing the value of feed and decreases with increasing the laser beam power, primarily for low feed rates. An exception to this phenomenon is only the tilt of the profile (*Rsk*). In this diagram, there is no clear relation between laser beam power and feed and this relation depends on combination of these two parameters. Remaining parameters confirm results described in literature, so decreasing the roughness. For analysis of 3–dimensional diagrams only parameters for which, according to Pareto chart, the significance for the confidence interval equal to p=0,05 occurred.

In Figure 4 compares the theoretical and the measured values roughness parameter *Rz*. Values of roughness obtained from regression model of two variables differ significantly from theoretical value. The theoretical value of roughness parameter *Rz* was calculated for arc projection according to the Formula 1:

$$Rz_t = \frac{f^2}{8r_\varepsilon} \tag{1}$$

where:
f– feed [mm/rev]
r_ε– insert nose radius [mm].

The higher laser beam power (to 1400 W), the values are closer to the theoretical one, due to among others reduction of surface hardening. The rapid increase of roughness is related to too deep heating in comparison with the depth of cut which causes modification of materials' ductility and results in heightening maximal peaks of profile due to plastic deformation of the workpiece. Differences in theoretical and measured values derive from not taking into consideration some phenomena, ex. elastic–plastic deformations and vibrations [13].

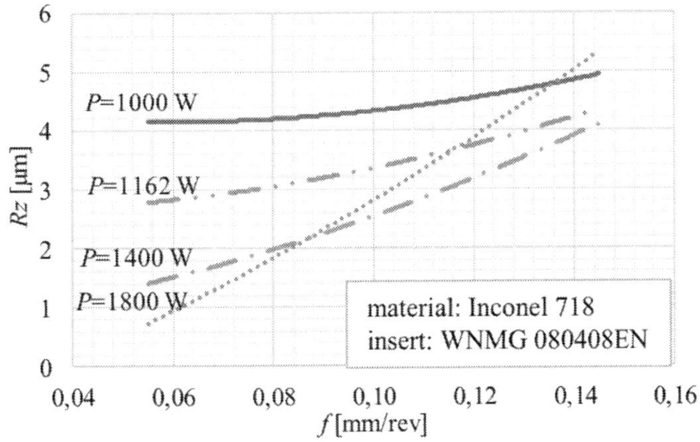

Figure 4. Theoretical surface roughness Rz_t roughness measured after laser assistant turning Inconel 718.

Profiles of usability of examined parameters are shown in Figure 5 with simultaneously multi–criteria optimization. Nevertheless, with increasing the laser beam power the value of average arithmetical profile deviation, due to simultaneously consideration of two roughness parameters (*Ra* and *Rsk*), the highest usability is obtained for medium laser beam power. Variable cutting speed, in combination with other parameters, influences the distribution of *Rsk* parameter. It is significant that the impact of cutting speed itself is reduced by other parameters which results from aforementioned Pareto charts. However, the comparison of parameters of lowest usability will worsen the roughness of machined surface even more than it results only from single–criteria analysis. For extremal values of cutting speed, *Rsk* values become negative which is not typical for turning because the roughness profile character changes from peak to valley.

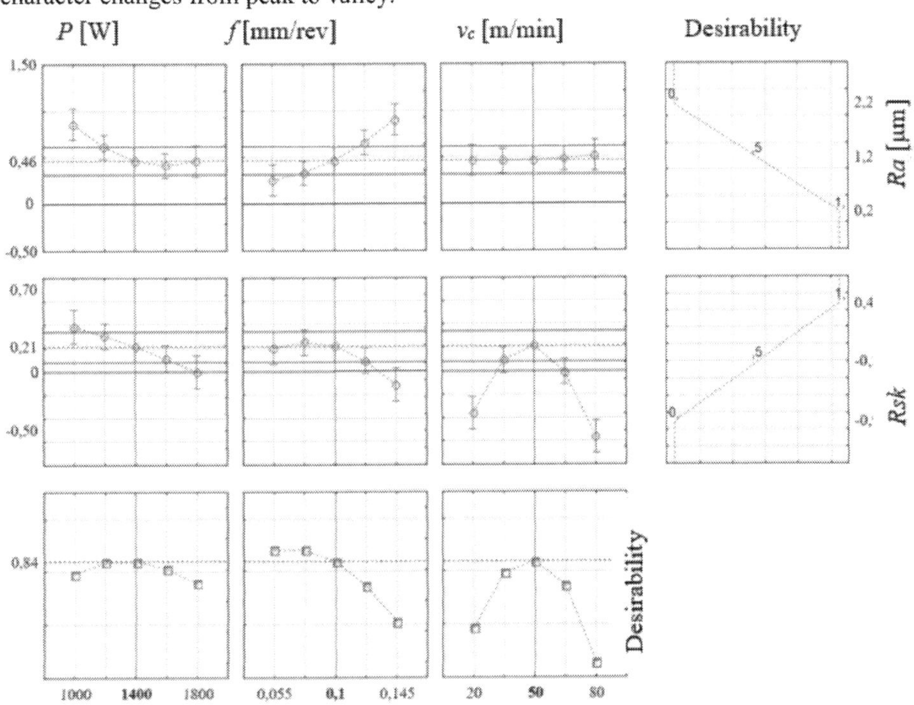

Figure 5. Results of usability function analysis.

4 Conclusion

On the basis of conducted research following conclusions were formulated:

- Increase of laser beam power allows to obtain lower values of machined surface roughness parameters. The significant restriction is the depth of workpiece heat softening.
- For low feed rates higher laser beam power can be applied (*P*= 1800 W) because roughness is convenient if low feed rate is applied. It results from low plastic deformation. This applies until the yield point is exceeded.
- In terms of roughness, the most effective feed rate is ranged between *f*= 0,055 mm/rev and *f*= 0,073 mm/rev, the most optimal laser beam power is equal to *P*= 1400 W and cutting speed v_c=50 m/min.

Acknowledgments

The presented research results, executed under the domestic project LIDER of No. 164/L-6/14/NCBR/2015 with grants for education allocated by the National Centre for Research and Development.

References

1. W. Grzesik, P. Niesłony, W. Habrat, J. Sieniawski, P. Laskowski, Tribology International **118**, 337 (2018)
2. D. Ulutan, T. Özel, International Journal of Machine Tools and Manufacture **51**, 250 (2004)
3. P. Kieruj, D. Przestacki, T. Chwalczuk, Archives of Mechanical Technology and Materials **36**, 30 (2017)
4. A.Thakur, S.Gangopadhyay, International Journal of Machine Tools and Manufacture **100**, 25 (2016)
5. T. Chwalczuk, P. Twardowski, P. Kieruj, P. Szablewski, Zeszyty Naukowe Politechniki Rzeszowskiej **295**, 307 (2017)
6. D.M., D'Addona, S.J. Raykar, M.M. Narke, Procedia CIRP **62**, 269 (2017)
7. D. Przestacki, T. Chwalczuk, MATEC Web of Conferences, **136** (2017)
8. D. Przestacki, M. Kukliński, A. Bartkowska, International Journal of Advanced Manufacturing Technology 93, 3111 (2017)
9. D. Przestacki, T. Chwalczuk, S. Wojciechowski, International Jiurnal of Advanced Manufacturing Technology **91**, 3887 (2017)
10. S. Wojciechowski, D. Przestacki, T. Chwalczuk, MATEC Web of Conferences, **136** (2017)
11. D. Przestacki, M. Jankowiak, Journal of Physics: Conference Series **483** (2014)
12. M. Anderson, R. Patwa, Y.C. Shin, International Journal of Machine Tools and Manufacture **46**, 1879 (2006)
13. T. Chwalczuk, M. Rybicki, D. Korzeniewski, D. Przestacki, Mechanik **10**, 1312 (2016)

Study on Stream Cooperation Connected Smart Manufacturing in Textile Timely Production System

Jae-Yong Lee[1,a], Dae-Hee Lee[1], Jae-Sung Song[2], Jun-Hee Park[3], Sung-Hun Ryu[4], and Jeung-Hyun Park[5]

[1]Korea Textile Machinery Convergence Research Institute, Mechatronics Team, 38542 Gyeongbuk, South Korea
[2]TNS Inc., CEO, 41026 Daegu, South Korea
[3]Korea Apparel Intelligence Technology Institute, 13217 Gyeonggi, South Korea
[4]YoungWoo Cnl Inc., 08380 Seoul, South Korea
[5]Bogang Engineering Inc., 42709 Daegu, South Korea

Abstract. In general, the flow of detailed data through the physical sensing of the mechanical system and the analysis of the production process is important as well as the development of the CPS(cyber-physical system) system in smartizing the production plant. In this study, stream is composed of three main fields in order to make fabric factory as smart factory. The three streams are divided into a knitting process, a dyeing process, and a sewing process. It can be made into a system that can monitor and manage the fabric production process as an integrated system by connecting the three separated streams. Physical sensors are developed and applied to the textile machines used in the three streams, and software programs for collecting and delivering data was performed together to construct a system for data collection per stream.

1 Introduction

The world textile market size is 73.8 million tons in 2010, and the industrial scale is composed of clothing > industrial > life style. In order to accommodate the marketing strategies of the apparel companies and ensure the sustainability of the order receipt in such market conditions, it is necessary to construct a flexible production system that enables immediate production according to orders and a production system for collaboration between streams to cope with small orders and large orders. The major production methods of apparel industry in each country are changing from finished production business to CMT(cut, make, trim). Therefore, it is necessary to cooperate with each other in the stream because the fabric producers require accurate delivery time and quality. As apparel brands increase inventory control, reordering, small quantity, mass production, timely delivery, and planning management, programming development has been carried out in connection with the development of software for linkage between planning and design work of apparel brands. In this study, it is required to develop a smart manufacturing system that can exchange ICT-based inter-stream production formation because it requires a high-productivity collaboration system for timely production/delivery and flexible production of hit products.[1-2]

[a] Corresponding author : jaeyonglee@hanmail.net

2 CPS in textile factory

Detailed functional design of equipment and system for the development of high productivity cyber-physical system(CPS) for smart factory, and function concept setting per unit system wear carried out.[3-4]

2.1 CPS system configuration

2.1.1 System Architecture

The design of cyber-physical system architecture for the development of a smart manufacturing system is shown in Fig. 1.

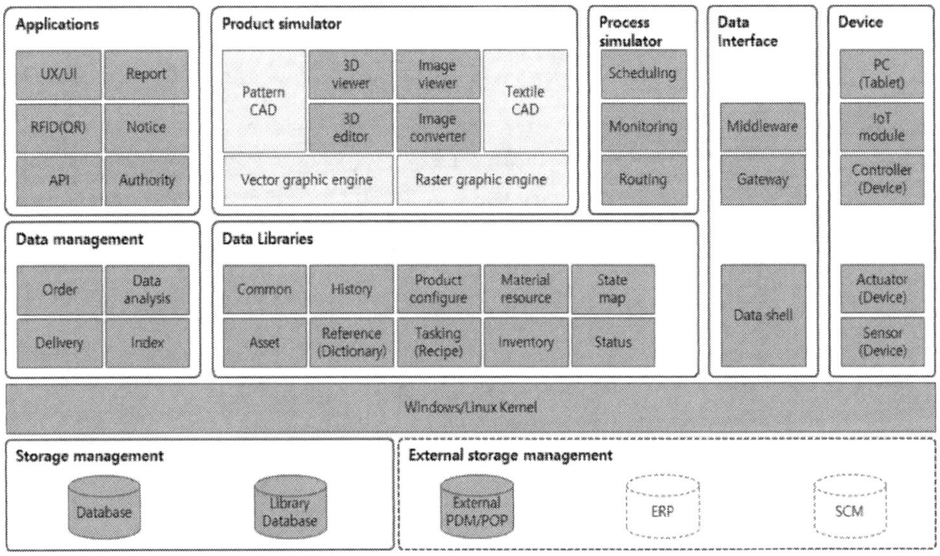

Figure 1. Architecture of CPS system configuration.

2.1.2 Process Simulator-modeling

The process of a smart factory is defined as a typical discrete event system (DES) because it consists of a number of resources (equipment, workers, work) and independent events (unit operations). Therefore, modelling is performed using Petri Net, which provides flexibility in finite resource modelling and model-based analytic performance evaluation among DES modelling tools.[5] Figure 2 shows the configuration sewing factory DES simulation.

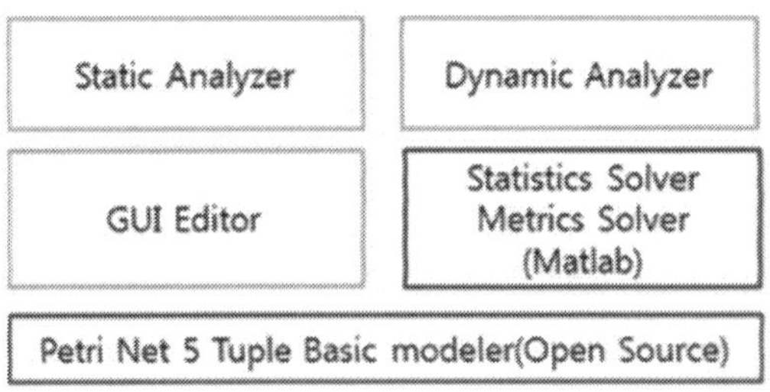

Figure 2. Sewing Factory DES Simulator.

2.1.3 Process Simulator-data packet

The data structure of process simulator is made through process modelling using Petri Net model. The data packet consists of 4M (Man, Machine, Material, Method) data of individual process for management of manufacturing process. Figure 3 shows 2-way channel information configuration for CPS-based production management.

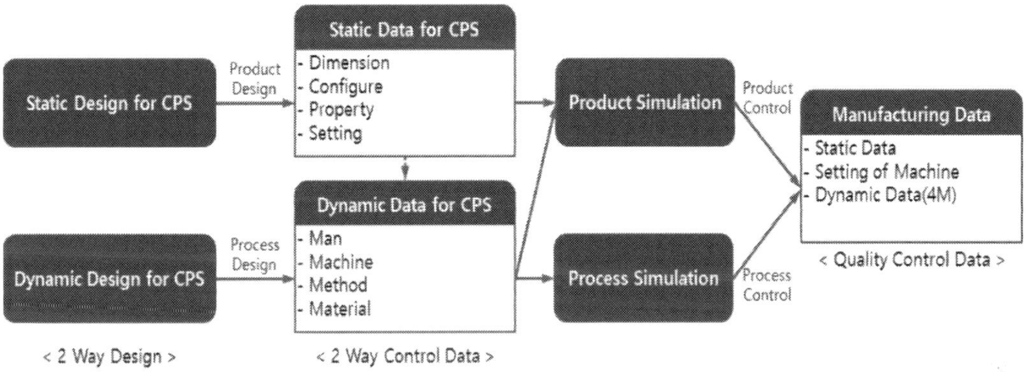

Figure 3. 2-way channel information configuration for CPS-based production management.

2.2 Knitting / Dyeing / Apparel Stream

The knitting / dyeing / apparel stream defines the sensor information detected in each machine, the respective machine status information, and the related information of the production fabric in association with the PDM through each POP terminal. Figure 4 shows three streams configuration of the connection through each configuration and data transfer.

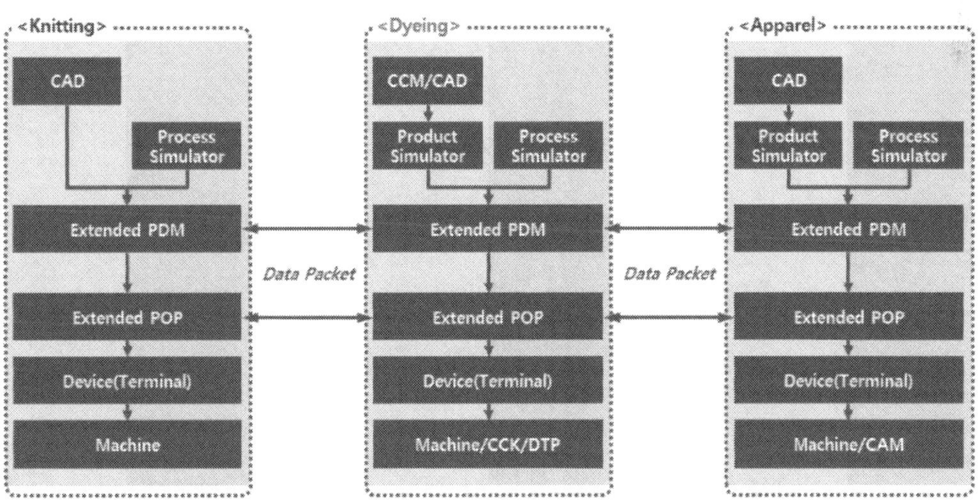

Figure 4. Three streams configuration of the connection through each configuration and data transfer.

3 Machine sensor data transferring and software programming

The smart process system was constructed by hardware parts sensor board and module development, communication and monitoring software programming of knitting machine in knitting stream. In dyeing stream, a software program for data collection was developed according to the dyeing process and a system for smart monitoring was constructed. In order to make the sewing machine smart in the apparel stream, the monitoring program and the system were constructed by the sensor development and application.[6]

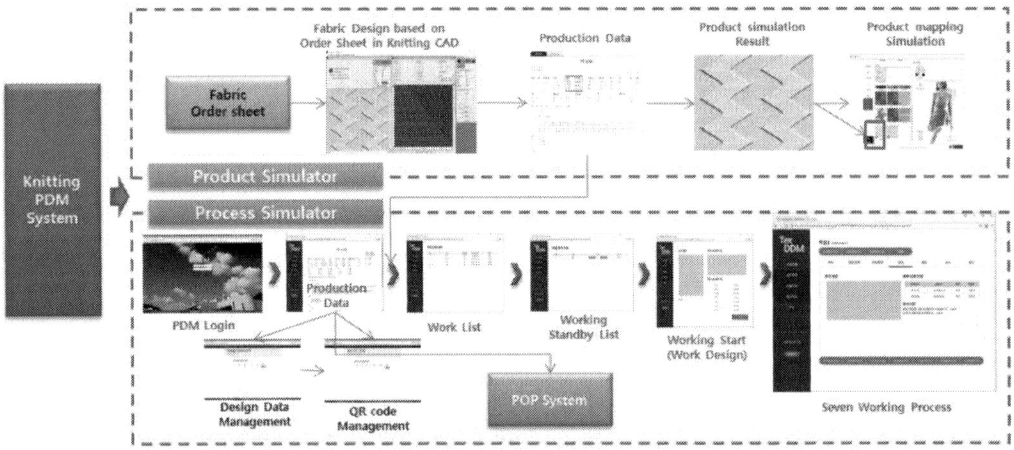

Figure 5. PDM function design and development in knitting stream.

Figure 5 shows the development of simulation technology using saved information after creating and storing knitting information (Meta Data) such as yarn information, bar information, and machine information based on seven bars in the knitting system.

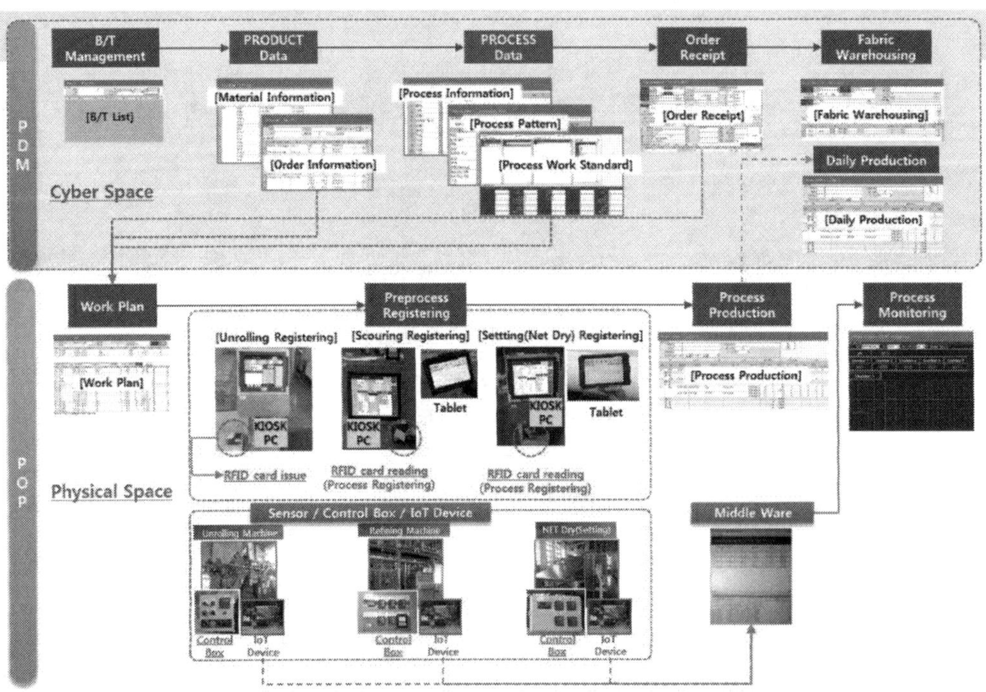

Figure 6. PDM and POP system design and development in dyeing stream.

Figure 6 shows PDM and POP programming and system development to implement CPS in dyeing stream. In order to collect work performance information (process terminal/sensing) in the preprocessing process, the system collects actual data and plans for collecting work performance information by process through the POP system using the terminal, RFID card, and sensing data linking method in the field.

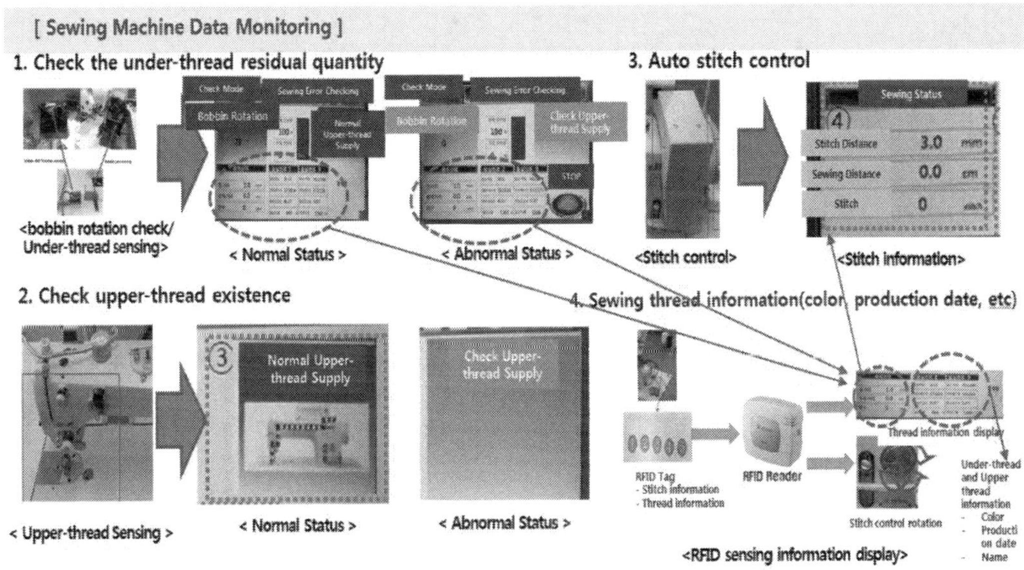

Figure 7. Sewing machine data monitoring and sensing system development in apparel stream.

In Figure 7, sensors such as laser sensor proximity sensor, and encoder were applied to collect mechanical information in the sewing machine to check the bobbin rotation state, under-thread volume, stitch number control and upper-thread supply status. The collected data is displayed on the monitor and transferred to the upper level POP and PDM to form an apparel stream.

4 Conclusions

In the textile industry, information integration and transmission among multiples streams is an important factor for smart factory applications. In addition, the current state of each machine and the process status of each stream should be applied to sensors and software programmed to collect data. In this study, we developed a system to apply smart factory in each stream by classifying streams into knitting, dyeing, and apparel fields. The smooth flow of data between streams is likely to transform the textile industry into a more innovative production environment.

The next study requires the development of a system that can manage all the streams from the apparel order to the finished product by monitoring each stream in the entire system by grouping the streams together.

Acknowledgement

This work was supported by the Ministry of Trade Industry and Energy (MOTIE). The project code is 10063057: Development of Textile Stream Connected Smart Manufacturing Innovation Technology by Timely Productive and Supplying for High Productive Textiles.

References

1. D. Lucke, C. Constantinescu and E. Westkamper, *Smart Factory – A Step towards the Next Generation of Manufacturing*, 115-118 (2008)
2. Rer.pol, Dipl.-Ing. Meike Tilebein, *The new revolution in textiles and fashion manufacturing : Industry 4.0 and its implications*, Creating a Smart Europe (2016)
3. L. Monostori, et al., *Cyber-Physical systems in manufacturing*, CIRP Annuals-Manufacturing Technology, **65**, 621-641 (2016)
4. Dan Catalin, et al., *Monitoring and Control Techniques of the installations on smart factory*, Tehnomus, 364-368 (2013)
5. S.Y. Moon, et al., *Simulation Modeling of Sewing Process for Evaluation of Production Schedule in Smart Factory*, ICIMSA (2016)
6. J.Y. Lee, et al., *Study on sensing and monitoring of sewing machine for textile stream smart manufacturing innovation,* MMVP, 295-297 (2017)

The Fabrication of Hierarchical Carbon Nanotube/carbon fibre/polyethylene Composites via Twin Screw Extruder

Chao Hu[1], Xinwen Liao[1], Qinghua Qin[1,a] and Gang Wang[2]

[1]*Research Schoole of Engineering, The Australian National University, Canberra, Australia*
[2]*Institute of Materials Science, Shanghai University, Shanghai, China*

Abstract: In this research, the hierarchical carbon nanotube/carbon fibre/polyethylene (CNT/CF/PE) composites were fabricated via the conventional twin-screw extrusion technique. For this multi-component composite, 1% wt CNTs were uniformly coated onto the surface of CF by using spray coating method. The effect of CNTs and CFs as reinforcing fillers on the mechanical properties was investigated through mechanical test and Scanning Electron Microscopy (SEM) characterization. It has been found that with the content increase of CFs (i.e. 0% wt, 5% wt, 10% wt, 15% wt, 20% wt, 25% wt), both of tensile strength and tensile modulus exhibited an upward trend. Moreover, CF/PE composites with CNT coatings have higher mechanical performance than the counterparts without CNT coatings. The SEM results demonstrated that with the spray coating of CNTs onto CFs, the surface roughness of CFs was increased, thus contributing to the improvement of interfacial bonding between the reinforcement (CFs) and the matrix (PE).

1 Introduction

The main advantages of using high performance fibres as reinforcement in composite manufacturing encompass higher tensile modulus, higher strength and stiffness, and lower density. However, apart from intrinsic properties of the reinforcement and the matrix, the entire performance of the resultant composite is also, to a large extent, modulated by the interfacial properties between these two phases. Excellent interfacial bonding contributes to the efficient load transfer from the matrix to the reinforcement, which largely avoids the occurrence of stress concentrations and enhances the overall mechanical performance[1]. Moreover, the interfacial strength of the fibre-matrix entities plays a significant role in fracture toughness, compressive strength and strain to failure, fatigue resistance, impact strength and damage initiation threshold[2].

With respect to the in-plane properties of laminated composite materials, the fibre reinforcement can provide sufficient strength for many practical applications, while for the out-of-plane (i.e. through thickness) properties, the matrix usually becomes the main restraining factor for the holistic enhancement of fibre-reinforced laminated composites[3]. Existing techniques such as z-pinning, stitching and braiding have been applied for the interfacial improvement[4]. However, these methods not only increase the additional manufacturing complexity and cost, but simultaneously compromise the in-plane strength. Therefore, fibre or matrix modifications via introducing nanoscale additives

[a] Corresponding author : qinghua.qin@anu.edu.au

open up a new avenue for the improvement of interfacial properties without increasing the overall volume of the composite. Numerous researches have been carried out in recent years in developing nano-engineered hierarchical composite materials where nanoscale additives (e.g. alumina, titania and silica) are finely integrated with conventional fibres[5-7].

Of all these nanofillers, the exploitation of carbon nanotubes (CNTs)—seamless cylinders of one or more layers of graphene with open or closed ends—is undoubtedly a big current interest in creating hierarchically reinforced composite structures[8]. Based on their unique spatial structures and remarkable strength and modulus, substantial improvements for the fibre-matrix interfacial strength and matrix dominated through-thickness performance have been achieved in glass fibre and carbon fibre reinforced composites, such as impact strength, fracture toughness, interfacial shear strength and fatigue life[9, 10]. In general, CNTs engineered hierarchical composites can be fabricated via two approaches: incorporation of CNTs into the matrix or direct adherence of CNTs onto the reinforcing fibre surface[11]. The first approach is normally limited to low concentrations of the nanofillers. With the loading increase of CNTs, the matrix will become increasingly viscose, leading to the difficulty of efficient dispersion of CNTs into the resin and sufficient filtration of the matrix into fabric reinforcement[12].Another drawback existing in the addition of CNTs into the matrix is the limited capacity of interfacial bonding between the fibre surface and the nanoscale reinforcement. Thus, more researchers have shifted the related study to directly incorporate CNTs into fibre tows or fabrics. Four different strategies can be identified for the direct integration with reinforcement phase: (1) epitaxial growth of CNTs from the fibre surface via chemical vapor deposition (CVD); (2) electrophoretic deposition of CNTs on the fibre surface; (3) chemical bonding between group-modified CNTs and fibres; (4) spray coating of fibres and fabrics by CNTs contained solutions[13-16].

However, some of these techniques have internal disadvantages for the CNTs deposition onto fibres and fabrics. Due to the high processing temperature and the reaction of the catalyst with fibre surface in the CVD process, both of the existing sizings around fibres during their manufacture and the tensile strength of the resultant composite are damaged, let alone the less flexibility in tuning the adhesion between CNTs and fibres[17]. The operational principle underlying the electrophoresis deposition technique also necessitates the possession of electric conductivity for fabric substrates, which to a certain extent limits the application of this method[18]. Furthermore, the complicated steps and cost involved in some of these techniques largely decrease the productivity and their potential for scale-up. Compared to these methods, the CNTs deposition by coating approach shows a great promise because of its easy operation and no harm to fibres (even able to heal surface defects and improve tensile properties)[19, 20]. It also does not need complex chemical modifications for pristine CNTs during the coating process, which are usually the prerequisite for other deposition techniques. By spray coating method, CNTs can be localized onto damage prone areas rather than deposited throughout the entire composite structure, enabling the fabrication of nano-engineered hierarchical composite systems with tunable CNTs distribution for improved material performance[21]. Consequently, the spray coating technology for CNTs deposition onto fibrous reinforcement has the most feasibility for large-scale industrial production.

In this research, carbon nanotubes with remarkable mechanical, electrical and thermal properties will be employed to create a hierarchical CNT-based high performance carbon fibre-reinforced composite. Since no report has been published concerning the effects of CNT coating on the extrusion of carbon fibre reinforced composites, the spray coating method for CNTs deposition onto carbon fibres will be investigated as well as the influence of the twin-screw extrusion on the resultant interfacial properties. The final nano-engineered composites will also be estimated via mechanical and morphological characterizations.

2 Experimental

2.1 Materials

The HDPE pellets used in this research were purchased from Qenos (Melbourne, Australia) with the tradename of Alkatane GE 7660. The mass density and average pellet size were 0.959 g/cm^3 and 4.6 mm, respectively. The reinforcer introduced into the composite was chopped polyacrylonitrile-derived CF from ACP Composites (Livermore, USA). Its average length was 3 mm. The commercial functionalized MWCNTs with hydroxyl groups with average diameter of 20 nm were manufactured by chemical vapor deposition technique.

2.2 Spray coating

Before spray coating, 1 mg/ml CNT/ethanol suspension was prepared by using an ultrasonic water bath (WUA-A09H, Thermoline Scientific, Austalia) to eliminate the agglomeration of CNTs. The ultrasonication was experienced for 10 min with the frequency of 35 Hz. Due to the light weight of chopped CFs, they were soaked with ethanol in advance to prevent them blowing during the spray coating process. After that, the ultrasonic treated CNT/ethanol suspension was loaded and spray deposited onto the surface of CFs with the 1% concentration by utilizing an air spray gun (Voilamart, Australia). The pressure was set at 20 psi with the flow rate of 0.2 ml/s from a 10 cm working distance. The CNT-coated CFs were dried for 24 h prior to the commencement of composite fabrication.

2.3 Extrusion and injection molding

The fillers (i.e. uncoated CFs or CNT-coated CFs) and matrix (i.e. HDPE) with varying ratios were fed into a laboratory scale miniature twin-screw extruder (WLG10, China), where the screw speed and barrel temperature were set at 60 rpm and 200 °C, respectively. The mixtures were compounded in the extruder for 5 mins and then transferred to the injection molding machine (WZS10D, China). During the injection molding process, the pressure was maintained at 980 bars for 10 s. The cylinder temperature was set at 200 °C and the mold temperature was 60 °C. Samples with standard dumbbell shape were fabricated for the tensile test. For all these samples, they can be categorized into two groups:

1. PE, 5% CF/PE, 10% CF/PE, 15% CF/PE, 20% CF/PE and 25% CF/PE
2. PE, 5% CNT/CF/PE, 10% CNT/CF/PE, 15% CNT/CF/PE, 20% CNT/CF/PE and 25% CNT/CF/PE

2.4 Tensile properties

According to the GB/T 1447 testing standard, tensile tests were performed on a universal testing machine (MTS CMT 5205, China) equipped with a load cell of 100 kN at a constant crosshead speed of 5 mm/min. At least 3 parallel samples for each composition were recorded for the following tensile strength and modulus analysis.

2.5 Scanning electron microscopy (SEM)

The surface morphology of uncoated CFs, CNT-coated CFs and the facture morphology of composites were observed by employing a scanning electron microscope (SEM, SU-1510, HITACHI, Japan) with the acceleration voltage of 15 kV. All samples were sputter-coated with gold before analysis so as to increase the electric conductivity and SEM resolution.

3 Results and discussions

3.1 Tensile properties of the composites

Table 1 listed the mechanical properties of the PE reinforced with various weight percentages of uncoated CFs and CNTs-coated CFs. In order to investigate the mechanical performance of the CNT-coated CF/PE composite and untreated CF/PE, pure PE was also tested as a control. The composite tensile strength increased along with growing amount of CFs and finally reached at the maximum value for 25 wt. % CFs content. Compared to the pure PE, the tensile strength had improved by 94.95% and 75.37% when it was reinforced with CNTs-coated and uncoated CNTs respectively. Meanwhile, for the same percentage content of CFs, CF/PE composites with CNT coatings had higher tensile strength than the counterparts without CNT coatings. Likewise, with the increase of CFs content, tensile modulus of the uncoated CF/PE composite increased constantly and reached the maximum value at 25 wt. % CNTs-coated CF content. Compared to the pure PE, the tensile modulus of PE composite with 25 wt. % uncoated CFs and CNTs-coated CFs had been improved by 110.11% and 1.15‰ respectively. Meanwhile, for the same content of CFs, the tensile modulus of CNTs-coated composites had higher tensile modulus than uncoated counterparts. Conversely, the elongation at break point corresponding to the maximum displacement decreased with the increase of CF content, which indicated that the composites became increasingly brittle with the content increase of CFs. Besides, the comparison of elongation at break point between CF/PE and CNT/CF/PE composites demonstrated that CF/PE with CNT-coating was more brittle than that of without CNT-coating.

Table 1. Tensile properties of pure PE and composites

Specimen	**Tensile Properties**		
	Tensile Strength (MPa)	**Tensile Modulus (MPa)**	**Elongation at break (%)**
PE	30.445	278.621	52.231%
5% CF/PE	31.023	398.963	51.175%
5% CNT/CF/PE	33.264	417.297	45.024%
10% CF/PE	36.215	503.415	32.421%
10% CNT/CF/PE	39.551	536.016	28.278%
15% CF/PE	42.177	540.048	19.270%
15% CNT/CF/PE	45.122	578.612	15.286%
20% CF/PE	47.495	550.552	14.667%
20% CNT/CF/PE	52.753	2723.125	5.871%
25% CF/PE	53.391	585.408	12.890%
1.25% CNT/CF/PE	59.354	3486.111	4.688%

The CNT-coated CF reinforced PE composite exhibited significant enhancement in tensile properties, which was higher than the uncoated counterparts. This is because the interface bonding between CFs and PEs was enhanced by the spray coating surface treatment on the CF, so that load can be better transferred from PE to CF, allowing composite materials to withstand greater loads. On the other hand, due to the significant mechanical strength of CNTs, it is generally believed that the incorporation of CNTs mixed into polymer matrix will result in a great improvement in strength. Further evidence for enhancement of CNT-coated CF interfacial adhesion will be presented in SEM studies.

3.2 Morphology of the composites

The morphological characterization of CNT/CF/PE composites was investigated through SEM imaging technique in this study. The information on orientation, distribution, surface conditions of CFs and CNTs as well as the interfacial adhesion between the PE matrix and reinforcements were correspondingly recorded. Figure 1 (a) and (b) shows the morphologies of CFs and CNTs at the magnification of 5KX and 100KX respectively. The surface of CF was extremely smooth, which indicated particularly poor interfacial adhesion between polymer matrix and CFs, and thus resulted in low mechanical performance of CF reinforced PE composites consequently. Therefore, CNTs has been widely utilized in various surface treatments such as spray coating, electrophoretic deposition, chemical grafting, containing sizing etc. to enhance the surface roughness of CFs.

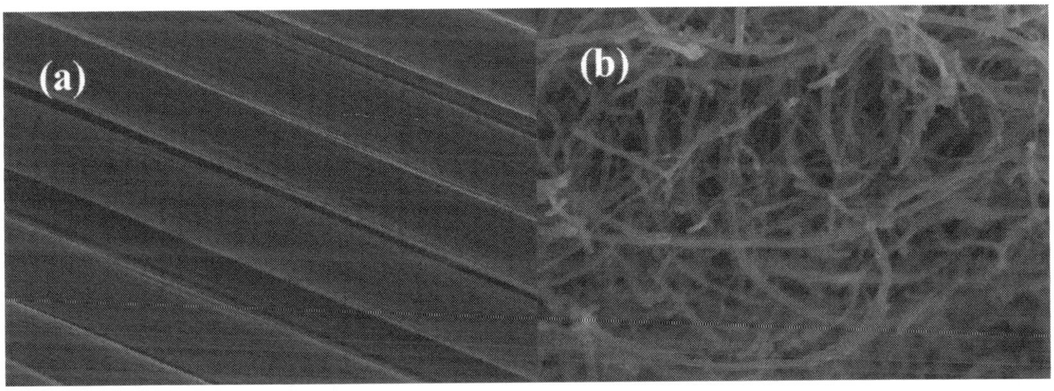

Figure 1. The morphological characterization of (a) CFs and (b) CNTs

In this study, spray coating surface treatment was applied to optimize the surface condition of CFs. Figure 2 (a) exhibited the morphology for CFs after coating with CNTs at the magnification of 100KX. It was obvious to see that the initial smooth homogeneous surface of CF became rougher after the non-uniformly CNTs layer wrapped onto the CF surface and thus CNFs coating on the CF surface was ascertained. The fracture surface of CNTs-coated CF/PE composites were also examined which was shown in Figure 2 (b). As it can be seen, with the roughness increase of CF surface after CNT coating, the PE matrix was attached onto CFs, eliminating the voids normally existing between the matrix and reinforcement. It contributes to the enhancement of interfacial bonding between these two phases, which was verified by the tensile test.

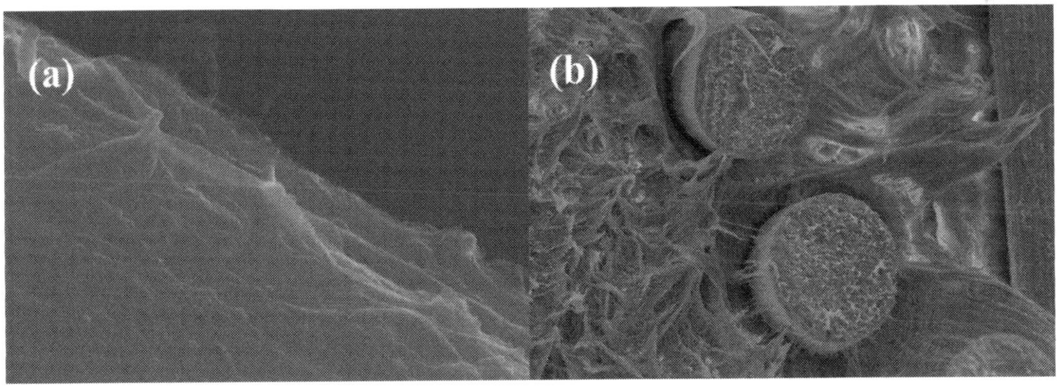

Figure 2. The morphological characterization of (a) CF surface after CNT coating and (b) fractural surface of CNT/CF/PE composites

4 Conclusions

In this research, the hierarchical CNT/CF/PE composites were first fabricated by combining the traditional spray coating method with twin-screw extrusion technique. Two groups of composites with or without CNT coatings were compared in terms of mechanical and morphological properties. The results showed that the tensile strength and modulus exhibited an upward tendency for these two groups with the rise of CF content. In particular, the incorporation of CNTs into CF/PE composites contributes to a greater improvement in tensile properties, indicating the enhancement of interfacial adhesion between the reinforcement and matrix. As the most promising technique for future scale-up production, the integration of spray coating with conventional extrusion method will pave the way for the following large-scale industrialization.

References

1. An, Q., et al., ACS Appl. Mat. Interfaces **5**(6): 2022-2032.(2013)
2. Godara, A., et al., Compos. Sci. Technol. , **70**(9): 1346-1352.(2010)
3. Chandrasekaran, V., et al., Carbon, **48**(13): 3692-3699.(2010)
4. Fan, Z., et al., Composites Part A **39**(3): 540-554.(2008)
5. Chang, L. and Z. Zhang, Wear, **260**(7): 869-878.(2006)
6. Zhao, S., et al., Compos. Sci. Technol. , **68**(14): 2965-2975.(2008)
7. Chen, C. and A. B. Morgan, Polymer, **50**(26): 6265-6273.(2009)
8. De Volder, M. F., et al., science, **339**(6119): 535-539.(2013)
9. An, F., et al., Mater. Des. , **33**: 197-202.(2012)
10. Li, M., et al., Carbon, **52**: 109-121.(2013)
11. Chou, T.-W., et al., Compos. Sci. Technol. , **70**(1): 1-19.(2010)
12. Kim, M., et al., Compos. Sci. Technol. , **69**(3): 335-342.(2009)
13. Díez-Pascual, A. M., et al., Curr. Opin. Solid State Mater. Sci. , **18**(2): 62-80.(2014)
14. Alsawat, M., et al., Carbon, **106**: 295-305.(2016)
15. Rider, A. N., et al., MRS Advances, **1**(12): 785-790.(2016)
16. Withers, G., et al., J. Compos. Mater. , **50**(6): 783-793.(2016)
17. Zhang, Q., et al., Compos. Sci. Technol. , **69**(5): 594-601.(2009)
18. Lee, S.-B., et al., Composites Part A **42**(4): 337-344.(2011)
19. Gao, S., et al., Acta Mater. , **55**(3): 1043-1052.(2007)
20. Siddiqui, N. A., et al., Composites Part A **41**(4): 539-548.(2010)
21. Zhang, H., et al., Composites Part A **70**: 102-110.(2015)

Parametric Study on Vibration and Harmonic Analysis of Moderately Thick Functionally Graded Plates Using FEM

Avadesh K. Sharma[1,a], M K Gaur[2] and R K Dwivedi[3]

[1]Associate Professor, Rajkiya Engineering College, Mainpuri, 205119
[2]Professor, Madhav Institute of Technology & Science, Gwalior
[3]Associate Professor, Maulana Azad National Institute of Technology, Bhopal

Abstract. Finite element method is used to investigate the free vibration and harmonic analysis of functionally graded plates. The material properties of the plates are assumed to vary continuously through their thickness direction according to a power-law distribution of the volume fractions of the plate constituents. The four noded shell 181 elements are used to analyse the functionally graded plates. The aim is to fill the void in the available literature with respect to the free vibration results of Functionally Graded plates. Convergence and Comparison studies with respect to the number of nodes has been carried out using FEM. The natural frequency, mode shape and harmonic analysis of FG plate has been determined using finite element package ANSYS.

1. Introduction

Composite materials have been widely used in aircraft and other engineering applications for many years because of their excellent strength-to-weight and stiffness-to-weight ratios. Functionally graded materials (FGM) are a class of composite materials that were first proposed by Bever and Duwez [1] in 1972. In a typical FGM plate the material properties continuously vary over the thickness direction by mixing two different materials [2], usually ceramic and metal. The gradual variation of properties avoids the delamination failure that are common in laminated composites. The computational modelling of FGM is an important tool to the understanding of the structures behavior, and has been the target of intense research, from micro to macro mechanics [3–6]. Researchers have also turned their attention to the vibration and dynamic response of functionally graded structures [7–9]. Sheng and Wang [10] investigated the effect of thermal load on vibration, buckling and dynamic stability of functionally graded cylindrical shells embedded in an elastic medium.A review of the main developments in FGM can be found in Birman and Byrd [11]. Sharma et. al. [12] presented the free vibration of shear-deformable antisymmetric angle-ply laminated rectangular plates having translational as well as rotational edge constraints. The Vibration attenuation using functionally graded material is studied by Saeed et. al. [13].

The vibration and harmonic responses of composite plates and FGMs have been extensively studied by a number of researchers [14-16]. However, the aim of the work is to study the effect of functionally graded materials on the vibration behaviour by doing modal and harmonic analysis of the models.

[a] Corresponding author : sharma.avadesh@gmail.com

2. Functionally Graded Material Properties

A functionally graded material plate as shown in fig. 1 is considered to be a plate of uniform thickness that is made of ceramic and metal. The material property is assumed to be graded through the thickness according to a Power-Law distribution that is

$$P(z) = (P_c - P_m)V_m + P_m \quad (1)$$

where *Pm*, *Pc*, *Vm* and *Vc* are the material properties and the volume fraction of the metal and ceramic, respectively, the compositions represent in relation to

$$V_c + V_m = 1 \quad (2)$$

The volume fraction of ceramic (V_C) can then be written as follows:

$$V_C = \left(\frac{z}{h} + \frac{1}{2}\right)^n \qquad (n \geq 0) \quad (3)$$

where the positive number n $(0 \leq n \leq \infty)$ is the power law or the volume fraction index. z is a distance parameter along the graded direction, while, h is the total length of the direction. To find out the results of material properties according to the power law distribution, this can be achieved by substituting the equations of material volume fractions Eq. (2) and Eq. (3) into Eq. (1).

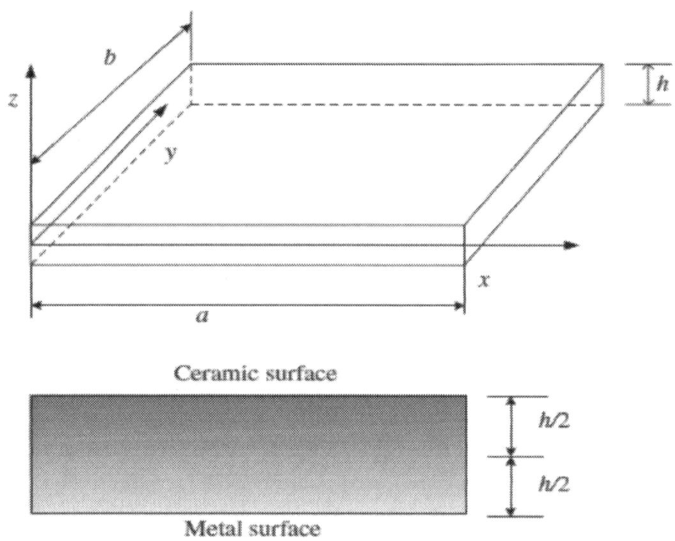

Figure 1. Geometry of Functionally graded plate

3. Results and Discussion

3.1 Modal Analysis

The variation of the frequency parameter with the boundary condition for functionally graded plates is described in Tables 1-2 respectively. It shows a comparison of the fundamental natural frequency parameters of functionally graded plates for volume fraction exponent n=0. It can be seen that the as the mesh size increases the results are converging and convergence are achived at the mesh size of 19x19. The results are compared with the Zhao et. Al. [17].

Table 1: Variation of the frequency parameter with the volume fraction exponent n =0 for square FG plates (a/h = 10) with CFFF

3x3	11.7174	22.4627	26.6912	33.8398	37.9197	38.1226	42.2459	45.4010	45.7544	51.0061
5x5	10.5809	20.7963	20.9147	29.6169	35.8369	36.0877	37.7263	37.7364	43.2174	44.1762
7x7	10.3100	19.9219	19.9483	28.2032	33.6206	33.8827	37.6205	37.6217	40.7289	40.8631
9x9	10.2299	19.7064	19.7360	27.8315	33.0467	33.3661	37.5852	37.5877	40.0541	40.1108
11x11	10.1758	19.5785	19.5823	27.6009	32.7228	33.0107	37.5606	37.5625	39.6554	39.6894
13x13	10.1512	19.5149	19.5256	27.4901	32.5559	32.8432	37.5486	37.5499	39.4632	39.4701
15x15	10.0340	19.2276	19.2295	27.0157	31.8887	32.2346	37.5341	37.5348	38.6379	38.6492
17x17	10.0006	19.1325	19.1369	26.8727	31.6978	32.0229	37.5260	37.5266	38.3941	38.3985
19x19	9.9918	19.1098	19.1111	26.8317	31.6556	31.9637	37.5215	37.5222	38.3361	38.3412
21x21	9.9830	19.0859	19.0865	26.7914	31.6021	31.9101	37.5171	37.5171	38.2631	38.2649
23x23	9.9666	19.0487	19.0518	26.7429	31.5258	31.8408	37.5089	37.5096	38.1824	38.1887
Zhao et. al [16]	9.6329	18.313	18.313	25.499						

Boundary conditions at different mesh size:

Table 2: Variation of the frequency parameter with the volume fraction exponent n =0 for square FG plates (a/h = 10) with

3x3	1.077363	2.617083	6.68115	6.6994	8.2656	9.61884	15.98373	16.70256	17.88	19.90044
5x5	1.059282	2.504943	6.292314	6.6389	7.95501	8.95734	15.30396	15.91632	17.56	17.79309
7x7	1.054494	2.489193	6.228243	6.6289	7.86429	8.80803	14.93163	15.90057	17.05	17.77923
9x9	1.0521	2.481381	6.200838	6.6238	7.83846	8.75952	14.81823	15.89175	16.91	17.68851
11x11	1.050777	2.476467	6.18786	6.6213	7.82019	8.73243	14.75334	15.88671	16.84	17.61984
13x13	1.049706	2.472498	6.17778	6.62	7.81074	8.71101	14.70483	15.88419	16.8	17.58204
15x15	1.046745	2.459205	6.142626	6.6182	7.77357	8.64045	14.55552	15.88104	16.62	17.43336
17x17	1.046052	2.457819	6.133995	6.6175	7.76412	8.62974	14.53158	15.87915	16.58	17.40375
19x19	1.045737	2.456937	6.131727	6.6169	7.76223	8.62533	14.52213	15.87789	16.57	17.39241
Zhao et. al [16]	1.0298	2.3907	6.0047	7.6356						

CCCC Boundary conditions at different mesh size:

Table 3: frequency response amplitude of FG square plate with different volume fraction exponent

B.C.	**Volume fraction exponent**	**Mode**	***Frequency (Hz)***	***Amplitude (m)***
CCCC	*n = 0*	*Mode 1*	1592.7	0.0000000146726
CCCC	*n = 1*	*Mode 1*	1324.5	0.0000000247805
CCCC	*n = 5*	*Mode 1*	858.76	0.0000000699683
CCCC	*n = 10*	*Mode 1*	811.53	0.0000000793071

3.2 Harmonic Analysis

To know FRF plot, harmonic analysis is to be done by providing the range of natural frequency of 0 Hz to 2000Hz and 100 substeps. It will generate FRF plot (linear) on graph of amplitude to frequency,

frequency (Hz) is taken on the x-axis and amplitude (m) on the y-axis. From this graph we come to know its resonance point. Also many details of system such as amount of displacement, by how much frequency by how much amount system excited. It can be said that the overall response of system can be known. It is clear from the Table 1 that as the volume fraction exponent increases the frequency response amplitude of FG plates increases.

These frequency response amplitudes can also be reduced by providing damping constant. In the study, constant damping ratio of 0.01 is taken. Effect of volume fraction exponent on the frequency response amplitude for different sets of FG plates is shown in Figure 2-5 respectively. It is clear from the figures that as the volume fraction exponent increases the frequency response amplitudes increases. Figure 6 shows the first four modes of functionality graded plates.

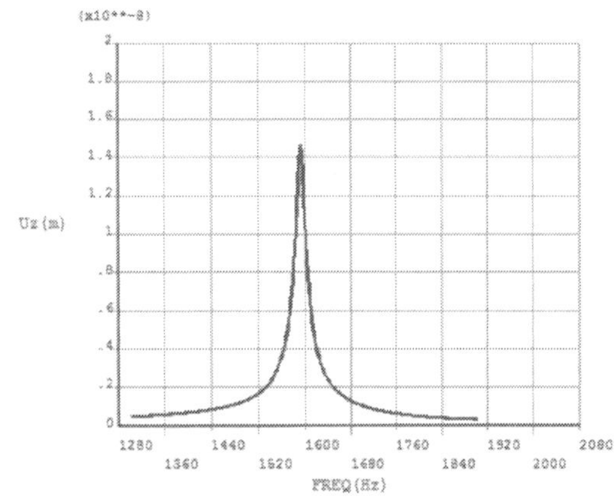

Figure 2. shows the FRF plots without damping FG plate at CCCC boundary condition at n=0.

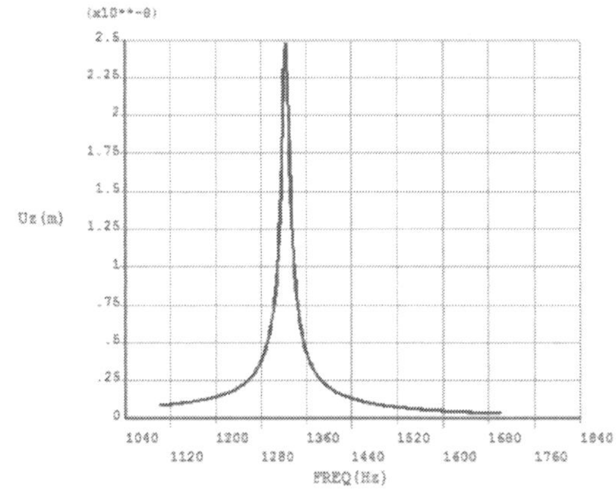

Figure 3. shows the FRF plots without damping FG plate at CCCC boundary condition at n=1.

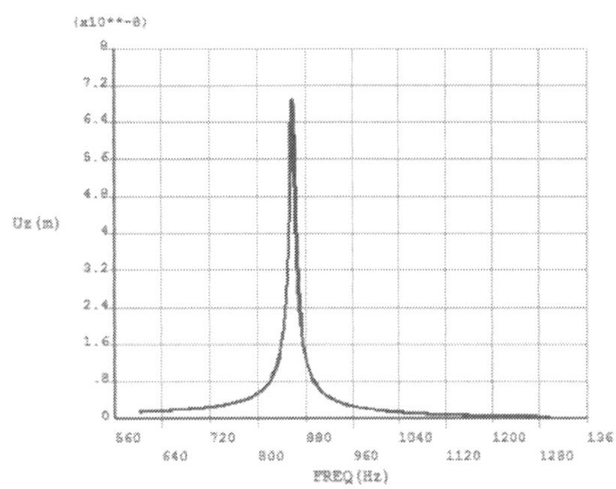

Figure 4. shows the FRF plots without damping FG plate at CCCC boundary condition at n=5.

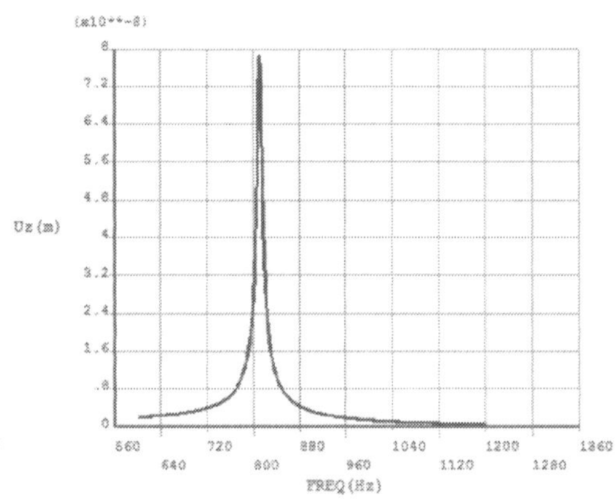

Figure 5. shows the FRF plots without damping FG plate at CCCC boundary condition at n=10.

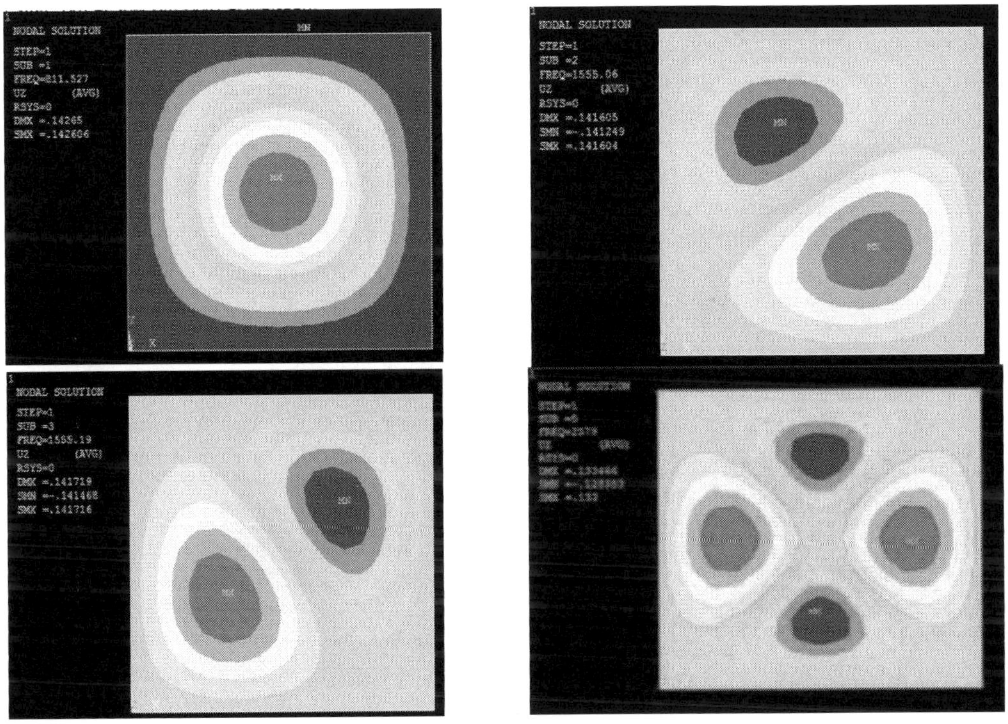

Figure 6. First four natural frequencies of functionally graded plates at volume fraction exponent n=10

4. Conclusion

In this study, vibration and harmonic analysis of FGM plates are analysed. Four functionally graded plates of different sets of materials are considered and their natural frequencies and frequency response amplitude at the fundamental mode are determined. Convergence tests and comparison studies have been carried out with the commercially available software (ANSYS). A four noded layered shell element (SHELL181) is used throughout the problem. The obtained results have illustrated a good agreement with those available in the literature for different volume fraction indices. It is concluded that as the volume fraction exponent increases the frequency response amplitude of FG plates increases and results also shows that for all the Functionally graded plates the frequencies decreases as the volume fraction exponent increases.

References

1. M.B. Bever and P.E. Duwez. Gradients in composite materials. Materials Science and Engineering, 10(0):1 – 8, 1972.
2. Y. Miyamoto, W.A. Kaysser, B.H. Rabin, A. Kawasaki, and R.G. Ford. Functionally Graded Materials: Design, Processing and Applications. Kluwer Academic Publishers, 1999.
3. F.J. Ferrante and L.L. Graham-Brady. Stochastic simulation of non-gaussian/non-stationary properties in a functionally graded plate. Computer Methods in Applied Mechanics and Engineering, 194(12-16):1675 – 1692, 2005.
4. H.M Yin, L.Z Sun, and G.H Paulino. Micromechanics-based elastic model for functionally graded materials with particle interactions. Acta Materialia, 52(12):3535 – 3543, 2004.
5. Zheng Zhong and Ertao Shang. Closed-form solutions of three-dimensional functionally graded plates. Mechanics of Advanced Materials and Structures, 15(5):355–363, 2008.

6. T. K. Nguyen, K. Sab, and G. Bonnet. Shear correction factors for functionally graded plates. Mechanics of Advanced Materials and Structures, 14(8):567–575, 2007.
7. H.L.Dai,X.Wang,Thermo-electro-elastic transient responses in piezoelectric hollow structures, International Journal of Solids and Structures 42 (2005) 1151–1171.
8. X.L.Huang, .H.Shen, Nonlinear vibration and dynamic response of functionally graded plates in thermal environments, International Journal of Solids and Structures 41 (2004)2403–2427.
9. J.Yang,S.H. Shen, Free vibration and parametric resonance of shear deformable functionally graded cylindrical panels, Journal of Sound and Vibration 261 (2003) 871–893.
10. G.G.Sheng, X. Wang, Thermal vibration, buckling and dynamic stability of functionally graded cylindrical shells embedded in an elastic medium, Journal of Reinforced Plastics and Composites 27 (2008)117–134.
11. Victor Birman and Larry W. Byrd. Modeling and analysis of functionally graded materials and structures. Applied Mechanics Reviews, 60(5):195–216, 2007.
12. Sharma A. K. & Mittal N.D., Free vibration analysis of moderately thick Anti-symmetric angle-ply laminated rectangular plates with elastic edge constraints, Mechanics of Advanced Materials and Structures, 21, 341–348, 2014.
13. Asiri Saeed, Hedia Hassan & EissaWael, Vibration attenuation using functionally graded material. World Academy of Science, Engineering and Technology, Vol 7, 2013.
14. Sharma A.K. & Mittal N.D., Review on stress and vibration analysis of composite plates. Journal of Applied Sciences, 10 (23), 3156-3166, 2010.
15. Sharma A.K. & Mittal N.D., Sharma A., Free vibration analysis of moderately thick antisymmetric cross-ply laminated rectangular plates with elastic edge constraints. International Journal of Mechanical Sciences, 53, 688 695, 2011.
16. Sharma A. K. & Mittal N. D., Free vibration analysis of laminated composite plates with elastically restrained edges using FEM. Central European Journal of Engineering. 3 (2), 306-315, 2013.
17. Zhao X., Lee Y.Y., Liew K.M., Free vibration analysis of functionally graded plates using the element-free kp-Ritz method, Journal of Sound and Vibration 319 918–939 (2009).

Study on Deformation Behavior of Ti6Al4V Parts Fabricated by Selective Laser Melting

Tao Yang[1], Tingting Liu[1], Wenhe Liao[1,a], Changdong Zhang[1], Shengqin Duan[1] and Xiangyuan Chen[1]
[1]*School of Mechanical Engineering, Nanjing University of Science and Technology, Nanjing, Jiangsu 210094,China*

Abstract. During the Selective laser melting process, the accumulation of the residual stress caused by the excessive temperature gradients and the rapid solidification is the main factor causing deformation; and the wetting quality of the meltpool also has an impact. How to control the deformation is the key problem to improve the quality of the parts. In this paper, the thin-wall specimens are divided into four distinct conditions, analyzing the meltpool date collected by the melting pool monitoring system and the simulation of temperature field, the relationship between the process parameters and the deformation under the same energy density is discussed. The results show that the temperature, the width and the intensity of the meltpool increase with the increasing of the laser power and the scanning speed, but the temperature gradients and the rate of the solidification of the meltpool are on the contrary, resulting in the decrease of the residual stress and the deformation of the part, which is help to improve the quality of the part.

1 Introduce

Titanium and associated alloys are favored by the aerospace field due to their low density, high strength, high temperature resistance and strong compatibility [1]. However, the parts are difficult cutting because of the high hardness, because it is costly to use traditional processing methods. Selective Laser Melting (SLM), a form of powder bed fusion (PBF) as defined in the standard terminology described in the ISO/ASTM 52900 standard, is a promising method that has the advantages of manufacturing flexibility, complex shape design, time and cost savings [2]. Forming the titanium alloy parts, especially the parts with complex geometric shapes, has obvious advantage. In recent years, it has received extensive attention and has a wide range of application in aerospace, biomedical and other fields [3].

However, the thermal stress is easily generated in the forming process due to the characteristics of melting fast and cooling fast in the SLM, which may cause the deformation of the parts and even cracks [4], and the defects have a significant influence on the dimensional accuracy and mechanical properties of the parts [5]. The Fraunhofer Institute of Laser Technology [6] found that the rate of temperature variation in the process has a great influence on the surface quality and p mechanical properties of the fabricated parts. Buchbinder et al. [7] and SiddiqueS et al. [8] studied that preheating the substrate may suppress cracks and reduce deformation effectively due to reduce the temperature

[a] Corresponding author : cnwho@njust.edu.cn

gradient. Karg et al. [9] found that reducing the cooling rate of the molten pool can reduce the susceptibility of Al-Cu alloys to cracks during SLM process.

The above results show that change the temperature field can control the solidification rate of the molten pool and the cooling rate of the specimen in the process Select a suitable temperature field can reduce the deformation of the specimens and improve the quality. In this paper, combining the data of the molten pool and the results of the experiment, with the help of the temperature field simulation, analyzing the influence of the process parameters on the solidification rate of the molten pool and the cooling rate of the specimen. The relationship between the solidification rate, cooling rate and the residual stress are investigated, in order to provide guidance for optimizing the process parameters and improving the quality of the titanium alloy parts fabricated by SLM.

2 Experiment method

2.1 Material and Equipment

The powder used in this experiment was gas atomized Ti6Al4V supplied by FALCONTECH (WuXi), The chemical composition is shown in Table 1. The Ti6Al4V powders were spherically shaped with a spheroidization ratio of almost 100% as seen in Fig.1(a). Granulometric parameters of the Ti6Al4V powders were evaluated by a laser particle size analyzer (Mastersize 3000), the D10, D50 and D90 of the sieved powder were 22.1 μm, 33.5μm and 50.1μm respectively as seen in Fig.1(b).

Table 1. Chemical component of Ti6Al4V powders (atomic fraction, %).

Al	V	Fe	C	O	N	H	Mo	Mn	Cu	Sn	Y	Zr	Ti
5.5~6.75	3.5~4.5	0.14	0.007	0.13	0.007	0.002	< 0.05	< 0.05	< 0.05	< 0.05	< 0.05	< 0.05	Bal

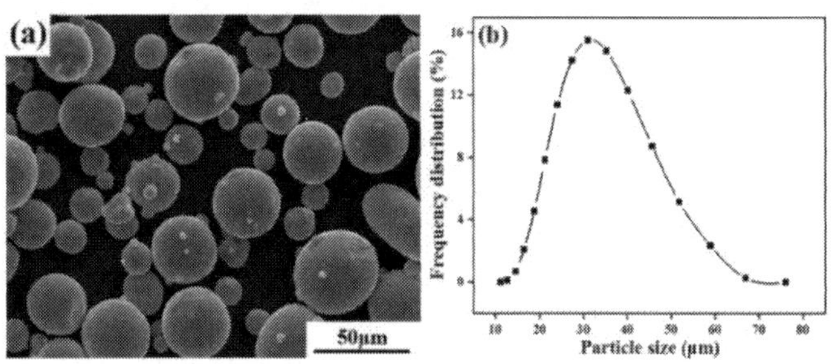

Figure 1. Typical particle morphology and particle size distribution curve of the Ti6Al4V powder: (a) SEM image of AlSi10Mg powder and (b) particle size distribution curve.

All specimens were fabricated using a Concept Laser M2. The M2 has an Yb-fiber laser with a laser power providing up to 400 W with a 60 μm laser spot size, the wave length range is 1064-1100 nm. All process were completed in an Argon atmosphere with an oxygen content below 0.1% [10].

2.2 Design of experiments

In this paper, four groups of experiments were designed to study the influence of laser power and scanning speed on the solidification time of the molten pool and the cooling time of the parts as the same linear energy density. The process parameters are shown in Table 2. The linear energy density (E) has been calculated using Eq. (1) describing the energy input per unit length during SLM, and the E is combination of effects of the SLM process parameters.

$$E = \frac{P}{V} \tag{1}$$

where P (W) denotes the laser power, V(mm/s) is the scanning speed.

Table 2. The process parameters of the specimens.

Number	Power P (W)	Velocity v (mm/s)	Scan strategy	Layer thickness, t (μm)	Linear energy E (J/cm)
1	120	1050	Single	30	114.3
2	180	1575			
3	250	2188			
4	300	2625			

In order to avoid the influence of the shape on the result, three different kinds of shape were designed such as circle, equilateral triangle and square, shown as Fig.2. The length of the square and the equilateral triangle is 10mm, and the diameter of the circular specimen is also 10mm, the wall thickness of all the specimens is 0.1mm, the height is 5mm. The specimens were scanned once and the laser scanning path is shown in Fig.3.

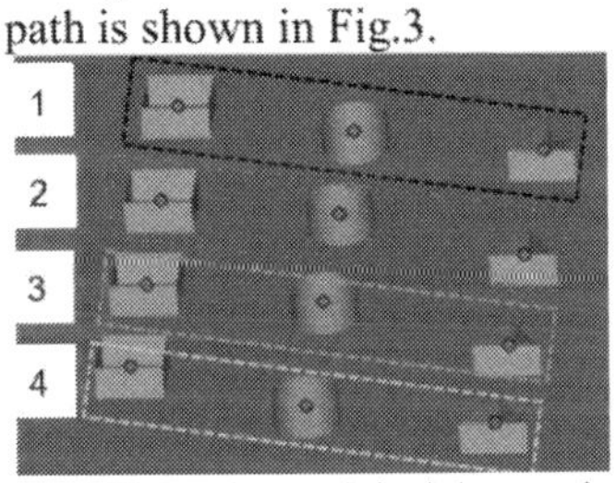

Figure 2. The model of the specimens.

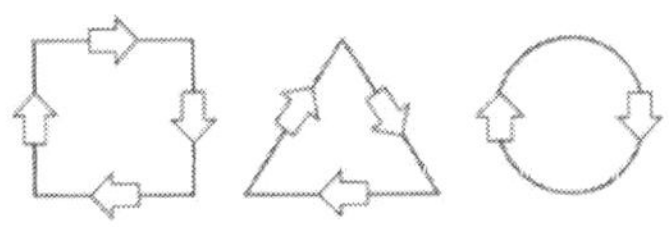

Figure 3. The scan path of the laser.

2.3 Experimental process

The date of the intensity of the molten pool and the area of the molten were collected and analyzed, the wall-thickness of the specimens were measured by using a vernier caliper, the roundness tolerance and the flatness tolerance of the specimens were measured by using a three-coordinate measuring machine (Global, Hexagon, Sweden). The simulation model shown in Fig.4 was built for the experiment to simulate the temperature field during the process. The size of the specimens is 1.6mm $\times$0.16mm$\times$0.12mm. The grid is a hexahedral grid, and the size is 0.04mm$\times$0.04mm$\times$0.04mm. The materials of the substrate and that the specimens is same, the substrate is divided into upper and lower parts. The lower part adopts a 0.4 mm larger grid and the upper part adopts a free grid to realize tetrahedral transition [11].

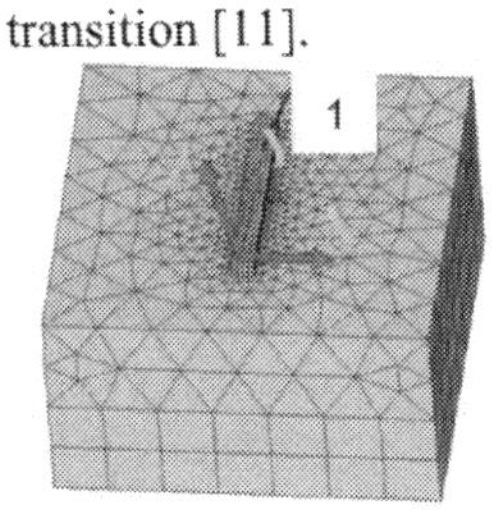

Figure 4. Finite element model.

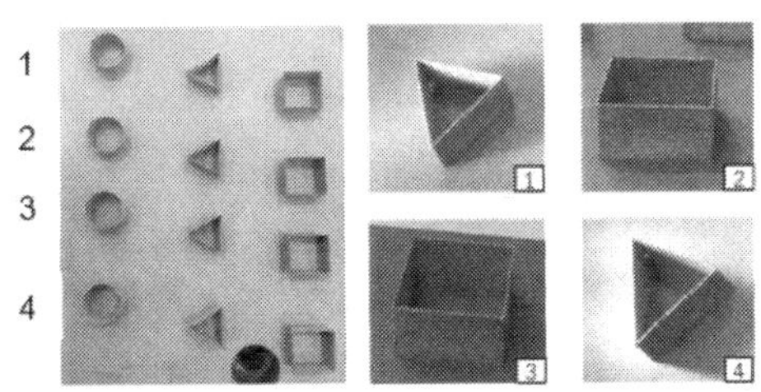

Figure 5. Fabricated specimens.

3 Results

The fabricated specimens shown as Fig.5. The square and triangular had a little deformation at the center of the wall, and the corners of the specimens were separated from the substrate. The circular specimens were of good quality and have no deformation.

3.1 The wall-thickness of the specimens

The wall-thickness of the specimens were measured at ten different positions selected randomly and averaged, the result was shown as Fig.6. The wall-thickness increases with the increasing of the laser power while the linear energy density is same. The wall-thickness value of the circular specimens is maximum and the wall-thickness value of the triangular specimens is minimum. Then simulated the temperature field of the molten pool, at the position of Y=0, the temperature curves with different process parameters are shown in Fig.7. The temperature higher than liquidus temperature is the molten pool, it shows that the highest temperature of the molten pool gradually increases as the laser power is increasing from 120W to 300W and the width of the molten pool has the same tendency.

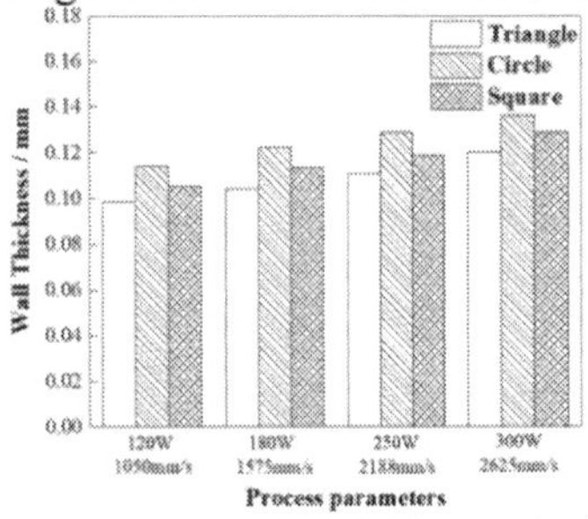

Figure 6 The average of the wall-thickness.

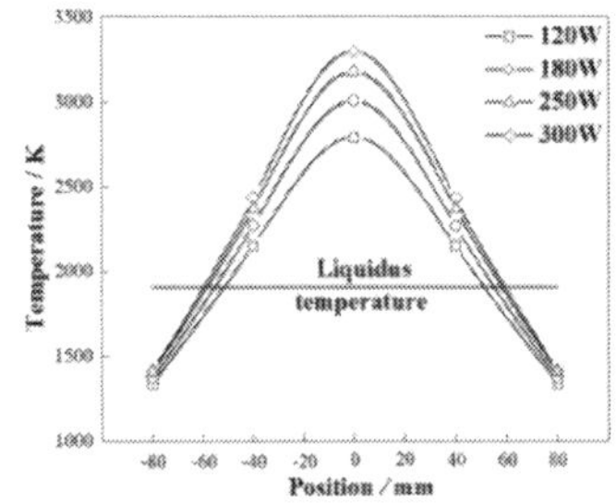

Figure 7 Temperature curves.

The data of the molten pool collected by QM-Meltpool is shown in Fig.8. As the power increasing, the area of the molten pool gradually increases shown as Fig.8(a) with then same tendency of the result of the wall-thickness (Fig.6) and the width simulation result of the molten pool (Fig.7). As the laser power increases, the temperature around the laser spot increases. There is more powder melted. The width of the molten pool and the wall-thickness of the specimens increase. When the linear energy density is same, the intensity of the molten pool increases due to the increasing of the area of the molten pool, shown as Fig.8(b).

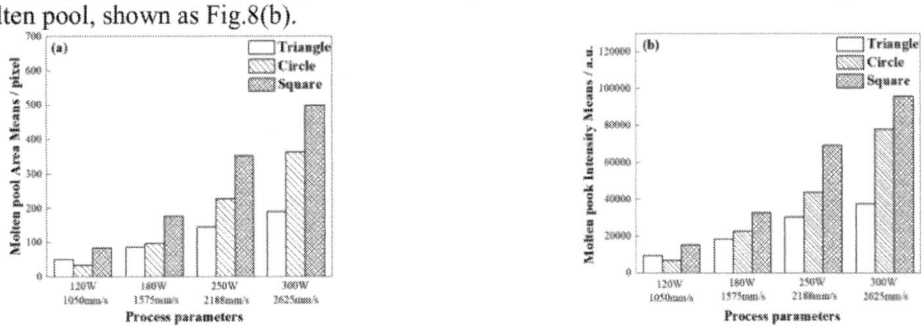

Figure 8 The data of the molten pool (a) the area of the molten pool and (b) the intensity of the molten pool.

3.2 The deformation of the specimens

The measurement results of the geometric tolerance are shown in Fig. 9. Under the condition that the online energy density is the same, the laser power increased from 120W to 300W, and the roundness average value shows a downward trend, indicating that the deformation of the test piece is reduced (Fig. 9(a)); likewise, the flatness of the triangular and square specimens also showed a downward trend (Fig. 9(b)). The above results show that under the condition of the same online energy density, the increase of laser power and scanning rate can effectively reduce the deformation degree of the test piece and improve the quality of the formed part.

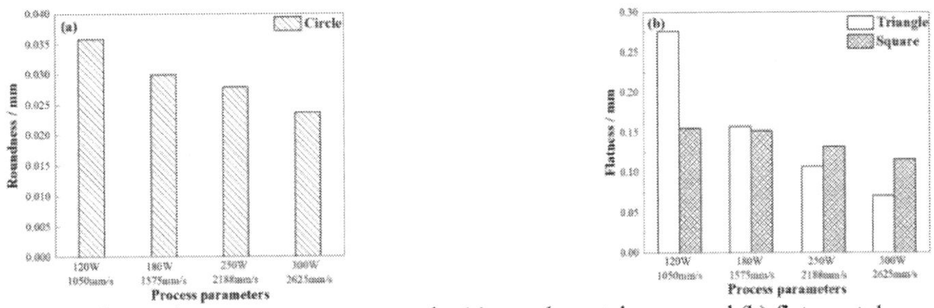

Figure 9 Shape tolerance measurement results (a) roundness tolerance and (b) flatness tolerance.

4 Discuss

The powder solidified rapidly after being melted, the wetting quality between the molten pool and the substrate, the solidification rate of the molten pool and the cooling rate of the specimens have a significant influence on the quality of the specimens, and all these influenced by the temperature of the molten pool.

4.1 The deformation of the specimens

In the SLM process, the law of the temperature changing at different position in the molten pool is same, taking a node in the second layer of this model randomly as (0,0), the temperature change process of this node with different process parameters as the laser spot scanning along the Y-axis shown as Fig.10.

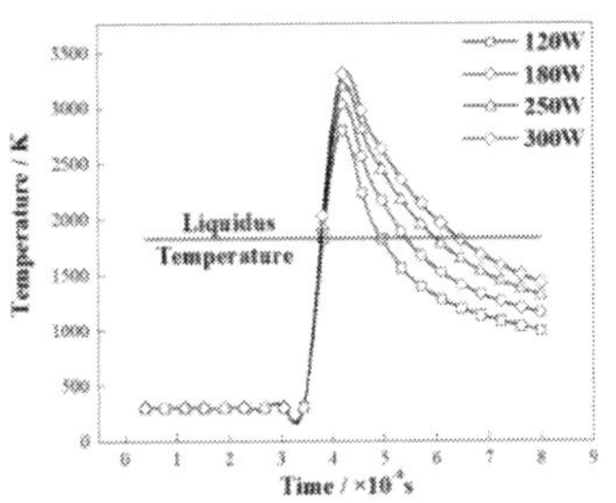

Figure 10 Temperature curves.

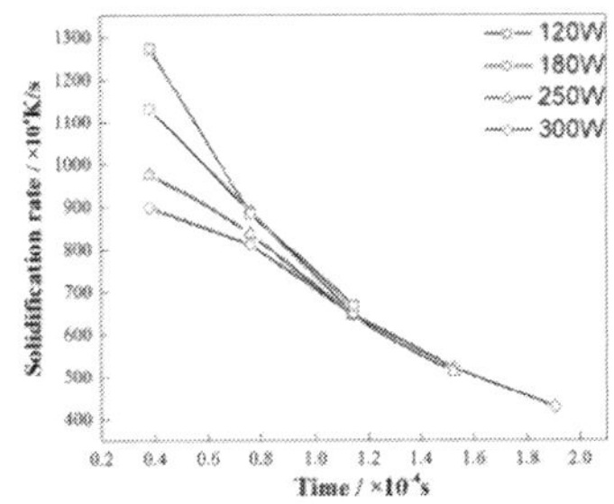

Figure 11 Simulation results about the solidification rate.

As the increasing of the laser power, the highest temperature of the (0,0) increases from the 2794 K to 3312 K as the laser power is increasing from 120W to 300W. The solidification time of droplet can be calculated using Eq. (2), the equation is as follow[12]:

$$\tau_{soli} = 2\left(\frac{a^2}{3\alpha}\right)\ln\left(\frac{T_0-T_t}{T_f-T_t}\right) \quad (2)$$

where a is the size of the initial droplet, α is the material thermal diffusion coefficient, T_0 is the temperature of the initial droplet, T_f is the liquidus temperature of material, T_t is the temperature of the substrate, and τ_{soli} is the solidification time of droplet.

According to the Eq. (2), the solidification time of the molten pool increases with the increasing of the initial size of the droplet and the initial temperature of the droplet. The solidification time of the molten pool increases with the increasing of the laser power due the increasing of the size of the initial droplet and the temperature of the initial droplet (Fig.6 and Fig.10). The longer solidification time is good for spreading and wetting with the last layer and can avoid the balling of the droplet due to the short solidification time as long as possible, which is good to improving the quality of the powder layer, and reducing the remelting phenomenon. This can reduce the thermal stress caused by remelting and reduce the possibility of the deformation.

4.2 The solidification shrinkage

Due to the thermal expansion of the material, volume condensation occurs during the solidification of the droplet, and adjacent regions of the droplet will produce different degrees of shrinkage due to different solidification rates. The shrinkage of the specimens can be calculated using Eq. (3).

$$\partial l = \delta \times l \times \partial t \tag{3}$$

where δ is the materials thermal expansion coefficient, l is the size of the specimens and ∂t is the amount of change in temperature per unit time. In this experiment, the same material is used, and the thermal expansion coefficient is same; the shape of the specimens is same. The amount of shrinkage is related to the solidification rate. In this solidification process of the droplet, the faster the solidification rate, the larger amount of shrinkage will be, and more obviously of the deformation will be. Taking the derivation with respect to the temperature, the results of the solidification rate is shown as Fig.11. The solidification rate shows obvious decreasing as the laser power increasing from 120W to 300W. When the laser power is higher, the amount of the shrinkage ∂l is smaller, the tendency is same as the change rules of deformation shown as Fig.6.

The specimens rapidly cooled after the droplet solidified. The temperature at different positions was uneven due to the different cooling rate, resulting in the temperature gradient. The difference in cooling rate between adjacent regions caused different degrees of shrinkage, and the portion with a large amount of shrinkage was constrained by the portion with a small amount of shrinkage, which caused by uneven temperature distribution and was also called thermal stress. As the scanning of the laser spot, the stress accumulated. Once the stress exceeded the yield strength of the materials, the deformations of the specimens appeared, even the cracks. The thermal stress can be calculated using Eq. (4).

$$\begin{aligned} \sigma_x &= \frac{\delta E \Delta T_x}{1-2u} \\ \sigma_y &= \frac{\delta E \Delta T_y}{1-2u} \\ \sigma_z &= \frac{\delta E \Delta T_z}{1-2u} \end{aligned} \tag{4}$$

where σ_x is the thermal stress in the direction of X, σ_y is the thermal stress in the direction of Y, σ_z is the thermal stress in the direction of Z, δ is the materials thermal expansion coefficient, E is the materials modulus of elasticity, u is the Poisson's ratio of materials, ΔT_x is the temperature gradient in the direction of X, ΔT_y is the temperature gradient in the direction of Y and ΔT_z is the temperature gradient in the direction of Z. The deformation caused by the thermal stress can by calculated using Eq. (5).

$$\begin{aligned} \varepsilon_x &= \frac{1}{E}\left[\sigma_x - u\left(\sigma_y + \sigma_z\right)\right] + \delta \Delta T_x \\ \varepsilon_y &= \frac{1}{E}\left[\sigma_y - u\left(\sigma_x + \sigma_z\right)\right] + \delta \Delta T_y \\ \varepsilon_z &= \frac{1}{E}\left[\sigma_z - u\left(\sigma_x + \sigma_y\right)\right] + \delta \Delta T_z \end{aligned} \tag{5}$$

where ε_x is the deformation in the direction of X, ε_y is the deformation in the direction of Y and ε_z is the deformation in the direction of Z. Then combined the Eq. (4) with the Eq. (5), the deformation was shown as Eq. (6).

$$\begin{aligned} \varepsilon_x &= 2\delta \Delta T_x \\ \varepsilon_y &= 2\delta \Delta T_y \\ \varepsilon_z &= 2\delta \Delta T_z \end{aligned} \tag{6}$$

It is known that the thermal stress was decided by the temperature gradient. The higher the temperature gradient, the larger the deformation amount caused by the thermal stress will be.

The molten pool simulation results are shown as Fig.12. The size of the molten pool increased obviously with the increase of the laser power.

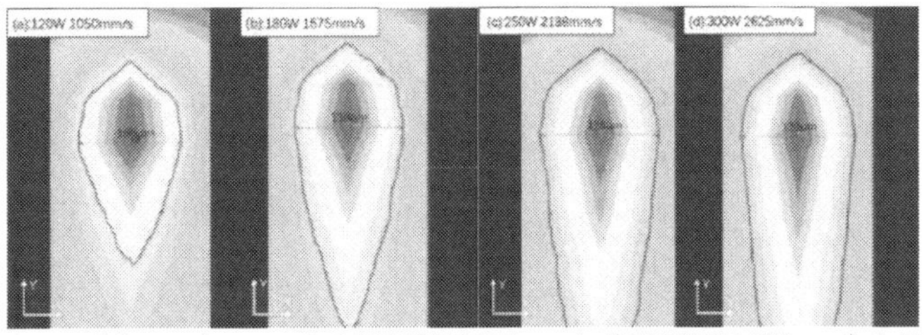

Figure 12 The dimensions of the molten pool, (a)120W,1050mm/s;(b)180W,1575mm/s;(c)250W,2188mm/s and (d)300W,2625mm/s.

When the laser spot scanning along with the direction of Y, taking the position at (0,0) with the highest temperature, then calculated the temperature gradient (TG) along X-direction and the TG along Y-direction respectively at the position of (40,0) and (0, -40), the results are shown as Fig.13.

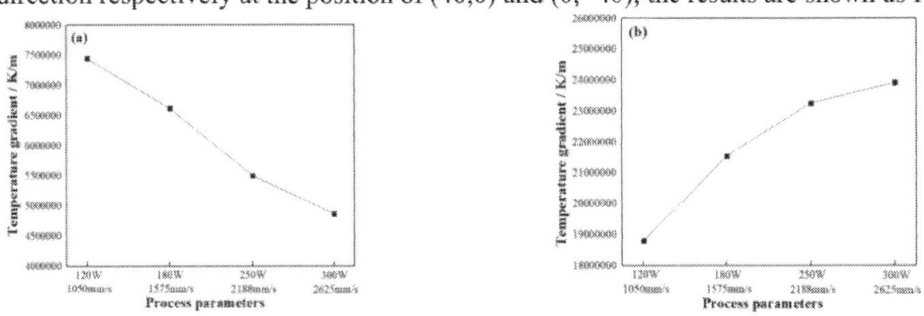

Figure 13 Influence of process parameters on the TG of the point (0, -40) and (40,0) as the laser beam reach the point (0,0), (a)the TG of the point (0,-40) along Y-direction and (b) the TG of the point (40,0) along X-direction.

When the linear energy density is same, ΔT_y decreases with the increasing of the laser power shown as Fig.13 (a) and ΔT_y shows the contrary tendency shown as Fig.13 (b). This means that ε_y caused by ΔT_y was smaller with the increase of the laser power as the linear energy density is same and ε_x caused by ΔT_x shown the contrary tendency. The ε_x on both side have equal value and opposite directions, and the deformation caused by ΔT_x at different side cancelled each other out. The ΔT_y has the significant influence on the deformation of the specimens and the deformation is smaller with higher laser power when the linear energy density is same.

5 Conclusions

In this work, the Ti6Al4V thin-wall specimens were fabricated by Selective Laser Melting (SLM) using various process parameters. The laser power had a significant influence on the deformation of the specimens. The main conclusions are summarized as follows:

(1) When the linear energy density is same, the solidification rate of the molten pool decreases with the increasing of the laser power, the shrinkage during the droplet solidification process is smaller with a higher laser power. The ΔT_y decreases with the increase of the laser power and the ε_y caused by the ΔT_y is smaller with a higher laser power and the ΔT_y has the significant influence on the deformation of the specimens. When the laser power increased from 120W to 300W, the roundness of the circular specimen reduced from 0.034mm to 0.024mm; the flatness of the triangular specimens and the square specimens reduced from 0.27mm and 0.16mm to 0.07mm and 0.11mm respectively.

(2) When the linear energy density is same, the intensity of the meltpool and the area of the molten pool increase with the increasing of the laser power.

(3) When the linear energy density is same, the solidification time of the molten pool increases with the increasing of the laser power, so as the temperature of meltpool, which is good for molten pool wetting with the substrate, improving the quality of the powder layer and reducing the balling phenomenon.

References

1. Z.T. Wu, X.T. Hu, T.D. Song, Multiaxial fatigue life prediction for titanium alloy TC4 under proportional and nonproportional loading. Int. J. Fatigue,**59(3)**:170-175(2014).
2. D. Herzog, V. Seyda, E. Wycisk, C. Emmelmann, Additive Manufacturing of Metals, Acta Mater.**117**:371-392(2016).
3. L.S. Bertol, W.K. Júnior, F.P.D. Silva, C. Aumund-Kopp. Medical design: Direct metal laser sintering of Ti–6Al–4V, Mater. Des. **31(8)** : 3982-3988(2010).
4. L.Thijs, F. Verhaeghe, T. Craeghs, J.V. Humbeeck, J.P. Kruth. A study of the microstructural evolution during selective laser melting of Ti–6Al–4V, Acta Mater. **58(9)**:3303-3312(2010).
5. B.Vrancken, V.Cain, R.Knutsen, J.V. Humbeeck. Residual stress via the contour method in compact tension specimens produced via selective laser melting, Scripta Mater. **87(87)**:29-32(2014).
6. Additive manufacturing of high-strength aluminum components, Annual report fraunhofer institute for technology(2010).
7. D. Buchbinder. Generative fertigung von Aluminiumbauteilen für die Serienproduktion, AluGenerativ Abschlussbericht.
8. S.Siddique, M.Imran, E.Wycisk, C.Emmelmann, F.Walther. Influence of process-induced microstructure and imperfections on mechanical properties of AlSi12 processed by selective laser melting, J. Mater. Process Tech. **221**:205-213(2015).
9. M.Karg, B. Ahuja, S. Kuryntsev, A. Gorunov, M.Schmidt. Processability of high strength Aluminium-Copper alloys AW-2022 and 2024 by Laser Beam Melting in Powder Bed. International Solid Freeform Symposium(2014).
10. K.Zhang, T.T.Liu, C.D.Zhang, W.H.liao. Study on Deformation Behavior in Selective Laser Melting Based on the Analysis of the Melt Pool Data. Chinese J. Lasers. **42(9)**: 0903007(2015).
11. D.N.Chen, T.T.Liu, W.H.liao, C.D.Zhang, K.Zhang. Temperature Field During Selective Laser Melting of Metal Powder Under Different Scanning Strategies, Chinese J. Lasers. **43(4)**:0403003(2016).
12. F. Gao, A.A. Sonin, Precise Deposition of Molten Microdrops: The Physics of Digital Microfabrication, Proc. R. Soc. London. Ser. A. **444** :533-554 (1994) .

A solid-analytical-based model for extracting cutter-workpiece engagement in 5-axis flank machining

Dongdong Li[1,a], Weimin Zhang[1,2], Tengfei Shang[1], Tianhong Gao[2], Shuo Liu[2] and Jürgen Fleischer[2,3]

[1] School of Mechanical Engineering, Tongji University, 201804 Shanghai, China.
[2] Sino-German Advanced Manufacturing Technology Center (CDHK), Tongji University, 201804 Shanghai, China.
[3] WBK Institute of Production Science, Karlsruhe Institute of Technology, 76131 Karlsruhe, Germany.

Abstract. 5-axis flank milling is applied extensively in aerospace, die-molds, and automotive industries because of high efficient material removal rate. Commercial CAM software can only simulate the tool path and collision at present, but cannot handle with cutting force during cutting process due to the variable geometrical cutter workpiece engagement (CWE) region of 5-axis milling. This paper presents a novel solid analytical model for extracting the CWE maps. The CWE is obtained analytically by performing Boolean operations between the cutter and in-process-workpiece (IPW) at any given cutter location (CL) point, instead of using the cutter and the removal volume. The proposed process simulation method could identify the CWE efficiently and precisely for general cutting tools. Finally, the CWE boundaries are mapped from a 3D space into a 2D plane defined by the immersion angle and the axial depth of a given cutter. The proposed solution can be easily integrated into the CAM software for predicting milling force and optimizing parameters in 5-axis milling.

1 Introduction

New processes in today's technology, e.g. electrochemical machining and 3D printing have been applied to manufacturing. However, parts machined with 5-axis milling still covers the main proportions in automotive, aircraft parts and molds. 5-axis milling is quite often used for surface machining of impellers, but it still has some defects that need to be optimized. The tool paths and CL file can be generated by commercial CAM. These solutions are only geometrical motion processes, namely, the kinematics properties of the machine tool and the cutting force between cutter and workpiece are not considered for the metal cutting process [1], so products are prone to quality problems. This article provides an optimization solution for processing, and the flow diagram is shown in Figure 1. Accurate modeling and extraction of the CWE area is an extremely important knot of the cutting force prediction model throughout the process.

Various approaches of CNC milling process have been described in previous research works. Kim et al. [2] presented the Z-map method to determine the cutter contact area, and cutting force acting on the engaged cutting edge elements are calculated using an empirical method. Lazoglu et al. [3] did a great study about ball end mill, meanwhile a boundary representation (B-rep) method was developed to find the cutter-workpiece engagement (CWE), which then is used to estimate milling force and

[a] Corresponding author : 1510289@tongji.edu.cn

feedrate optimization. Erdim et al. [4] proposed a new approach to shape representation called composite adaptively sampled distance fields. This function yields the minimum Euclidean distance from a point to the closest in the boundary of the set. Actually, these could be categorized into three major approaches: discrete representations (based on z-buffer, dexels or voxels), solid modeling (B-rep, CSG), and point based methods [5].

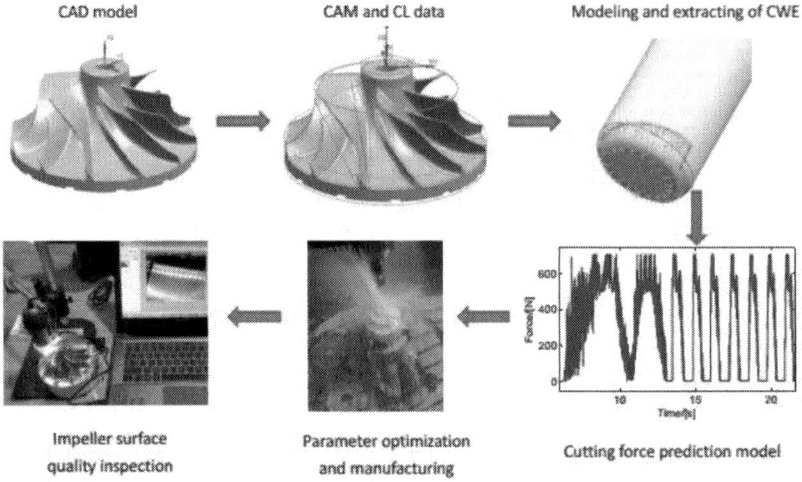

Figure 1. Optimization of 5-axis milling process

According to the Ref. [5,6], the solid model method of CWE extraction has the most accurate results among these methods, though it still has the challenge of computational complexity and cost. This technical trend must be mentioned, that is, the solid modeling solution is greatly promising in industry 4.0 and can be easily integrated into the CAD/CAM softwares.

2 CWE model for flank milling

In order to develop CAM, Chiou et al. [7] presented a solution for the determination of instantaneous cutter swept envelope, which provides a feasible tool envelope concept for solid modeling methods. Multiple researchers perform modeling, simulation and optimization of 5-axis NC machining by a ball end mill or a taper ball end mill [2-6,8-10]. However, there are very few CWE extraction studies about 5-axis flank milling in recent years. So a solid analytical model for CWE map in 5-axis flank milling is proposed in this paper. Generally speaking, a bull nose cutter is usually selected for flank milling considering the characteristics of surfaces.

2.1 Workpiece coordinate system and tool coordinate system

To accurately describe the cutter movement in 5-axis milling, it is needed to build the workpiece coordinate system (WCS) and tool coordinate system (TCS). On one hand, the CAD model, toolpath generation, and CL file are generally described in the WCS. On the other hand, the CWE and engagement angle are firstly preferred to be illustrated in the TCS. Therefore, the coordinate conversion between the fixed WCS and the instantaneous TCS must be clearly defined before milling process. The toolpath and CL file generated by CAM are shown in Figure 2, the standard format is usually used to describe the cutter trajectories in 5-axis milling, for example, GOTO/x, y, z, i, j, k in each line of CL file that represents the tool tip position and tool axis orientation, respectively.

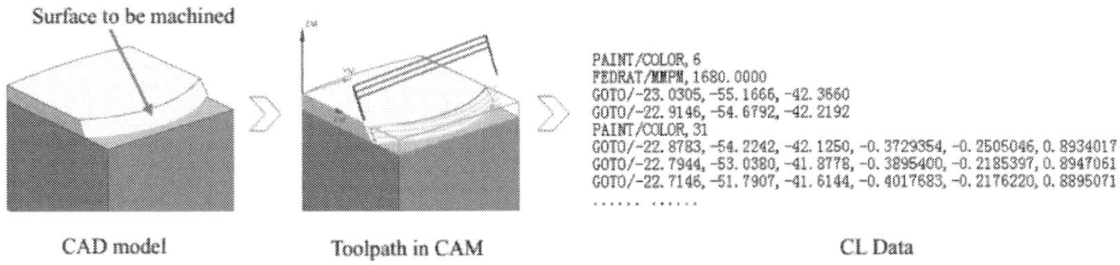

Figure 2. Generation of the toolpath and CL data in CAM

Schematic diagram of both coordinate systems are shown in Figure 3. The cutter contact (CC) points (x_c, y_c, z_c) can be calculated from the CL points and cutter geometry [11]. Then the projection point (x_p, y_p, z_p) of CC points can be calculated by the following equation:

$$\begin{cases} x_p = x_c - r \cdot i \\ y_p = y_c - r \cdot j \\ z_p = z_c - r \cdot k \end{cases} \tag{1}$$

Where r is the radius of cutter end part.

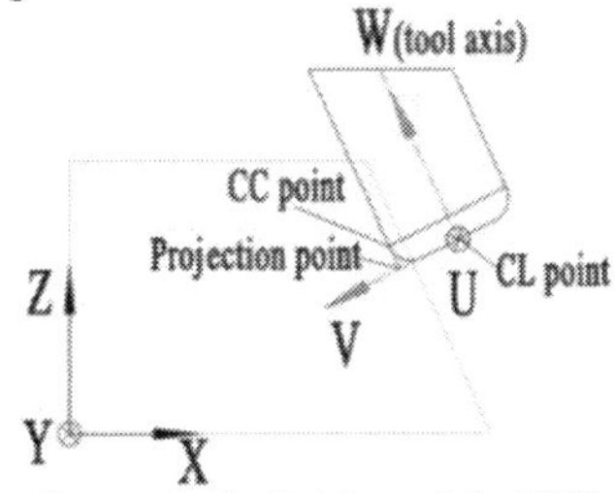

Figure 3. Definition of the TCS

The projection point and the CL point (x_t, y_t, z_t) are together connected to obtain the V-axis unit vector:

$$V = \frac{\left[(x_t - x_p),(y_t - y_p),(z_t - z_p)\right]}{\sqrt{(x_t - x_p)^2 + (y_t - y_p)^2 + (z_t - z_p)^2}} \tag{2}$$

As a vector cross product of the V-axis and W-axis unit vector, the U-axis unit vector can be expressed as:

$$U = V \times W = \begin{bmatrix} i & j & k \\ V_i & V_j & V_k \\ W_i & W_j & W_k \end{bmatrix} \tag{3}$$

In the 5-axis flank milling, the coordinate conversion between WCS and TCS contains two parts: the translation transformation and rotation transformation. Homogeneous matrix transformation of 5-axis coordinates has already been studied extensively. Thus, the matrix equation can be represented as follows:

$$T = \begin{bmatrix} U_i & V_i & W_i & x_t \\ U_j & V_j & W_j & y_t \\ U_k & V_k & W_k & z_t \\ 0 & 0 & 0 & 1 \end{bmatrix} \tag{4}$$

Obviously, the mathematical equation between the WCS and TCS can be written as:

$$[x, y, z, 1]^T = T \cdot [u, v, w, 1]^T \tag{5}$$

2.2 Surface normal vector of the cutter

The surface normal vector of a cutter plays a vital role in calculating the feasible CWE area. According to the geometry of a bull-nose cutter, the whole cutter surface can be divided into a cylindrical section and an arc section. An arbitrary point $M(u_1, v_1, w_1)$ and its surface normal $N(n_i, n_j, n_k)$ of the cutter in the TCS can be expressed as follows:

For the cylindrical section ($w_1 > r$),

$$\begin{cases} M(\phi, w_1) = (\frac{D}{2}\sin\phi, \frac{D}{2}\cos\phi, w_1) \\ N(\phi, w_1) = (\frac{D}{2}\sin\phi, \frac{D}{2}\cos\phi, 0) \end{cases} \tag{6}$$

For the arc section ($0 \le w_1 \le r$),

$$\begin{cases} M(\phi, w_1) = ((\frac{D}{2} - r + \sqrt{r^2 - (r - w_1)^2})\sin\phi, (\frac{D}{2} - r + \sqrt{r^2 - (r - w_1)^2})\cos\phi, w_1) \\ N(\phi, w_1) = (\sqrt{r^2 - (r - w_1)^2}\sin\phi, \sqrt{r^2 - (r - w_1)^2}\cos\phi, w_1 - r) \end{cases} \tag{7}$$

Where ϕ denotes the immersion (engagement) angle of cutting edge at the CL point, and $\phi \in [0, 2\pi]$.

2.3 Feasible contact surfaces (FCS) on a bull-nose cutter

Different from the 3-axis milling process, the feed vector $F_{a,b}$ of the cutter is no longer constant in 5-axis flank milling due to the additional two rotary motion, here the first subscript a represents the a-th cut, and the second subscript b represents the b-th cutting location in the a-th cut. Now, two adjacent CL points in the TCS are marked as $M^{TCS}(u_{a,b}, v_{a,b}, w_{a,b})$ and $M^{TCS}(u_{a,b+1}, v_{a,b+1}, w_{a,b+1})$, respectively. Through the Eq. (5) in Section 2.1, there must be two corresponding adjacent points $M^{WCS}(x_{a,b}, y_{a,b}, z_{a,b})$ and $M^{WCS}(x_{a,b+1}, y_{a,b+1}, z_{a,b+1})$ in the WCS. Thus, the instantaneous feed vector (velocity vector) of point M could be described as follows:

$$F_{a,b} = [(x_{a,b+1} - x_{a,b}), (y_{a,b+1} - y_{a,b}), (z_{a,b+1} - z_{a,b})] \tag{8}$$

Once obtaining the normal and feed direction of the arbitrary point on the cutter surface, the cutter could be divided into three areas: the forward area, backward area, and envelope boundary, as the FCS area of a cutter is shown in Figure 4. The different areas are defined as the following Eq. (9).

$$\begin{cases} F_{a,b} \cdot N_{a,b} > 0 & Forward\_area \\ F_{a,b} \cdot N_{a,b} = 0 & Envelope\_boundary \\ F_{a,b} \cdot N_{a,b} < 0 & Backward\_area \end{cases} \tag{9}$$

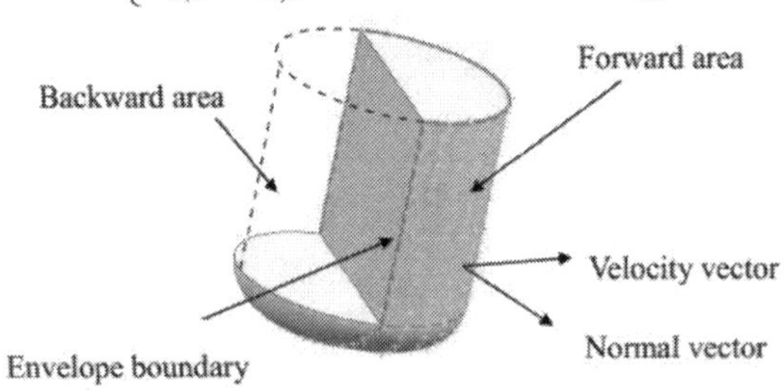

Figure 4. Feasible contact surface area of a cutter

The cutter may contact the in-process workpiece through the forward area and envelope boundaries. The combination of forward area and envelope boundaries determine the feasible contact surface of cutter. Finally the CWE areas are generated when the FCS engages the IPW during the material removal process. As the CWE are the subsets of the FCS, namely, CWE⊆ FCS. The entire CWE extraction process is described clearly.

3 Simulation results

3.1 Loop module for 5-axis flank milling

The cutter swept volume is generated by the cutter movement along the toolpath that can be obtained from the CAM module. As the loop diagram of the CWE model shown in Figure 5, when the cutter swept volume is calculated, the material removal volume can be simulated through a boolean subtraction with the raw stock. Then the CWE area can be extracted by the proposed method in Section 2.

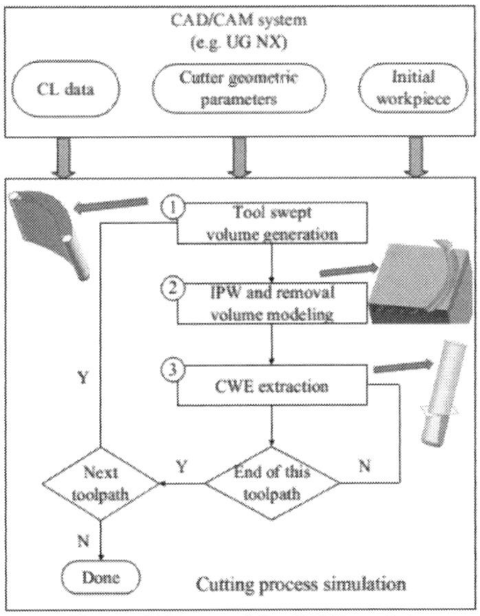

Figure 5. Loop of the CWE model

To compute the CWE area at a given CL point, the cutter body is intersected with some slicing planes which results in arcs or rings at each height increment in the TCS. These arcs/rings are then intersected with the IPW. The intersection results in several engagement curves. The starting and the end coordinates of these curves define the engagement points. Ultimately, engagement points are used to calculate the entrance and exit angles with respect to the plane of feed and cross-feed direction in the TCS. When these steps are done, the entire process should be updated including the IPW, cutter, and CL points. In each cycle firstly the cutter body at the current CL(i) point and the swept volume is subtracted from the IPW. Next step, the CL point counter for the loop is incremented by one and the cutter moves to the CL(i+1) point. As a result, the IPW is updated and becomes ready for the CWE computation of the remaining CL points. Finally, the cutter body is created to be used again in the next iteration.

3.2 Compute the engagement angle

The CWE area and engagement angle computation form the main part of the proposed model. It also consumes most of the computational cost. The cutter body is sliced into a certain number of discs as is shown in cyan-yellow planes in Figure 6, in which they are shown as the removal volume after boolean operation between the cutter and workpiece. During this process, desired number of planes are created perpendicular to the tool axis at each plane height. To create these planes, normal direction and reference direction, which are the same with the cutter, are required.

These planes are perpendicular to the normal direction of the cutter at a given CL point and start from the tool tip point extending until the specified height of the cutter with very small height increments. These planes will be applied in combination with the cutter body for creating the cutter discs. Then, the slicing planes and the cutter body are intersected by the "intersect surface" function. Each intersection operation will result in a disc/arc lying on the corresponding plane, as the CWE calculation is shown in Figure 6.

The discs/arcs are later intersected with the copy of IPW. The IPW is duplicated to preserve the original IPW to be used in the following iterations [9]. The intersection results in an arc for single engagement or multiple arcs for multiple engagement that are in Figure 6 and Figure 7. So the solid analytical model is capable of detecting the frequently occurring multiple engagement conditions in 5-axis flank milling.

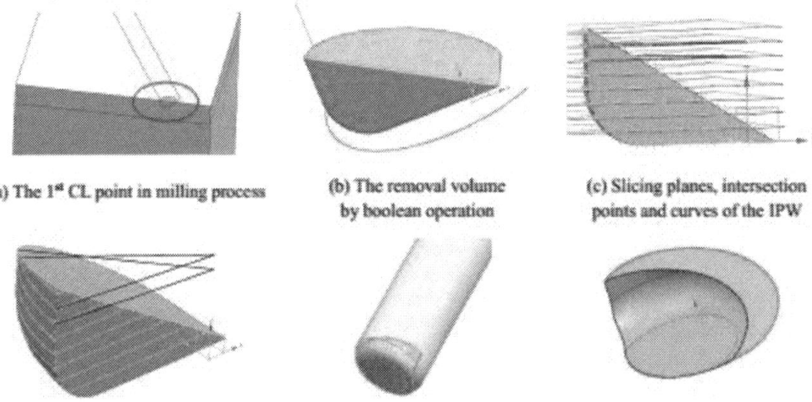

(a) The 1st CL point in milling process (b) The removal volume by boolean operation (c) Slicing planes, intersection points and curves of the IPW

(d) Calculation of the CWE and engagement angle (e) The CWE of the 1st cutting, 11th CL point (f) The removal volume of the 11th CL point

Figure 6. Calculation of the CWE and engagement angle

The coordinate values of the start and end points from each arc are extracted and stored. They are called engagement points and are used in angle calculation. The angle between vector pointing to the current engagement point from the cutting center and the cross-feed direction vector is calculated. The entrance and exit angles are computed accurately by utilizing the inverse transformation of Eq. (5) to view them in the tool axis direction. This allows for the engagement angles to be measured with respect to the cross-feed direction. The calculated entrance and exit angles for each disc/arc are saved as CWE map to be used as a boundary condition for the cutting force prediction, as the CWE engagement angles are shown in Figure 7 and Table 1.

Table 1. CWE output of the 1st CL point (entrance (En) and exit (Ex) angles)

CL point	Arc No.	En angle/degree	Ex angle/degree
01	1	8.997	107.393
01	2	6.952	93.554
01	3	6.534	84.637
01	4	6.364	76.345
01	5	6.335	67.799
01	6	6.367	58.524
01	7	6.399	48.208
01	8	6.431	35.834
01	9	6.463	17.253

Additionally, in order to assess the performance of the proposed method, the output engagement angles of CWE model, i.e. Table 1, could be acted as an input in a cutting force model in research [1,8,12]. The engagement angles are utilized as the upper and lower boundaries (ϕ_{en}, ϕ_{ex}) in Eq. (10)

to estimate the cutting force. Here $dF_{x,y,z}(\phi)$ is the elemental cutting force in X, Y and Z directions as a function of angular position of a cutter. Validation can be proved by comparing the simulated force data with the experimental force data that can be measured by a dynamometer. The simulation and experiment should be under the same condition of process parameters.

$$F_{x,y,z} = \int_{\phi_{en}}^{\phi_{ex}} dF_{x,y,z}(\phi)\,d\phi \tag{10}$$

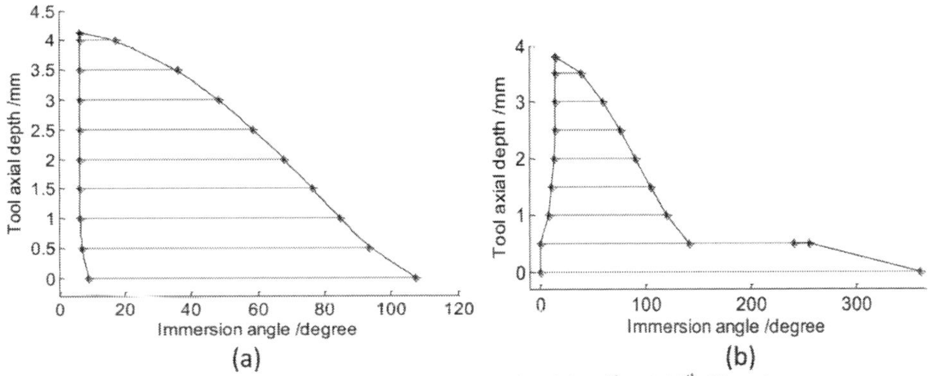

Figure 7. The CWE and engagement angle of the 1st and 11th CL points

4 Conclusion

The aim of this paper is introducing a solid analytical model to extract the CWE in 5-axis flank milling. The CL data and cutter geometry parameters are firstly generated from CAM, then the boolean subtraction operations are performed to obtain the possible CWE between the cutter swept volume and the raw workpiece, and the CWE and engagement angles are extracted finally. The proposed model can be integrated into commercial CAM applied to feedrate scheduling, tool orientation optimization and prediction of cutting force in 5-axis milling.

Acknowledgments

The research is supported by the National key R&D project-the Construction, Reference Implementation and Verification Platform of Reconfigurable Intelligent Production System, China (Grant No. 2017YFE0101400).

References

1. Ferry W. B, Altintas Y. J MANUF SCI E-T ASME, **130** (1), 339-353 (2008).
2. Kim G. M, Cho P. J, Chu C. N. INT J MACH TOOL MANU, **40** (2), 277-291 (2000).
3. Lazoglu I, Boz Y, Erdim H. CIRP ANN-MANUF TECHN, **60** (1), 117-120 (2011).
4. Erdim H, Sullivan A. Procedia Cirp, **8**, 438-443 (2013).
5. Altintas Y, Kersting P, Biermann D, et al. CIRP ANN-MANUF TECHN, **63** (2), 585-605 (2014).
6. Aras E, Albedah A. INT J ADV MANUF TECH, **73** (9-12), 1351-1362 (2014).
7. Chiou C. J, Lee Y. S. INT J MACH TOOL MANU, **42** (14), 1497-1507 (2002).
8. Zhang X, Yu T, Wang W. INT J ADV MANUF TECH, **74** (9-12), 1611-1624 (2014).
9. Yigit I. E, Lazoglu I, S. E. L. K. Procedia Cirp, **31**, 179-184 (2015).
10. Ju G, Song Q, Liu Z, et al. INT J ADV MANUF TECH, **80** (5-8), 1297-1310 (2015).
11. Si H, Wang L, Zhang J, et al. INT J ADV MANUF TECH (9), 1-13 (2017).
12. Boz Y, Khavidaki S. E. L, Erdim H, et al. *in Machining of Complex Sculptured Surfaces*. Springer, 67–125 (2012).

Vibration analysis of rotating rings with complex support stiffnesses

Fuchun Yang[1,a], Yue Zhang[1] and Hailong Li[1]

[1]*School of Mechanical Engineering, Shandong University, Jinan 250061, China*

Abstract. Vibration characteristics of rotating rings with complex support stiffnesses are studied. The complex stiffnesses of the rotating ring include discrete stiffnesses and partially distributed stiffnesses. The governing equations are established by Hamilton's principle. The governing equations are cast in matrix differential operators and discretized using Galerkin's method. The eigenvalue problem is dealt with state space matrix and the natural frequencies and vibration modes are obtained. The properties of natural frequencies and vibration modes of rotating rings are studied. The results illustrate that frequency separation and frequency veering happen with the increase of rotation speed. The vibration modes are not dominated by only one nodal diameter while dominated by several nodal diameters because the discrete and partially distributed stiffnesses disrupt the axisymmetry of rotating rings. The influences of several parameters to vibration properties of rotating rings are also investigated.

1 Introduction

Vibration properties of rotating rings are investigated by researchers for different structures, for instance, rotors, bearings, gears, etc. This work is motivated by star-type planetary gear sets, where rotating gears often have a thin rim [1], especially in aerospace applications. The rotating rings are supported by both discrete stiffnesses of mesh pairs and partially distributed stiffnesses of support bearings and these complex supports may cause quite different vibration properties compared to free rotating rings.

Vibrations of rotating rings have been investigated by many researchers. Bert and Chen [2] studied the in-plane and out-plane vibration of rotating rings on an elastic foundation. Free vibrations of rings on rigid radial supports were investigated by Rao and Sundararajan [3], where the ring was separated into several parts with supports located at the ends of each segment. Carrier [4] derived the governing equations for rotating rings in a rotating basis. Endo et al. [5] investigated different strain measures in the derivation of the equation of motion. Bickford and Reddy [6] included shear deformation and rotary inertia in their derivation. Reddy and Bickford [7] studied the vibration of rotating rings with equally spaced discrete stiffnesses. Huang and Soedel [8] derived a model using Lagrangian strain. They calculated natural frequencies for varying rotation speeds and find the dynamic response. Nackenhorst and Brinkmeier [9] used two observer systems, that is, Lagrangian and Eulerian systems, to derive rotating ring models. Wu and Parker [10] studied the vibration of rings on a general elastic foundation. There are fewer studies on the vibration of rotating rings with discrete supports. Canchi

[a] Corresponding author: yfc26@163.com

and Parker [11] studied the parametric instability of rotating rings connected to moving discrete springs with time-varying stiffness. Cooley and Parker [12] studied the vibration of high-speed rotating rings coupled to space-fixed stiffnesses.

As mentioned above, the support mode of bearings has not excited the attention of researchers and there has been no study on the vibration of rotating rings with discrete stiffnesses and partially distributed stiffnesses. In this work, the work will mainly focus on vibration properties of rotating rings with complex support stiffnesses.

2 Vibration model of rotating ring

The vibration model of a rotating ring with complex support stiffnesses is shown in Figure1. The rotating speed is ω. The ring-fixed basis is {**i**, **j**} and the {$\mathbf{e}_1$, $\mathbf{e}_2$} basis is a local basis on the middle surface of the ring, where $u(\theta, t)$ and $v(\theta, t)$ are radial and tangential deformations, respectively. M is a material point on the middle surface and θ is the relative angle to the basis **i**, where $0\leq\theta\leq2\pi$. R is the radii of the middle surface. The density of the ring is ρ, its elastic modulus is E, its section area is A. k_{m1} and k_{m2} are discrete stiffnesses in tangential direction. k_b is the partially distributed stiffnesses between θ_{b1} and θ_{b2}.

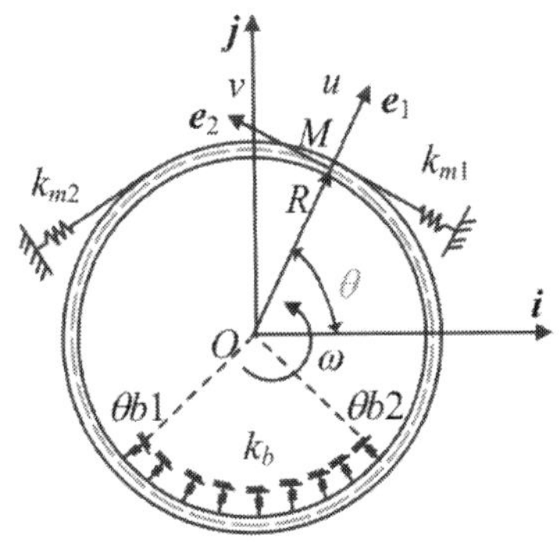

Figure 1. vibration model of a rotating ring with complex support stiffnesses.

By applying the kinetics principles, the kinetic energy of the ring is obtained as

$$K_E=\frac{1}{2}\rho AR_1\int_0^{2\pi}\left[(\dot{u}-\omega v)^2+(\dot{v}+\omega u+R\omega)\right]\mathrm{d}\theta \tag{1}$$

The strain energy of the ring is derived using the engineering strain as

$$S_e=\frac{1}{2}\iint_{A\theta}\sigma\varepsilon dAR_1\mathrm{d}\theta=\frac{1}{2}\int_0^{2\pi}\left[\frac{EA}{R_1}(u+\partial v/\partial\theta)^2+\frac{EI}{R_1^3}\left(\partial v/\partial\theta-\partial^2u/\partial\theta^2\right)^2\right]\mathrm{d}\theta \tag{2}$$

The potential energy of discrete and partially distributed stiffnesses is

$$S_s=\sum_{i=1}^{Nb}\frac{1}{2}k_b\left[u\delta\left(\theta-\theta_{bi}\right)\right]^2+\sum_{i=1}^{Np}\frac{1}{2}k_{mi}\left[v\delta\left(\theta-\theta_{pi}\right)\right]^2 \tag{3}$$

Accordingly, the governing equations can be derived by Hamilton's principle as

$$\begin{cases}\rho AR\dfrac{\partial^2u}{\partial t^2}-2\rho AR\omega\dfrac{\partial v}{\partial t}-\rho AR\omega^2u+\dfrac{EA}{R}\left(u+\dfrac{\partial v}{\partial\theta}\right)-\dfrac{EI}{R^3}\left(\dfrac{\partial^3v}{\partial\theta^3}-\dfrac{\partial^4u}{\partial\theta^4}\right)+\sum\limits_{i=1}^{Nb}k_bu\delta\left(\theta-\theta_{bi}\right)=0 & (4)\\ \rho AR\dfrac{\partial^2v}{\partial t^2}+2\rho AR\omega\dfrac{\partial u}{\partial t}-\rho AR\omega^2v-\dfrac{EA}{R}\left(\dfrac{\partial u}{\partial\theta}+\dfrac{\partial^2v}{\partial\theta^2}\right)-\dfrac{EI}{R^3}\left(\dfrac{\partial^2v}{\partial\theta^2}-\dfrac{\partial^3u}{\partial\theta^3}\right)+\sum\limits_{i=1}^{Np}k_{mi}v\delta\left(\theta-\theta_{pi}\right)=0 & (5)\end{cases}$$

To investigate the general properties of rotating rings, Equation (1) is normalized by the following variables

$$\omega_0=\frac{1}{R_1^2}\sqrt{\frac{EI}{\rho A}},\ \bar{u},\bar{v}=\frac{u,v}{R_1},\bar{t}=\omega_0 t,\bar{\omega},\bar{\Omega}=\frac{\omega,\Omega}{\omega_0},\bar{k}_{mp},\bar{k}_b=\frac{k_{mp},k_b R_1^3}{EI},\gamma=\frac{EAR_1^2}{EI} \tag{6}$$

The dimensionless equations are obtained by substitution of the above normalized variables to Equation (4) and (5), the equations are written as

$$\begin{cases}\dfrac{\partial^2 u}{\partial t^2}-2\omega\dfrac{\partial v}{\partial t}-\omega^2 u+\gamma\left(u+\dfrac{\partial v}{\partial\theta}\right)-\left(\dfrac{\partial^3 v}{\partial\theta^3}-\dfrac{\partial^4 u}{\partial\theta^4}\right)+\sum_{i=1}^{Nb}k_b u\delta(\theta-\theta_b)=0 & (7)\\[2ex] \dfrac{\partial^2 v}{\partial t^2}+2\omega\dfrac{\partial u}{\partial t}-\omega^2 v-\gamma\left(\dfrac{\partial u}{\partial\theta}+\dfrac{\partial^2 v}{\partial\theta^2}\right)-\left(\dfrac{\partial^2 v}{\partial\theta^2}-\dfrac{\partial^3 u}{\partial\theta^3}\right)+\sum_{i=1}^{Np}k_{mp} v\delta(\theta-\theta_p)=0 & (8)\end{cases}$$

In Equation (7) and (8) the bar on variables are removed for convenience and Equation (7) and (8) are rewritten in matrix form by partial differential operators as

$$\mathbf{M}\ddot{\mathbf{x}}+\omega\mathbf{G}\dot{\mathbf{x}}+\mathbf{K}\mathbf{x}=\mathbf{0} \tag{9}$$

where $\mathbf{x}=[u\ v]^T$ and **M**, **G**, **K** are the corresponding matrix. In order to analyse the vibration of the rotating ring, Equation (5) is discretized using Galerkin's method by choosing basis functions,

$$\mathbf{x}=\left[\sum_{n=-N}^{N}c_n e^{jn\theta}\quad \sum_{m=-M}^{M}d_m e^{jm\theta}\right]^T e^{\lambda t} \tag{10}$$

Then the eigenvalue equations about c_n and d_m are obtained and it is solved following the state space eigenvalue problem in reference [13].

3 Results

Since this work is motivated by planetary gear sets, the parameters of a gear ring in a planetary gear system are given in Table 1.

Table 1. Parameters of a rotating ring.

Ring radius R (mm)	80
Cross section area A (mm^2)	250
Area moment of inertia I (mm^4)	2083
Elastic modulus E (GPa)	206
Density ρ (kg/m^3)	7800
Mean mesh stiffness k_{mp} (N/m)	5×10^8
Support stiffness k_b (N/m)	1×10^8

For each nodal diameter there are positive and negative natural frequencies and modes that correspond to forward and backward waves, respectively. In the following analysis only the positive frequency properties are illustrated since the positive and negative frequencies have the same magnitude but opposite sign.

Figure 2a shows the natural frequencies of a rotating ring with complex support stiffnesses calculated by the Galerkin method. As shown in Figure 2a, for the stationary system the eigenvalues have degeneracy two, which is typical in axisymmetric spinning systems, such as shafts, disk-spindle system. For non-zero rotation speeds, the eigenvalues separate, becoming district two frequencies which corresponding to forward and backward waves, respectively. The natural frequencies increase with the rotating speed. And in high frequency areas, frequency veering occurs drastically which means the vibration mode exchange at these areas drastically, but for the lower order frequencies there is no frequency veering and the rotation speeds affect the frequencies slightly. In addition, there are no critical speeds and instability for the system since the frequencies are purely imaginary and never reach to zero in the analysis speed range.

It can be seen from Figure 2b that c_n and d_m are comprised of several components of lower orders and there is no district dominated component. This means that since the complex support stiffnesses with discrete and partially distributed stiffnesses disrupt the axisymmetry of the rotating ring, the vibration modes are not dominated by only one nodal diameter anymore as free rings does, but dominated by several nodal diameters. And in the lower order modes, the partially distributed stiffnesses affect the vibration mode heavily.

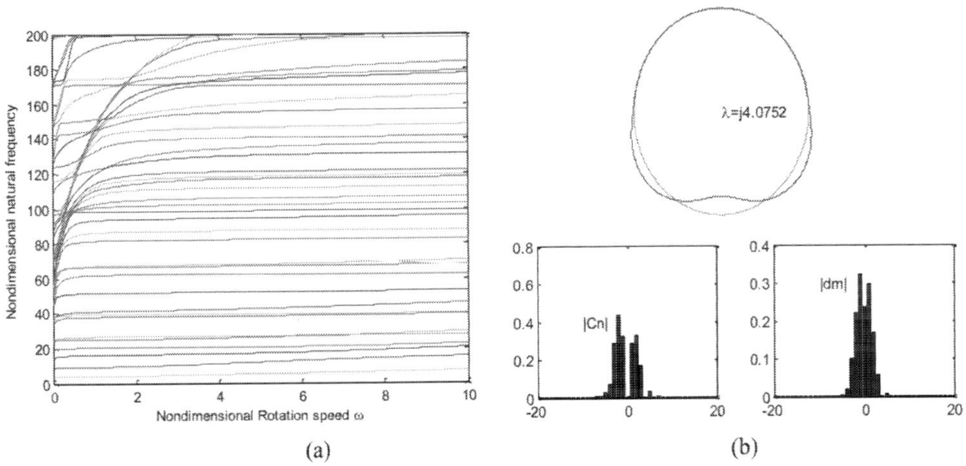

Figure 2. (a) Natural frequencies vs. rotating speed; (b) representative vibration mode and its components at speed ω=2 (dash line-original position, solid line-vibration mode).

Figure 3a illustrates the influence of partially distributed stiffnesses to the natural frequencies. It can be seen that almost all the frequencies are affected by the partial distributed stiffnesses, but the intermediate or higher frequencies are affected heavily while the lower frequencies are affected slightly. And in the analysis range of support stiffness in Figure 3a, frequency veering also happens with the increase of the stiffnesses.

Figure 3b shows the influence of discrete stiffnesses to the natural frequencies. As seen in Figure 3b a majority of frequencies are affected slightly by the discrete stiffness and only the intermediate frequencies and partial high frequencies are affected heavily. And frequency veering also exists in certain areas with the increase of discrete stiffnesses.

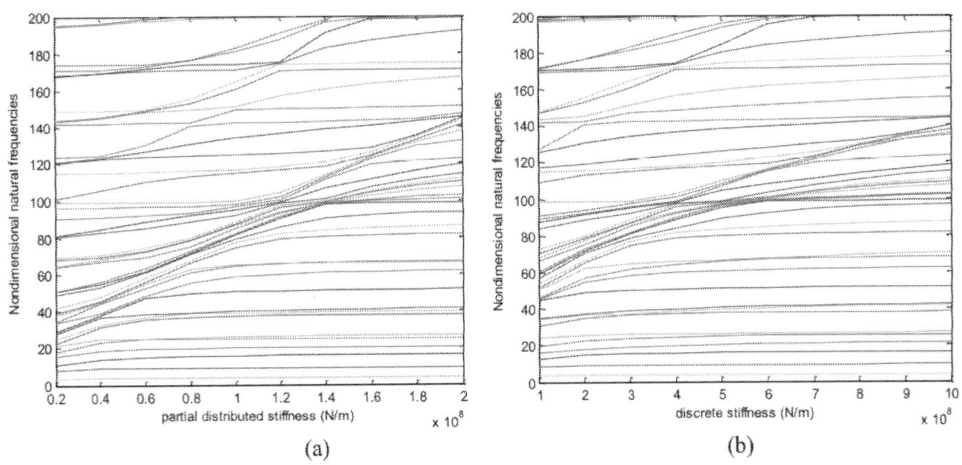

Figure 3. Influence of stiffnesses to natural frequencies (a) partial distributed stiffness and (b) discrete stiffness to natural frequencies at ω=2.

4 Conclusions

This paper established the governing equation of a rotating ring with complex support stiffnesses by Hamilton's principle. The governing equation is treated in matrix form and the eigenvalue problem is solved using Galerkin's method. The complex support stiffnesses change the vibration properties compared to free rotating rings. The phenomenon of frequency veering occurs with the increase of rotation speed. The influences of partially distributed and discrete stiffnesses to natural frequencies and naturel mode were also investigated. The results show that frequency separation and frequency veering happen with the increase of rotation speed. The vibration modes are not dominated by only one nodal diameter while dominated by several nodal diameters because the discrete and partially distributed stiffnesses disrupt the axisymmetry of rotating rings. The changes of the discrete or partially distributed stiffnesses mainly affect the intermediate frequencies and with the increase to stiffnesses the frequency veering also happens.

Acknowledgements

This work was supported by the National Natural Science Foundation of China (Grant numbers 51775309) and Young Scholars Program of Shandong University.

References

1. A. Kapelevich, Y. Shekhtman. Intl. Conf. on Gear, Munich, Germany, (2017)
2. C. Bert, L. Chen. J. Sound Vib., **122** (1988)
3. S. Rao, V. Sundararajan. J. Appl. Mech., **36** (1969)
4. G. Carrier. Q. Appl. Math., **3** (1945)
5. M. Endo, K. Hatamura, M. Sakata, O. Taniguchi. J. Sound Vib., **92** (1984)
6. W. Bickford, E. Reddy. J. Sound Vib., **101** (1985)
7. E. Reddy, W. Bickford. J. Sound Vib., **103** (1985)
8. E. Huang, W. Soedel. J. Sound Vib., **115** (1987)
9. U. Nackenhorst, M. Brinkmeier. Arch. Appl. Mech., **78** (2008)
10. X. Wu, R. Parker. J. Sound Vib., **295** (2006)
11. S. Canchi, R. Parker. J. Vib. Acous., **128** (2006)
12. C. Cooley, R. Parker. J. Sound Vib., **333** (2014)
13. L. Meirovitch. AIAA Journal, **12** (1974).

Vibration of thin cylindrical shells on an elastic foundation coupled with multiple discrete stiffnesses

Fuchun Yang [1,2,a] and Wenlei Qiu [1]

[1] *School of Mechanical Engineering, Shandong University, Jinan 250061, China*
[2] *Key Laboratory of High-efficiency and Clean Mechanical Manufacture at Shandong University, Ministry of Education, Jinan 250061, China*

Abstract. Vibration properties of thin cylindrical shells on an elastic foundation coupled with multiple discrete stiffnesses were investigated. The discrete stiffnesses were modelled as external forces. Hamilton's principle was applied to deduce the governing equations. To study the natural vibration properties, the wave-like solutions were applied. Then the governing equations were discretized in matrix form to obtain the eigenvalues. The properties of natural frequencies and vibration modes of thin cylindrical shells were studied. The results illustrate that each natural frequency of cylindrical shells is mainly dominated by one vibration mode, that is, flexural, longitudinal and shear modes. The natural frequency will get closer with the increase of circumferential wave numbers. The influences of several parameters to vibration properties of cylindrical shells were also investigated.

1 Introduction

Vibrations of cylindrical shells have received great attention for over a century in various engineering structures, such as rotor systems of gas turbine engines, high-speed centrifugal separators, rotating satellite structures, and planetary gear sets. This work is motivated by multiple stage planetary gear systems. The ring gears in this structure usually mesh with several planet gears and are supported on elastic foundations.

Many researchers have investigated the vibration of cylindrical shells in different boundary conditions. In the reference [1], the propagation of harmonic waves in thin-walled circular cylindrical shells made of orthotropic and isotropic materials is analysed in the framework of the Sanders linear shell theory. Forsberg [2] studied the influence of general boundary conditions on the modal characteristics of thin cylindrical shells. Weingarten [3] analysed the free vibration of thin cylindrical shells. Warburton [4] studied the natural frequencies and modes for different end conditions. Chung [5] presented a general analytical method for evaluating the free vibration characteristics of a cylindrical shell with classical boundary conditions. Lam and Wu [6] used the first-order shear deformation theory to presented analytical solutions for thick laminated composite cylindrical shells. Ng and Lam [7] studied the dynamic stability of simply-supported, isotropic cylindrical panels under combined static and periodic axial forces. In addition, some researchers studied the vibration of rotating cylindrical shells. Kim and Bolton [8] considered the effects of rotation on wave propagation of

[a] Corresponding author: yfc26@163.com

circular cylindrical shells. Omer [9] dealt with the free vibration analysis of rotating laminated cylindrical shells. Padovan [10] obtained the complete set of natural frequencies and corresponding mode shapes for circular cylindrical shells. Huang and Soedel [11] studied the forced response of a special cylindrical shell.

Although the researchers have studied the vibration of cylindrical shells on classical boundary conditions or rotating conditions, there is no study on the vibration of cylindrical shells with multiple discrete stiffnesses as shown in Figure 1. This work will establish the governing equations of thin cylindrical shells on an elastic foundation with multiple discrete stiffnesses and study their vibration properties.

2 Analytical model

The vibration model of a cylindrical shell with multiple-stage discrete stiffnesses is shown in Figure1. The fixed coordinate is {**i**, **j**, **k**} and the {$\mathbf{e}_1$, $\mathbf{e}_2$, $\mathbf{e}_3$} basis is a local basis on the middle surface of the shell, u, v and w are axial, tangential and radial deformations, respectively. M is a material point on the middle surface and θ is the relative angle to the basis **i**, where $0\leq\theta\leq 2\pi$. R is the radii of the middle surface. The density of the ring is ρ, its elastic modulus is E, Poisson's ratio is μ, the length is L and the thickness is h. k_{ip} is discrete stiffness in the ith stage with index p. The thin cylindrical shell is supported on the elastic foundation with stiffness k_r and k_θ in radial and tangential directions, respectively. The angles between radial directions of the cylindrical shell and each discrete stiffness are β.

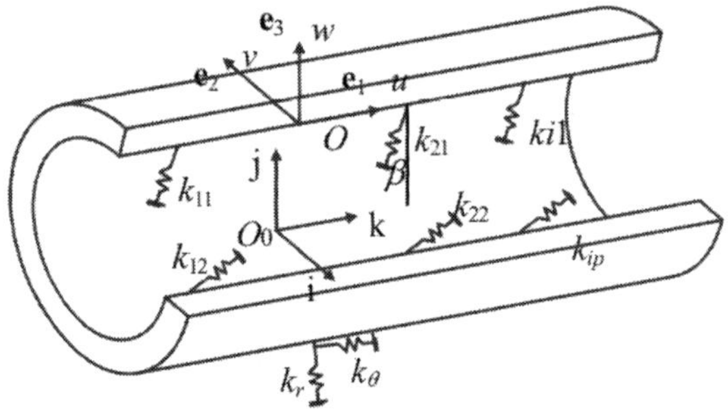

Figure 1. Model of thin cylindrical shell on an elastic foundation coupled with multiple discrete stiffnesses.

By the classical shell theory and the principles of kinetics, the kinetic energy, strain energy of the thin cylindrical shell and the work of external forces caused by discrete stiffnesses were obtained, and the Hamilton's principle was applied to establish the governing equation of thin cylindrical shells as following,

$$\rho h\ddot{u}-\frac{Eh}{R^2(1-\mu^2)}\left(R^2\frac{\partial^2 u}{\partial z^2}+\frac{1-\mu}{2}\frac{\partial^2 u}{\partial\theta^2}+\frac{R(1+\mu)}{2}\frac{\partial^2 v}{\partial z\partial\theta}+\mu R\frac{\partial w}{\partial z}\right)=0 \tag{1}$$

$$\rho h\ddot{v}-\frac{Eh}{R^2(1-\mu^2)}\left[\frac{R(1+\mu)}{2}\frac{\partial^2 u}{\partial z\partial\theta}+\frac{R^2(1-\mu)}{2}\frac{\partial^2 v}{\partial z^2}+\frac{\partial^2 v}{\partial\theta^2}+\frac{\partial w}{\partial\theta}+\frac{h^2}{12}\left(\frac{1-\mu}{2}\frac{\partial^2 v}{\partial z^2}+\frac{\partial^2 v}{R^2\partial\theta^2}-\frac{\partial^3 w}{\partial z^2\partial\theta}\right.\right.$$
$$\left.\left.-\frac{\partial^3 w}{R^2\partial\theta^3}\right)\right]+k_\theta v+\sum_i^{M_n}\sum_p^{N_s}k_{ip}\left(v(z_{mi},\theta_{ip},t)\sin^2\beta+w(z_{mi},\theta_{ip},t)\sin\beta\cos\beta\right)\delta(z-z_{mi})\delta(\theta-\theta_{ip})=0 \tag{2}$$

$$\rho h\ddot{w}+\frac{Eh}{R^2(1-\mu^2)}\left[\mu R\frac{\partial u}{\partial z}+\frac{\partial v}{\partial\theta}+w+\frac{h^2}{12}\left(R^2\frac{\partial^4 w}{\partial z^4}+2\frac{\partial^4 w}{\partial z^2\partial\theta^2}+\frac{\partial^4 w}{R^2\partial\theta^4}-\frac{\partial^3 v}{\partial z^2\partial\theta}-\frac{\partial^3 v}{R^2\partial\theta^3}\right)\right]+k_r w$$
$$+\sum_i^{M_n}\sum_p^{N_s}k_{ip}\left(v(z_{mi},\theta_{ip},t)\sin\beta\cos\beta+w(z_{mi},\theta_{ip},t)\cos^2\beta\right)\delta(z-z_{mi})\delta(\theta-\theta_{ip})=0 \tag{3}$$

where δ is Dirac delta function, i is the group index of discrete stiffnesses on the same circular, p is the index number in a group, M_n and N_s are the number of group and the number of discrete stiffnesses in one group, respectively. z_{mi} and θ_{ip} are the positions of discrete stiffnesses in the axial direction and circumferential direction, respectively. To investigate the vibration properties of thin cylindrical shells, the set of displacements were assumed to have the wave-like forms,

$$u_{mn}(z,\theta,t) = A_{mn}\cos\frac{m\pi z}{L}e^{(j\omega_{mn}t-jn\theta)} \tag{4}$$

$$v_{mn}(z,\theta,t) = jB_{mn}\sin\frac{m\pi z}{L}e^{(j\omega_{mn}t-jn\theta)} \tag{5}$$

$$w_{mn}(z,\theta,t) = C_{mn}\sin\frac{m\pi z}{L}e^{(j\omega_{mn}t-jn\theta)} \tag{6}$$

where m and n represent the axial half wave number and circumferential wave number, respectively. Equation (4)-(6) were substituted into Equation (1)-(3) and explored the orthogonality properties of $\cos m\pi z/L$, $\sin m\pi z/L$, $e^{jn\theta}$ to yield the eigenvalue equations. The equations were written in matrix form as

$$\mathbf{M}_{mn}\ddot{\mathbf{x}}_{mn} + \left(\mathbf{K}_{mn} + \mathbf{K}^{b}_{mn} + \mathbf{K}^{m}_{mn}\right)\mathbf{x}_{mn} = \mathbf{0} \tag{7}$$

where $\mathbf{x}$=[A_{mn}, B_{mn}, C_{mn}] and $\mathbf{M}_{mn}$ is mass matrix, $\mathbf{K}_{mn}$, $\mathbf{K}^{b}_{mn}$, $\mathbf{K}^{m}_{mn}$ are corresponding system stiffnesses, foundation stiffnesses and discrete stiffness matrix, respectively. Equation (7) is an equation about the variables A_{mn}, B_{mn}, C_{mn}, and the eigenvalue problem can be solved by standard eigenvalue method.

3 Numerical results

Since this study is motivated by multiple stage planetary gear systems, the parameters of a thin cylindrical shell which denotes a ring gear in the planetary gear system are given as ρ=7800kg/m^3, E=206GPa, μ=0.3, k_r=k_θ=1×10^{8}N/m, k_{ip}=5×10^{8}N/m, R=160mm, h=10mm, L=300mm, β=1.22, z_{mi}=L/6, L/2, 5L/6, θ_{ip}=0, 2π/3, 4π/3, i=1, 2, 3, p=1, 2, 3, M_n=N_s=3, the discrete stiffnesses are axially and circumferentially equispaced and the phases between adjacent axial group of the discrete stiffnesses are set to π/4.

The eigenvalue equation obtained from Equation (7) is six order equation, that is, there are six natural frequencies associated with the (m, n)th wave-like solution. However, the negative and positive solutions appear as pair with the same magnitude but opposite sign. Thus, for each positive or negative n, the three positive natural frequencies are chosen and the negative natural frequencies are discarded. There are then only three distinct natural frequencies associated with each (m, n) wave mode, whether $n > 0$ or $n<0$. And each frequency is primarily associated with a particular wave type, that is, shear, or flexural, longitudinal. In what follows, we present the results of numerical investigations, which were carried out by a range of m from 1 to 5 and n from 1 to 10.

Figure 2a illustrates the natural frequency of the thin cylindrical shells on an elastic foundation with multiple group of discrete stiffnesses. It can be seen that the natural frequencies will increase with the increasing of m or n and the natural frequencies can be classified into three groups. And the three groups of frequencies correspond to flexural, longitudinal and shear modes, respectively. For each (m, n)th order, the vibration of the cylindrical shell is comprised of three vibration modes, but one of the three modes is dominant and the other two modes are ignorable. By analysing the vibration mode corresponding to each frequency in Figure 2a, it can be found that the lower frequencies mainly associate with flexural mode as shown in Figure 3a, the intermediate frequencies mainly associate with longitudinal mode as shown in Figure 3b, and the higher frequencies mainly associate with shear mode as shown in Figure 3c. Figure 2b shows the natural frequencies of shear mode with axial half wave number m equals from 1 to 5 and circumferential wave number n equals from 1 to 10. It can be observed that the natural frequencies of the cylindrical shell increase with n at each m nonlinearly and the natural frequencies get closer with the increase of n and m.

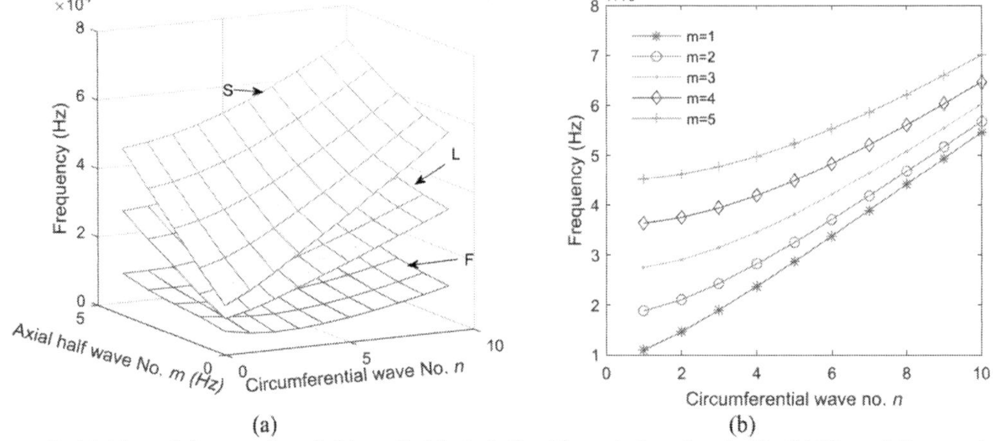

Figure 2. (a) Natural frequencies of thin cylindrical shell with m=1~5 and n=1~10; (b) Natural frequencies of shear mode.

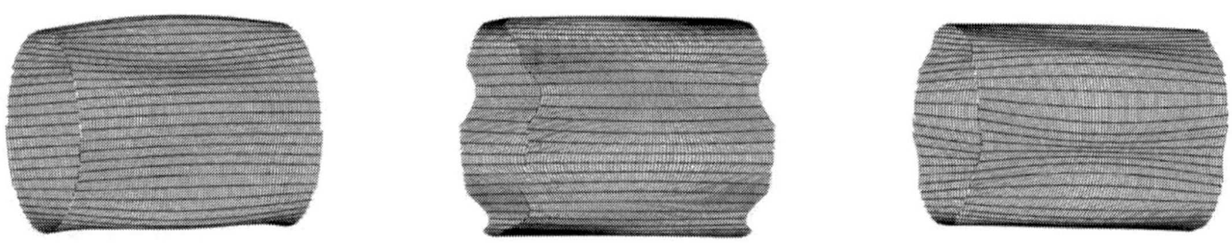

Figure 3. Vibration modes (a) flexural mode; (b) longitudinal mode; (c) shear mode.

The influences of discrete stiffnesses and support stiffnesses to the natural frequencies are illustrated in Figure 4. It can be seen from Figure 4 that with the increase of two types of stiffnesses the natural frequencies of the three modes will all increase. However, Figure 4a shows that the discrete stiffnesses mainly affect the natural frequencies of shear mode, i.e., the high order natural frequencies, and the natural frequencies of the other two modes change very slightly. Figure 4b shows that the support stiffnesses mainly affect the natural frequencies of flexural mode, i.e., the low order natural frequencies. And the intermediate natural frequencies of longitudinal mode almost have no change with the increase of support stiffness.

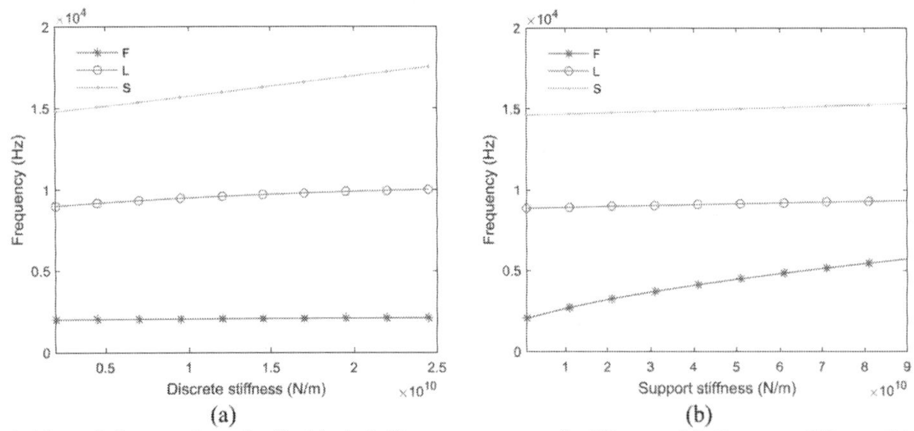

Figure 4. Natural frequencies of cylindrical shell vs. two types of stiffnesses (a) discrete stiffness; (b) support stiffness where m=1 and n=2.

4 Conclusions

Vibration properties of thin cylindrical shells with multiple-stage discrete stiffnesses and foundation stiffnesses were investigated in this paper. The governing equations were obtained by the Hamilton's principle. In order to get the uniform equations, the discrete stiffnesses were modelled as external forces. The wave-like solutions were applied to study the natural vibration properties and the governing equations were discretized in matrix form to obtain the eigenvalues. The properties of natural frequencies and vibration modes of cylindrical shells were studied. The influences of discrete stiffnesses and foundation stiffnesses to natural frequencies of thin cylindrical shells were also investigated. The results show that each natural frequency is mainly dominated by one vibration mode, that is, flexural, longitudinal and shear modes. The natural frequency will get closer with the increase of circumferential wave numbers. The discrete stiffnesses mainly affect the higher order natural frequencies and the support stiffnesses mainly affect the lower order natural frequencies.

Acknowledgements

This work was supported by the National Natural Science Foundation of China (Grant numbers 51775309) and Young Scholars Program of Shandong University.

References

1. T. Chen, C. Bert. Nucl. Eng. Des., **42** (1977)
2. K. Forsberg. AIAA J **2** (1964)
3. V. Weingarten. Aero. Corp. Sys. R&P Div., El Segundo, California, Report No. **169** (1962)
4. G. Warburton. J. Mech. Eng. Sci., **7** (1965)
5. Chung H. J. Sound Vib., **74** (1981)
6. K. Lam, Q. Wu. J. Compo., **31** (2000)
7. T. Ng, K. Lam, J. Reddy. Intl J. Solids Struct., **36** (1999)
8. Y. Kim, J. Bolton. J. Sound Vib., **275** (2004)
9. C. Omer. Thin Wall. Struct., **25** (2007).
10. J. Padovan. J. Sound Vib., **31** (1973)
11. S. Huang, W. Soedel. J. Sound Vib., **118** (1987)

The effect of third principal stress in the measurement of residual stresses by hole-drilling method

David Halabuk[1,a] and Tomas Navrat[2]

[1]*Faculty of Mechanical Engineering, Brno University of Technology, Technicka 2896/2, 616 69 Brno, Czech Republic*

Abstract. One of the most popular and widely used technique for measuring residual stresses is the hole-drilling method. By this method, it is possible to evaluate only biaxial residual stresses located in plane parallel to the surface, but some processes produce a triaxial stress state. For this reason, the evaluation of triaxial stress state by the method used for biaxial state was assessed in this paper. A hole-drilling experiment was simulated by the finite element method for two different stress states. The first stress state considered constant residual stresses in all directions. The second one considered constant residual stresses in a plane parallel to the surface and the residual stress in a direction perpendicular to the surface was equal to zero on the surface and increased with depth. Both states were simulated for various ratios of stress in a direction perpendicular to the stresses in plane. The obtained results show that residual stress in a perpendicular direction affects the evaluation of residual stresses in plane. If the residual stress in the perpendicular direction is high compared to stresses in plane, the error produced by the evaluation of triaxial stress state by the method for biaxial stress state can also be high.

1 Introduction

Residual stresses are usually created in a component as a consequence of manufacturing, heat treatment or loading and can have positive or negative effects. On the one hand, they can cause a change of the component's shape after machining or cracks (Fig. 1) [1]. On the other hand, they can also suppress the creation of cracks or decrease the total stress in a component during loading. The positive or negative effect depends on a sign and magnitude of residual stresses. For this reason, it is important to know the correct value of residual stresses in components. The hole-drilling method is one of the most popular and widely used techniques for measuring residual stresses near the surface of component. This method is based on drilling a small hole to the measured component which causes a redistribution of residual stresses around it (Fig. 2). After that, relaxed strains are measured by a strain gauge rosette and residual stresses are evaluated from these measured strains [2].

[a] Corresponding author : david.halabuk@vutbr.cz

Figure 1. Beam split due to relaxation of residual stresses [1].

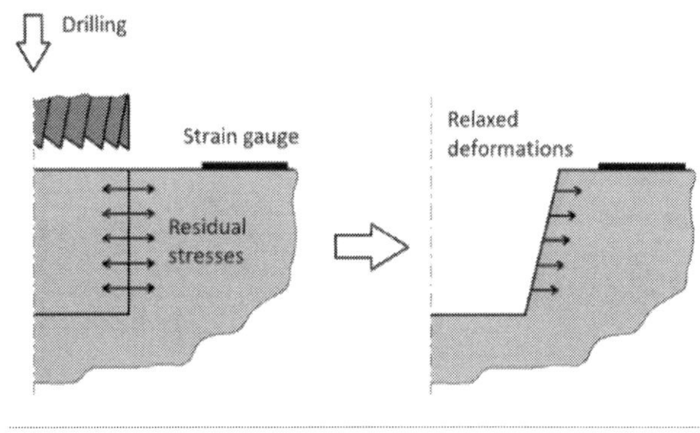

Figure 2. The principle of hole-drilling method.

The accuracy of residual stress evaluation depends on many factors. If the hole is drilled eccentrically, shape or diameter of drilled hole is different from considered ones, or there are extensive plastic deformations the error is produced during evaluation [3, 4]. It is important to known which factors influence the evaluation of residual stresses and how strong this influence is.

There are several methods for evaluating the measured strains. Some of them are convenient for uniform residual stress distribution in depth direction, others for non-uniform ones. Moreover, some methods deal with eccentrically drilled hole or with plastic deformations which occurred around drilled hole when residual stresses were close enough to yield strength of material before hole drilling. However, all methods are able to evaluate only biaxial stress states when two principal stresses (σ_x, σ_y) are located in a plane parallel to the surface of the component. They assume that third stress (σ_z) situated in perpendicular direction to the surface is equal to zero or has a negligible effect on the evaluation of residual stresses σ_x, σ_y.

As shown in [5, 6], processing such as shot peening, welding, or quenching can produce triaxial residual stress states. In most cases, residual stress σ_z is equal to zero at the surface of component but an increase in σ_z can be significant under the surface and thus σ_z can affect the evaluation of residual stresses. It is a problem to evaluate triaxial residual stress states because strain gauge rosettes do not provide necessary amount of information. If we want to deal with residual stress in a direction perpendicular to the surface, we also need to measure a strian or a displacement in this direction. However, in practical applications this displacement is hard to measure. Wern et al. [7] modified an integral method, usually used for biaxial states, for the evaluation of triaxial residual states by additional information about the displacement in the direction perpendicular to the surface. However, there is no known publication in which this method is applied in practice. Moreover, the influence of residual stress σ_z on the evaluation of residual stresses in plane parallel to the surface is also unknown. Since in practical application, as mentioned above, triaxial residual stress state can be created it is important to know this influence. This is the reason why the evaluation of triaxial residual stress states by a method taking into account biaxial ones and the calculation of errors which are produced by this procedure are presented in this paper.

2 Methods

A hole drilling experiment was simulated by the finite element method. For this purpose, the ANSYS software was used. By simulating the hole drilling experiment, it is possible to exclude the uncertainty of material characteristics, measured residual stresses, technological factors or errors caused by measurement process.

A model of thick solid body with a uniform mesh around the drilled hole was created. Linear elements were used for the mesh; the total number of elements was over 2.5 million. Drilling process was simulated by a sequential removal of layers of mesh in place where a hole should be created. Fig. 3 depicts the finite element mesh of a model with a drilled hole.

The linear model of material with an elastic modulus of 210 GPa and Poisson's ratio of 0.3 was used. The drilled hole had a diameter of 2 mm and a depth of 1 mm. The drilling process was simulated by 0.05 depth increment.

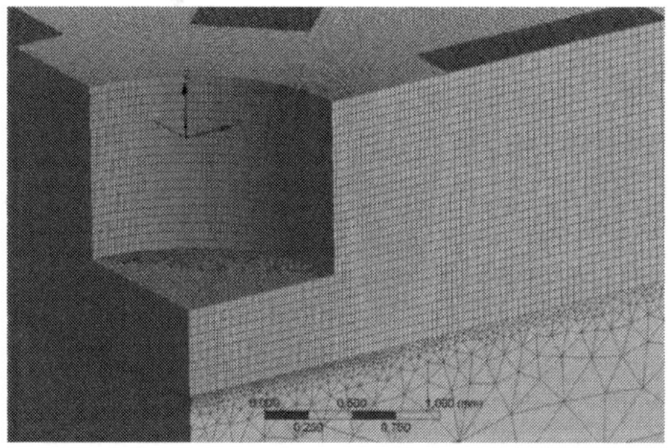

Figure 3. The mesh of finite element method model.

Relaxed strains, after removing the 'drilled-out' elements, were obtained from averaging nodal strains on surface bodies which represented grids of a strain gage rosette. These surface bodies had dimensions of a real strain gauge rosette 1-RY61-1,5/120S (HBM) and were bonded to the surface of measured solid body. Because of symmetry, only a quarter of the whole model was simulated. Two residual stress states were simulated (Fig. 4).

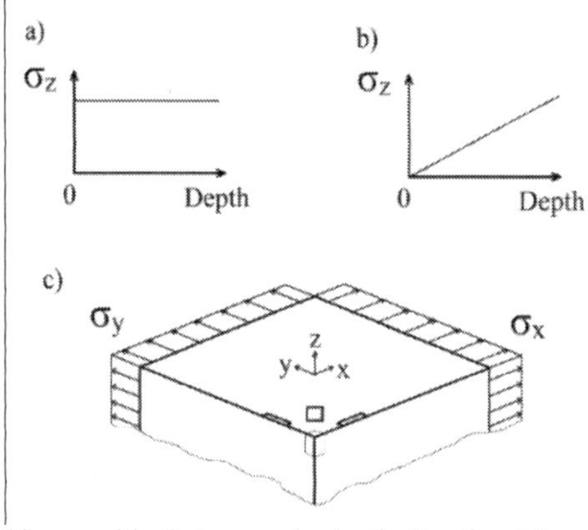

Figure 4. Stress state with a uniform residual stress σ_z in depth direction (a) and stress state with σ_z equal to zero on the surface and a linear increase with depth (b); stresses σ_x and σ_y are uniform for both stress states (c).

The first stress state considers a uniform residual stress σ_z whose value differs from 0 to 250 MPa for analyses. Residual stresses σ_x and σ_y were also uniform with a constant value of 500 MPa for all analyses. The stress state was induced by a command which inputs the values of stresses into every element of the mesh at the beginning of the analysis. For the first state, seven analyses with the σ_z to σ_x ratio equal to 0, 0.05, 0.1, 0.2, 0.3, 0.4 and 0.5 were calculated.

The second stress state had σ_z equal to zero on the surface and its value increased with depth. Values of residual stresses for the second stress state depending on the depth direction are shown in Fig. 5. Individual analyses for this state are labelled by the σ_z to σ_x ratio in the depth of 1 mm. This stress state is closer to reality than the first one because σ_z is usually equal to zero on the surface. Residual stresses σ_x, σ_y are, as for the first stress state, uniform with the value of 500 MPa. This stress state was induced by a non-uniform temperature distribution in combination with volume forces.

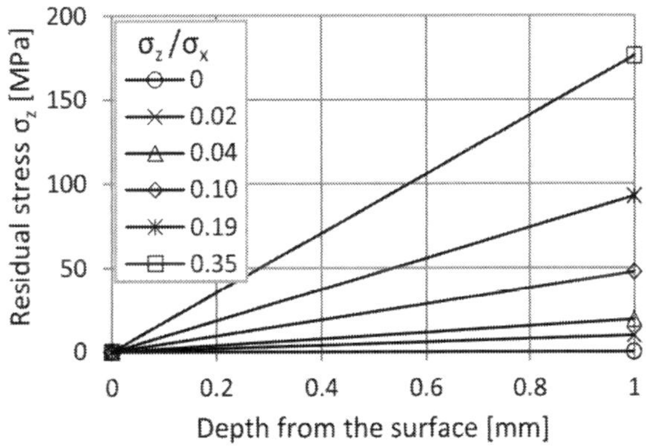

Figure 5. Values of residual stresses σ_z depending on the depth direction for the second stress state.

Strains obtained from simulations were evaluated by the EVAL-7 software and residual stresses were calculated by the ASTM E837-13 Extended Uniform calculation method. Thus calculated residual stresses were compared to the simulated ones and a relative error was calculated as follows.

$$\delta = \frac{\sigma_e - \sigma_s}{\sigma_s} \cdot 100\ [\%] \tag{1}$$

where δ is the relative error, σ_e and σ_s are evaluated and simulated residual stresses in the x-direction respectively. Because simulated stresses in the x and y-directions were the same, also relative error was the same in these directions except for a negligible numerical error. For this reason, relative errors were plotted only for residual stress σ_x.

3 Results and discussion

Fig. 6 illustrates the distribution of normal strain in the x-direction after hole drilling for the first stress state and $\sigma_z/\sigma_x = 0$ (a) and $\sigma_z/\sigma_x=0.5$ (b). It can be noticed that the distribution in component is slightly different between states with different σ_z to σ_x ratios, but more importantly, the difference on the surface in the area of strain gauge grid situated in the x-direction can be observed. If there is a difference of strains on the surface of the component, the value of strain measured by the strain gauge rosette is also different, which affects the evaluation of residual stresses.

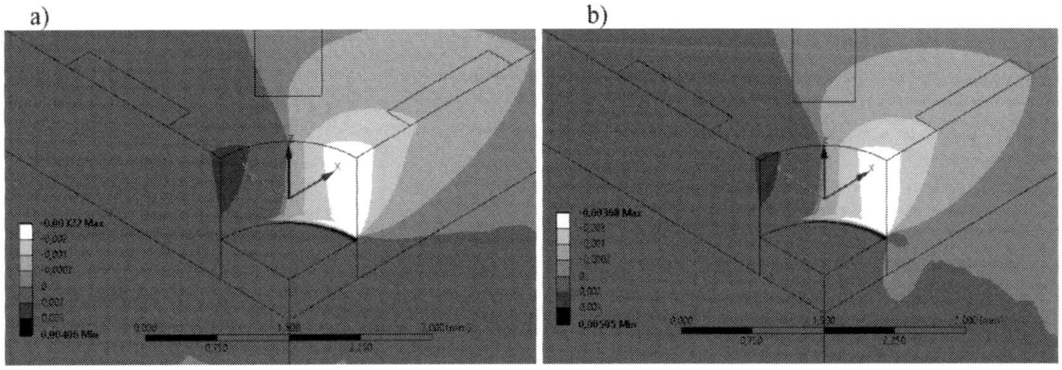

Figure 6. The strain distribution in the x-direction for the first stress state with $\sigma_z/\sigma_x = 0$ (a) and $\sigma_z/\sigma_x = 0.5$ (b).

Values of the measured strains for various σ_z to σ_x ratios are plotted in Fig. 7. The presence of residual stress σ_z caused that the strains measured by the strain gauge rosette are higher for all depth increments than those with σ_z equal to zero. The higher the σ_z to σ_x ratio, the higher the measured strains.

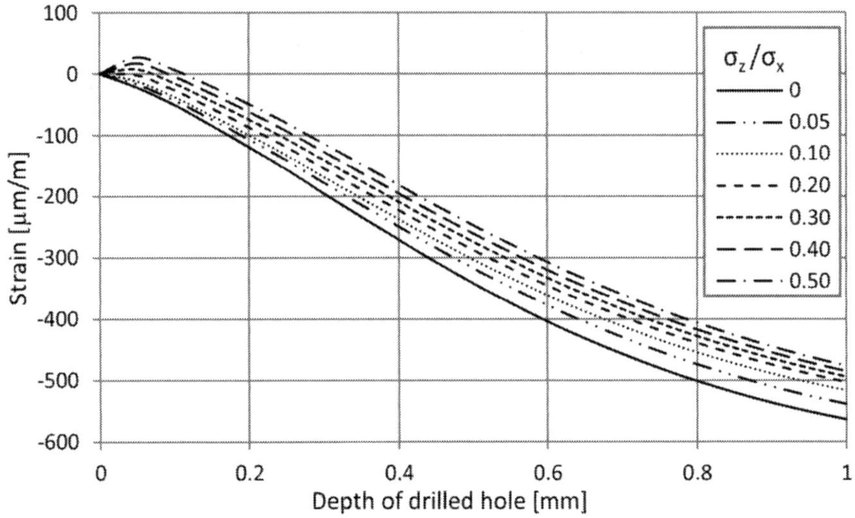

Figure 7. Strains obtained from strain gauge grid situated in *x*-direction for state with uniform σ_z in depth direction

For a state with non-uniform residual stress σ_z there is also upward shift of curves (Fig. 8). However, in this case, due to linear increase of σ_z from zero on the surface, the shift starts from zero in the zero depth and increase with depth. In addition, the shift for same σ_z to σ_x ratios in the depth of 1 mm is lower for this stress state than for the stress state with uniform σ_z.

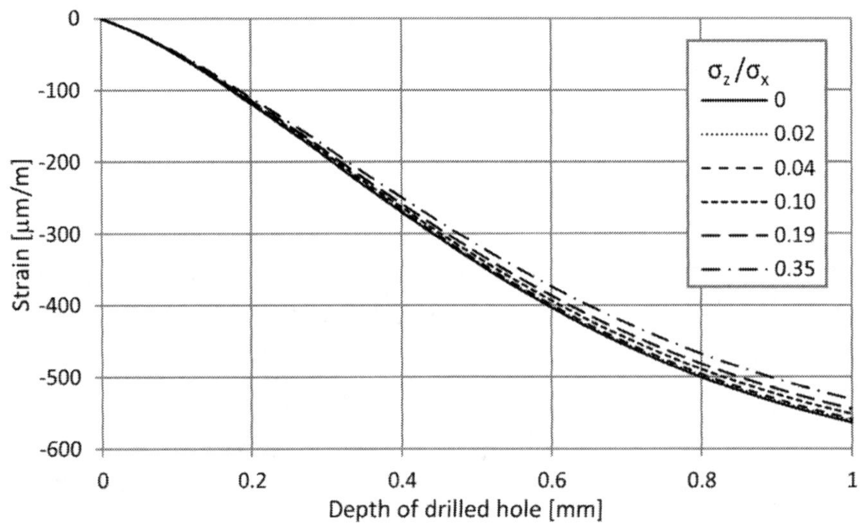

Figure 8. Strains obtained from strain gauge grid situated in *x*-direction for state with σ_z equal to zero on the surface and increasing with increasing depth

Relative errors of evaluated residual stresses depending on the σ_z to σ_x ratio for a state with uniform and non-uniform residual stresses σ_z are illustrated in Fig. 9. Higher values of measured strains for all depth increments for the stress state with a uniform residual stress σ_z caused that relative errors are high already for small σ_z to σ_x ratio. For higher σ_z to σ_x ratios a relative error is high and the presence of σ_z has a significant influence on the evaluation of residual stresses. If σ_z is not constant with depth, but increases linearly from zero on the surface, the values of relative errors are considerably lower. For small σ_z to σ_x ratios the effect of σ_z on the evaluation of residual stresses is negligible. However, if σ_z is high enough compared to stresses in the plane parallel to the surface, it could negatively affect the evaluation of residual stresses.

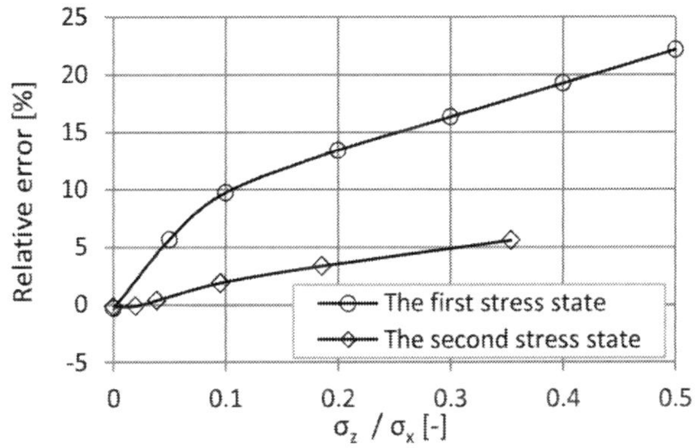

Figure 9. Relative error of evaluated residual stresses for the stress state with uniform σ_z (the first stress state) and σ_z equal to zero on the surface and increasing with increasing depth (the second stress state).

4 Conclusion

The presence of residual stresses in the direction perpendicular to the surface causes that strains measured by the strain gauge rosette are different than strains in biaxial stress state. For this reason, the evaluation of triaxial stress states by a method used for biaxial stress states produces additional errors. The value of these errors depends on the σ_z to σ_x ratio. Simulations of the state with a uniform residual stress σ_z show that the relative error of evaluated stresses could be high already for small σ_z to σ_x ratio. However, in practice, a state with residual stress σ_z equal to zero on the surface and with an increase in the depth direction is more realistic. This state produces considerably lower relative errors and for small ratios the relative error is negligible. Nevertheless, if residual stress σ_z is high compared to residual stresses σ_x or σ_y, the relative errors of evaluated residual stresses in the plane parallel to the surface can increase to significant values. The value of the σ_z to σ_x ratio which is acceptable depends on the required accuracy of residual stress evaluation.

Acknowledgment

This work is an output of projects FV 18 – 18 and NETME CENTRE PLUS (LO1202) created with financial support from the Ministry of Education, Youth and Sports under the „National Sustainability Programme I".

References

1. F. Campus, *Effects of residual stresses on the behaviour of structures* (1954)
2. G. S. Schajer, *Practical Residual Stress Measurement Methods* (John Wiley, 2013)
3. D. Halabuk and T. Navrat, *EAN 2017 - 55th Conference on Experimental Stress Analysis 2017*, 8-14 (2017)
4. T. Navrat, D. Halabuk, M. Vlk, P. Vosynek, Defect Diffus. Forum **382**, 208-212 (2018)
5. G.A. Webster and A.N. Ezeilo, Int. J. Fatigue **23**, 375-383 (2001)
6. S. Hossain, C.E. Truman, D.J. Smith, M.R. Daymond, Int. J. Mech. Sci. **43**, 235-243 (2006)
7. H. Wern, R.Cavelius, D. Schläfer, Strain **33**, 39-46 (1997)

The improvement on thermal performance of herbal ball compress

Wasan Kamsanam [1,a] and Rachaneewan Aungkurabrut [1]

[1] *Mechanical Engineering Department, School of Engineering, University of Phayao, 56000, Thailand*

Abstract. Herbal ball compress is an equipment usually employed in Thai traditional medical treatment to relieve muscular pain. The herbal ball needs to be prepared by steam heating prior to the application in order to extract essential oils from herbal ingredients. This study aims at improving thermal performance of herbal ball so that it can maintain temperature at a suitable level for a particular period of time. In this article, the investigation to facilitate the herbal ball compress usage is performed in various schemes. Adding some materials of high specific heat capacity is one approach. In addition, employing electric powered heat sources such as cable heater, Peltier sheet, ceramic resistor and directly supplied steam are also included in the test. The experimental results revealed that the herbal ball with small water balloons inside can prolong the time duration at a desired temperature range. Direct steam supply system was built by modifying a household electric pan. The experiment demonstrated that the temperature of herbal ball was capably controlled at a desired level through the treatment. It is supposed that the appropriate method could benefit thermal performance of herbal ball compress in regard to temperature control, treatment duration, economical aspect as well as safety circumstance.

1 Introduction

Herbal ball compress has been widely used as a traditional therapeutics in Thai community for centuries [1]. Nowadays, usage of herbal ball compress either at home for personal remedies or at spa services has been spread around the globe. It was approved that muscle sprains, muscular and joint pain can be effectively relieved by hot herbal compress treatment [2-4]. The herbal compress treatment is thus considered as an elaborative practice contributed to its aromatherapy, thermal therapy and herbal therapy. As its name implies, herbal ball compress is a spherical shape cotton pouch with a handle on top. There are variety of herbs in the ball for example Plai or Cassumunar ginger (Zingiber cassumunar Roxb. or Zingiber montanum (J. Koenig) Link ex A. Dietr.), turmeric (Curcuma longa L.), and camphor [5]. To use the herbal ball compress, it needs to be steamed for 10-15 minutes [5] in order to allow heat transferring to the herbs and stimulate the isolation of essential active substances and volatile oils from herbal ingredients.

The effectiveness of thermal modality like herbal ball compress may depend on various factors. Duration of application and therapeutic temperature are among those influencing parameters. It has been suggested that the appropriate temperature for the use of thermotherapy to relieve muscular spasm is between 40 to 45°C [6]. In addition, in order to achieve the useful physiologic effect, heating

[a] Corresponding author : tao_wasan@hotmail.com

modalities should be applied for 20 to 30 minutes [7, 8]. In the therapeutic treatment, hot herbal ball is pressed with a rolling motion on the affected area for a while and taken off for muscular relaxation and moved to the next areas. Such procedure is repeated for 20 to 30 minutes to achieve effective result. It is supposed that the useful area of the herbal ball compress is located at the lower part while the rest which covers larger surface is responsible to heat loss. Thus, herbal ball compress cannot hold the suitable temperature as required.

Though the unique use of herbal ball compress offers several potential health benefits, a major consideration is convective heat loss which results in steadily decrease of temperature. Generally, two or three herbal balls compress are used in a single treatment: one is in use while the others are heated in steamer. Replacing with the hot and fresh herbal ball when the one in use starts to cool is a common approach, even though this is not an economical solution. Several researchers in Thailand put effort into keeping the herbal ball compress temperature at a desired level for longer time span. Electric herbal ball was patented in 2008 [9] and produced for commercial purpose. Before the use, herbal powder in a cotton bag is wrapped around a spherical ball with an electric heater installed inside. Operating temperature can be set and monitored from a separated controller unit. It should be noted that this device was designed to use with herbal powder whereas the capability to apply with dry herbs in ordinary form has not been reported. Thermal energy from chemical reaction is also exploited to heat herbal ball compress as exhibited in [10]. This invention is, however, not widely used due to a sense of complexity. In addition, information regarding temperature controlling has not been revealed.

According to the previous study and the invented products as reviewed above, there is a gap to improve the herbal ball compress in term of thermal performance. The aim of this study is to efficiently control the herbal ball compress temperature at a suitable level and, also, prolong its applying duration to fully achieve the clinical effects of the heat treatment. The modification on herbal ball compress is associated with adding material of high specific heat capacity and supplying heat sources by various schemes directly into the ball. The experimental results of the modified herbal ball compress will be compared with the traditional one. It is supposed that the redesigned herbal ball compress would fulfil the deficiency of original product. Moreover, it is an opportunity to upgrade local brands to be in favoured worldwide.

2 Material and method

In this study, thermal performance of traditional herbal ball compress is compared with the ones which have been modified in various arrangements. The main ingredients of herbal ball compress are variety of dried herbs as has been described in the introduction section. The prepared herbs are wrapped by cotton cloth in spherical shape with a handle on top. Diameter and weight of the herbal balls compress in the test are about 22 cm and 200 g, respectively. All tested material used in the current investigation are produced by Baan Tham Community Enterprise in Dok Khamtai District, Phayao Province, Thailand. Thermal performance of herbal ball compress will be observed through temperature profile with time evolution. Prior to perform the temperature measurement, determination of moisture content (MC) is conducted.

It is expected that MC in herbal ball compress may have effects on the experimental results. Thus, this parameter is evaluated corresponding to the Standard Test Methods for Direct Moisture Content Measurement of Wood and Wood-Base Materials (ASTM D4442-92) in Method A—Oven-Drying (Primary) [11]. To determine the MC, a force convection oven is used to dry herbal ball compress. Hot air temperature in the oven chamber will be controlled to maintain at 103±2°C throughout the drying process and the evaporated moisture is able to be vented via the device's ventilation system. Moisture content based on dry basis can be calculated as follows:

$$MC\ (\%) = \frac{(A-B)}{B} \times 100 \tag{1}$$

where A is original mass and B is oven-dry mass. In equation (1), the oven-dry mass is obtain at the endpoint of drying process. The endpoint is assumed when the mass reduction within 3 hours of specimen is equal or less than two times the sensitivity of selected balance. An electric balance used in the current experiment has the sensitivity of 0.001 g. Hence, the end point is achieve when the mass loss of herbal ball compress is less than 0.002 g in a three hours drying period.

Regarding experimental procedure, each herbal ball compress will be heated from room temperature until reaching around 90°C and keeping at this level for 10 – 15 minutes to allow essential oil from herbs ingredient to be extracted. Then, heating is terminated to let the herbal ball temperature decrease. Time duration when the herbal ball decreases from 45°C to 40°C is recorded for subsequent analysis. In order to observe temperature variation of the herbal ball compress, four K type thermocouple probes are installed at different locations as indicated in Figure 1. The temperature measurement points are chosen corresponding to the usable area where the herbal ball is being compressed and rolled over patient's body. Herbal ball compress filled with materials of high specific heat capacity such as water balloons and white pebbles (river rock) are included in the test to compare with the traditional herbal ball. There are two sets of herbal balls filled with water balloon: one herbal ball is added with a water balloon of 120 g and another is added with six water balloons of 20 g each. For the herbal ball filled with rock, the mass of white pebbles is prepared approximately 120 g. Besides adding extra materials, heat source supplied into herbal ball by different methods are also taken for the comparison. The heating suppliers chosen in this investigation are 60 W Peltier sheet, cable heater of 65 W rated power, 9 pieces of 5 W ceramic resistor and steam generated from an 800 W electric steamer. To supply steam into the herbal ball directly, a high heat resistance silicone tube of 3/4 inch nominal diameter is connected to the cover lid of steamer while another end of the tube is inserted into the herbal ball compress. Mass flow of steam can be controlled by a regulator installed on the steam tube.

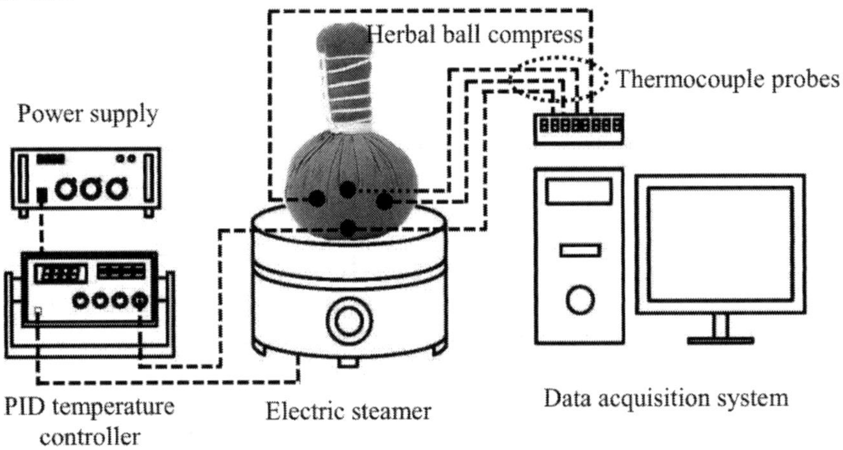

Figure 1. Illustration of experimental setup and a herbal ball in the experiment.

In this experiment, measured variables contain errors or uncertainties which can be categorised into two groups: precision uncertainty (*P*) and bias uncertainty (*B*). Regarding the determination for moisture of content from equation (1), its overall uncertainty (U_{MC}) is evaluated at 95 percent confidential level by the following combination of a precision contribution to the uncertainty of MC (P_{MC}) and a bias contribution to the uncertainty of MC (B_{MC}) [12]:

$$U_{MC}=\sqrt{P_{MC}^2+B_{MC}^2} \quad (2)$$

The precision contribution (P_{MC}) and the bias contribution (B_{MC}) can be found by the error propagation equation of Kline and McClintock [13] as shown in equation (3) and (4), respectively.

$$P_{MC}^2=\left(\frac{\partial MC}{\partial A}\right)^2 P_A^2+\left(\frac{\partial MC}{\partial B}\right)^2 P_B^2 \quad (3)$$

$$B_{MC}^2 = \left(\frac{\partial MC}{\partial A}\right)^2 B_A^2 + \left(\frac{\partial MC}{\partial B}\right)^2 B_B^2 \tag{4}$$

The precision limit P_A and P_B can each be calculated as 2 times the standard deviation of unsteadiness of a set of mass measurements of the original mass (A) and oven-dry mass (B) of the herbal ball compress, respectively. Regarding the bias limit B_A and B_B, it cannot be directly determined by any statistical technique. However, it is estimated from a half of the readability of the digital balance [14]. According to the calculation, the overall uncertainty of moisture of content (U_{MC}) is about ±0.5%. With the same methodology, the uncertainty of temperature measurement is evaluated at ±0.53°C.

3 Experimental results and discussion

In this study, the effort to prolong the temperature of herbal ball compress at an appropriate level is carried out by several techniques. It is supposed that material of high specific heat capacity could accomplish this intention. Water is a high specific heat capacity substance due to hydrogen bonding among water molecules. Thus, the quantity of water in the herb represented by percentage of moister content could have significant effects on thermal characteristic of the herbal ball compress.

Figure 2 shows the relationship between the time duration when the temperature of herbal ball compress decreases from 45°C to 40°C as a function of dry basis moisture content. The temperature range from 45°C to 40°C is in consideration here as it is the effective temperature for therapeutic treatment. Equation developed by data regression and R-squared are also depicted. Dashed line in the figure is drawn from the developed equation. It can be seen from the graph that at low percentage of moisture content the time duration when temperature of herbal ball compress decrease from 45°C to 40°C is shorter than that of the higher moisture content. Regarding specific heat property, it may represent the ability of material to store thermal energy and maintain its temperature. At temperature around 45°C, specific heat capacity (c_p) of water is 4.18 kJ/kg·K whereas the average c_p of wood based material is in the range of 0.77 to 1.78 kJ/kg·K [15, 16]. An increase in moisture content improves the bulk specific heat capacity of herbal ball compress because the specific heat of the water is greater than that of herbs. Thus, at the specified temperature range, the herbal ball compress with higher moisture content could extend the time duration comparing to that of the drier one. From this observation, it is supposed that herbal ball compress with water balloon or white pebble rocks inside could improve its thermal property. The investigation regarding this matter is revealed in Figure 3.

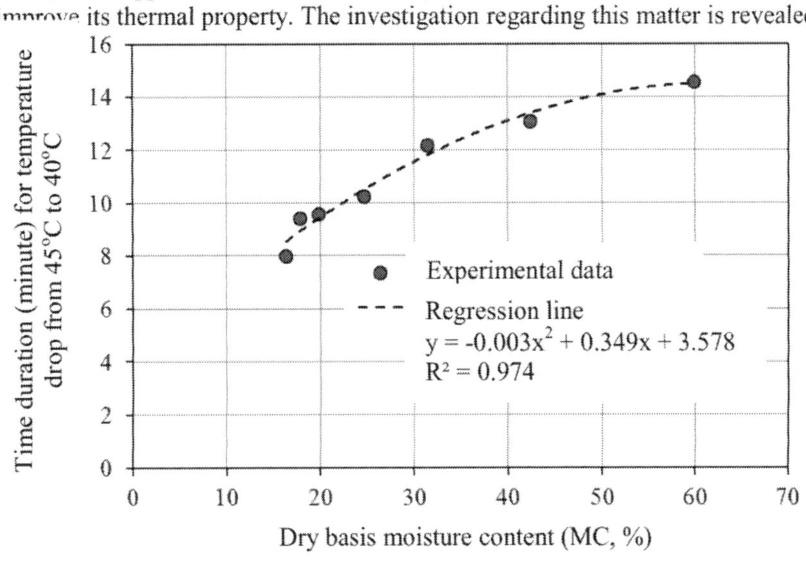

Figure 2. Dry basis moisture content (MC, %) with time duration for temperature drop from 45°C to 40°C.

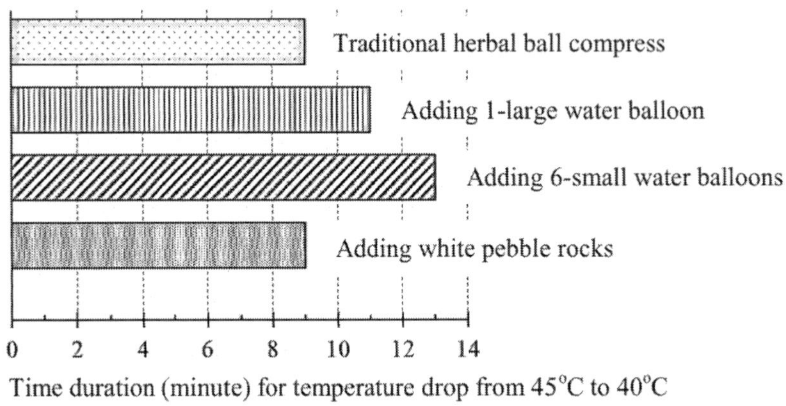

Figure 3. Time duration for temperature drop from 45°C to 40°C of herbal ball added with different materials.

From Figure 3, the bar graph represents time duration while temperature decreases from 45°C to 40°C on the traditional herbal ball comparing to the ones with added different materials. Water balloon inside the herbal ball can extend the time duration due to the higher specific heat capacity of water enhancing heat storing of the herbal ball. Six small water balloons provide the longer time than that with one large water balloon. This could be due to scattered location of small water balloons which is capable of absorbing more heat energy than one large water balloon. Regarding the pebble rocks, the result is similar to that of the traditional herbal ball as its mean specific heat capacity is only about 0.8 kJ/kg·K [17].

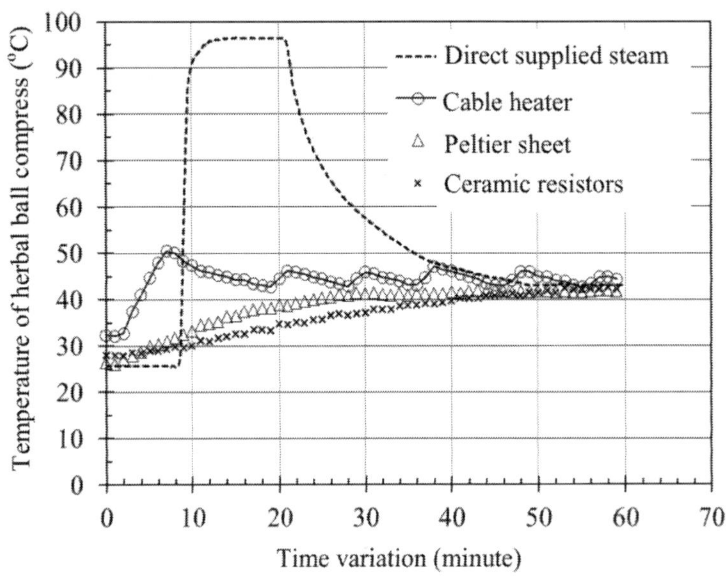

Figure 4. Temperature of herbal ball compress with time variation for various heating methods.

Although adding several small water balloons in the herbal ball could prolong the temperature to be within the desired level, it does not meet the suggested duration. In order to allow heat to increase the regional blood flow and to release volatile oils from herbs to help exerting anti-inflammatory action on the affected muscles, hot herbal compression should be applied around 20 - 30 minutes [7]. To achieve this time span, another options is installing heating source into the herbal ball. As shown in Figure 4, using cable heater, Peltier sheet and ceramic resistors as heating elements could raise herbal ball temperature to 45°C. However, it is not sufficient to facilitate the release of volatile oils from herbal ingredients which required at up to 90°C. Increasing rating power of the heating elements may damage the herbal ball as the heating will concentrate only on the elements' surface and herb ingredients may possibly be burned. For supplying steam directly into the herbal ball, the temperature profile shows that this method achieves thermal requirement for therapeutic heat treatment.

4 Conclusions

It was approved that herbal ball compress which is a blend of herbs could be used to relieve muscle sprains, joint pain and discomfort. To achieve therapeutic purpose, the herbal ball needs to be heated for 15 - 20 minutes to extract essential oils from herb ingredients. In addition, hot herbal compression at 40 – 45°C for 20 – 30 minutes is recommended to receive the effective result. In traditional treatment procedure, two or more herbal balls compress are required as the temperature steadily decreases and it is difficult to control the temperature to be in the appropriate level. Herbal ball compress with added water balloons could extend time duration for the required temperature range. However, it is not long enough to meet the therapeutic suggestion. Heating modalities by cable heater, Peltier sheet and ceramic resistors are chosen for the test. According to their capabilities, they may not be suitable to be used as heat sources for herbal ball compress. Direct supplied steam is likely to be the suitable practice as the temperature of herbal ball compress can be controlled at a required condition. Moreover, a conventional electric steamer for household use with a little modification can be applied in this method. This will benefits the initial investment as only steam tube and a few accessories are additionally required.

Acknowledgement

The authors gratefully acknowledge the contributions of Mr.Pramote Ploydeelert, Mr.Wichian Saensong and Mr.Donlawat Hanyut for their helps on setting up a part of the experimental apparatus and a preliminary testing on the system capability. The authors also wish to express their appreciation to the reviewers and editors for their very useful comments and contributions.

References

1. V. Chokevivat, A. Chuthaputti. *6GCHP*, pp.1-25 (2005)
2. J. Boonruab, N. Nimpitakpong, W. Damjuti, JEBIM, **23**, pp. 1-8 (2018)
3. P. Sukonthasarn, THPJ, **1**(1), pp. 23-35 (2004)
4. Institute of Thai Traditional Medicine, *Thai Traditional Medicine and Health Care* (2003)
5. T. Dhippayom, C. Kongkaew, N. Chaiyakunapruk, P. Dilokthornsakul, R. Sruamsiri, S. Saokaew, A. Chuthaputti, eCAM, **2015**, 942378 (2015)
6. J. F. Lehmann, *Therapeutic heat and cold*, pp. 439-447 (1990)
7. P. S. Tepperman, M. Devlin, CFP, **32**, pp. 1110 (1986)
8. M. Y. Kim, J. H. Kim, J. U. Lee, Y. M. Kim, J. A. Lee, N. M. Yoon, B. Y. Hwang, K. J. Kim, H. M. Lee, B. Kim, J Phys Ther Sci, **23**, 5, pp. 797-801 (2011)
9. B. Khemjareun, PAT. 4209, Thailand (2008)
10. S. Upalee, S. Liawruangrath, PAT. 13508, Thailand (2018)
11. ASTM, *ASTM D4442* (2007)
12. J. H. Kim, T. W. Simon, R. Viskanta, J Heat Transfer, **115**, 1, pp. 5-6 (1993)
13. S.J. Kline, F.A. McClintock, Mech Eng, **75**, pp. 3-8 (1953)
14. R.H. Dieck, W.G. Steele, G. Osolsobe, *ASME PTC 19.1-2005* (2005)
15. Ł. Czajkowski, W. Olek, J. Weres, R. Guzenda, Wood Sci Technol, **50**, 3, pp. 537-545 (2016)
16. M. Gupta, J. Yang, C. Roy, FUEL, **82**, 8, pp. 919-927 (2003)
17. E. C. Robertson, B. S. Hemingway, *USGS* (1995)

Mechanical Vibration Mechanism Motion Simulation of Ultrasonic Composite Vibration Polishing Equipment

Wan Hongqiang[a], Han Peiying, Ge Shuai, Li Fancong and Zhang Simiao

School of Mechanical Engineering, Xi'an Technology University, Xi'an 710021, China

Abstract. In order to design a kind of mechanical vibration part with stable motion in ultrasonic composite vibration polishing equipment. firstly, the main working principle of ultrasonic composite vibration polishing equipment is briefly introduced. Secondly, the mechanical vibration structure of the design is analyzed theoretically, again, using ADAMS software to simulate the crankshaft connecting rod piston mechanism. When the crankshaft speed is 750r/min, the stroke of the piston is 90mm, which is similar to the theoretical analysis value. Finally, using ANSYS Workbench software to analyze the crankshaft statics. The analysis shows that the design of the mechanical vibration structure is reasonable, which provides a theoretical basis for the study of the mechanical vibration structure of ultrasonic composite vibration polishing equipment.

1 Introduction

With the development of other disciplines such as aerospace, VLSI and optical instruments, the quality requirements for the surface of the key parts needing precision machining are more and more high, and the characteristic dimensions of the substrates used in the IC circuits have reached the 45nm[1] with the large scale development trend. Traditional parts polishing relies on the experience of the workers to complete, but its processing cycle is long and low efficiency, and its processing quality is not stable [2], the current mainstream polishing method is composite vibration polishing, mainly including ultrasonic electrochemical composite polishing, ultrasonic-chemical composite polishing, ultrasonic auxiliary mechanical vibration composite polishing, ultrasonic fluid jet composite polishing, ultrasonic magnetic abrasive composite polishing and other polishing methods [3-7], but these polishing methods for the current requirements still have deficiencies. So domestic and foreign research institutions are still actively exploring new methods which have high-quality, high efficiency and low-cost surface processing.

In this paper, a composite surface machining technique combining ultrasonic machining, vibration processing and free abrasive machining is presented. Because the stability of mechanical vibration structure of ultrasonic compound vibration polishing equipment directly affects the polishing quality of workpiece surface, the mechanical vibration part of ultrasonic compound vibration polishing equipment is analyzed in this paper.

[a] Corresponding author : 413162931@qq.com; Tel:08613363941134

2 Working principle of ultrasonic composite vibrating polishing equipment

Ultrasonic composite Vibration polishing working principle as shown in Figure 1., the workpiece is driven up and down by the mechanical vibration mechanism, and the workpiece is polished by the high speed relative motion of the abrasive and workpiece, and the energy assisted by the cavitation effect of ultrasonic vibration is used to remove the workpiece surface material by the aid of ultrasonic vibration, to achieve the workpiece surface polishing purposes.

3 Structural design of mechanical vibration part

The mechanical vibration part of ultrasonic compound vibrating polishing equipment is mainly connected with crankshaft connecting rod by crankshaft, connecting the piston through crank connecting rod, connecting clamp under the piston, and with the driving of crankshaft, the fixture is vibrating up and down in vertical direction. The mechanical vibration mechanism is the evolution of the crank-connecting rod structure, so the crank-piston mechanism can be simplified to crank-slider mechanism for theoretical analysis, and its structure diagram is shown in Figure 2.

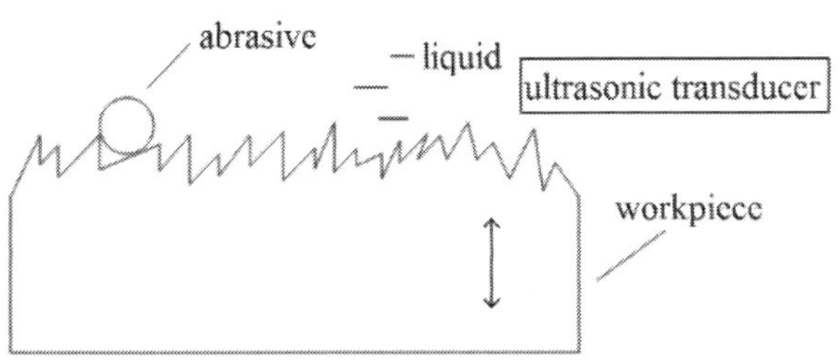

Figure 1. Schematic diagram of ultrasonic composite polishing process.

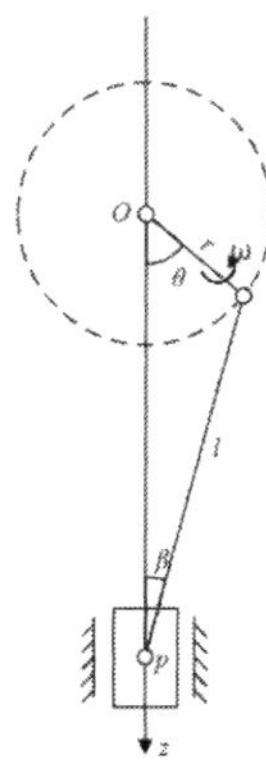

Figure 2. Mechanism sketch of crank slider.

Wherein, "r" is the crank length, the "l" is the connecting rod length, the point P is the fixture particle, "O" is the crank rotation center, and "θ" is the angle that the crank rotates from the initial position, the "β" is the angle that the connecting rod rotates from the initial position, and "z" is the displacement of the jig. In this paper, the relation between the displacement, velocity, acceleration and θ of the slider is mainly studied.

The displacement expression (1) of the slider is based on the triangular relation, which is based on the 2 schematic diagram of the crank-slider structure, namely.

$$z = r\cos\theta + \sqrt{l^2 - r^2\sin^2\theta} \tag{1}$$

$$\theta = \omega t \tag{2}$$

$$\omega = \frac{2\pi}{T} = 2\pi n \tag{3}$$

In the formula, ω--the angular velocity of the crank,

t--rotation time

T--rotation cycle,

n--the speed of the driving motor of mechanical vibrating mechanism.

The speed of the slider can be expressed by (4).

$$v = \frac{dz}{dt} = \frac{dz}{d\theta}\bullet\frac{d\theta}{dt} = \omega\bullet\frac{dz}{d\theta} \tag{4}$$

In the type, "v" is the speed of the slider.

Substituting (1), (2) into (4), we can get the slider-type velocity expression.

$$v = -\omega r \bullet \sin\theta \bullet \left[1 + \frac{r\cos\theta}{\sqrt{l^2 - r^2\sin^2\theta}}\right] \tag{5}$$

Because the acceleration of the slider can be expressed by the following expression.

$$a = \frac{dv}{dt} = \frac{dv}{d\theta} \bullet \frac{d\theta}{dt} = \omega\frac{dv}{d\theta} \tag{6}$$

In the formula, "a" is the acceleration of the slider.

Therefore, substituting (5) into (6) can be used to get the acceleration expression of the slider (7).

$$a = -\omega^2 r\left[\cos\theta + \frac{r(l^2\cos 2\theta + \mathrm{r}^2\sin^4\theta)}{(l^2 - \mathrm{r}^2\sin^2\theta)^{3/2}}\right] \tag{7}$$

The rotating speed of mechanical vibrating mechanism is 750r/min, the length of crank is r=45mm, the length of connecting rod is l=150mm. When the clamp reaches the maximum displacement, the maximum displacement is z=195mm, the corresponding velocity and the acceleration value are v=0m/s, a=3.6x10^5mm/s^2 respectively, when the rotation angle, the jig achieves the smallest displacement, its minimum displacement z=105mm, the corresponding velocity and the acceleration value respectively are v=0m/s, a=-1.97x10^5mm/s^2. The stroke of the slider is 90mm.

4 Kinematic Simulation and dimension optimization of crank mechanism

In the ultrasonic compound vibrating polishing equipment, the mechanical vibration part plays a vital role, the polishing of workpiece is the result of the interaction between the mechanical vibration part and the ultrasonic vibration part, so it is necessary to carry on the motion simulation to the mechanical vibration part. For the convenience of analysis, this paper simulates the mechanical vibration part by using the software of Adams. The model of crankshaft linkage piston mechanism is established in Soildworks software, which is saved as Parasoild format and imported into Adams software, and the model of crankshaft linkage piston mechanism after import is shown in Figure 3.

The model of the crankshaft connecting rod piston is applied to constrain and drive. The model unit is set to MMKS units, and the gravity is negative in the Y axis, with a value of 9.806.

4.1 Add constraint

-The title is set in bold 14-point Arial, flush left, unjustified. The first letter of the title should be capitalised with the rest in lower case. You should leave 35 mm of space above the title and 6 mm after the title.

-Add a rotation pair. The rotation pair between the ground and the crankshaft, the rotating pair between the connecting rod and the crankshaft, the rotating pair between the connecting rod and the piston pin are added respectively.

-Adds a mobile pair. Add a moving pair between the ground and the piston.

-Add a fixed constraint. Adding a fixed pair between the connecting rod cover and the bolt, a fixed pair between the connecting rod lid and the connecting rod, a fixed pair between the piston pin and the piston.

-Adds a mobile pair. Add the piston relative to the ground of the moving pair.

4.2 Add Driver

Add a constraint at one end of the crankshaft and the rotational speed is set to 750r/min.

Add constraints and driven models as shown in Figure 4.

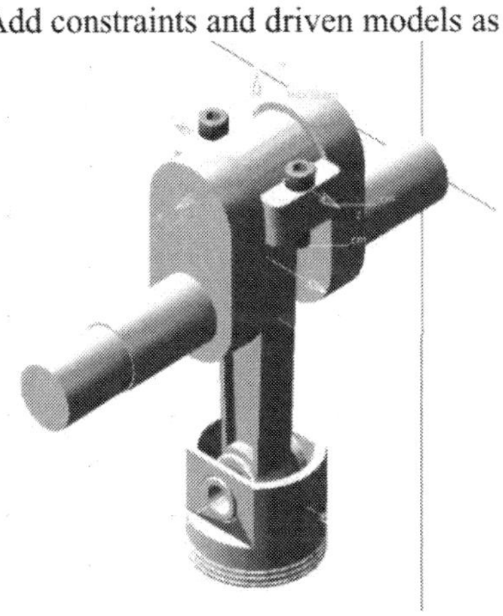

Figure 3. The model of the piston mechanism of the crankshaft connecting rod in ADAMS.

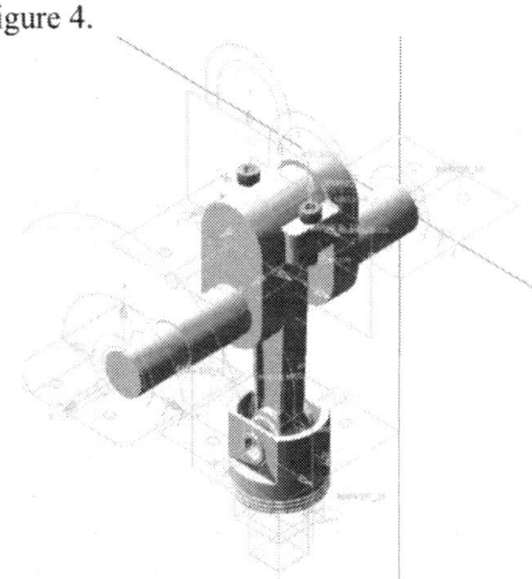

Figure 4. Impose constraints and drives on the model.

The displacement, velocity and acceleration of the piston are simulated. The simulation time is two cycles 0.16s, and in order to make the simulation data more precise, the simulation steps are 100 steps. The simulation results are shown in Figure 5 -7. respectively.

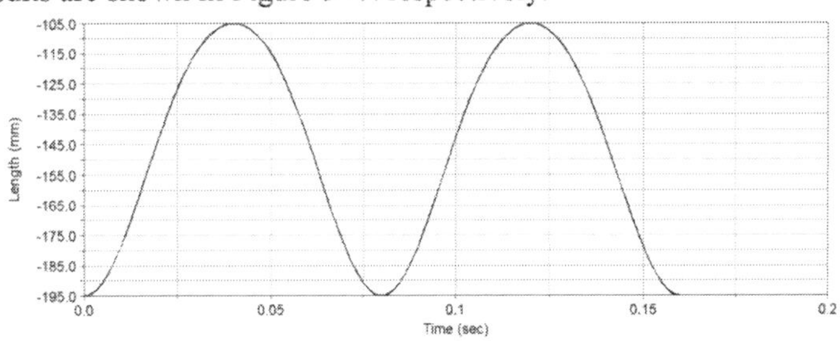

Figure 5. Displacement diagram of the piston.

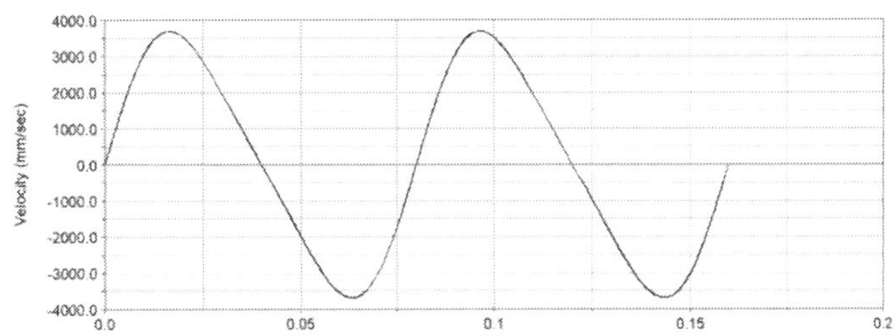

Figure 6. Velocity diagram of the piston.

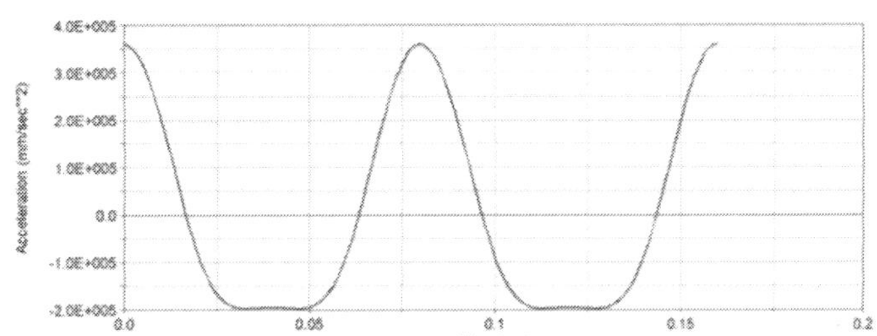

Figure 7. Acceleration diagram of the piston.

As can be seen from Figure 9, the displacement and velocity curves of the piston are sinusoidal distributions, and the change interval of displacement is [-195, -105]mm, and the change interval of velocity is [-3686.4, 3686.4]mm/s. Therefore, we can calculate the piston stroke of 90mm, which is consistent with the theoretical calculation value, which shows that the above theoretical analysis is correct. The acceleration change interval of the piston is $[-1.97\times10^5, 3.6\times10^5]$mm/s^2.The displacement, velocity and acceleration of piston are stable with time curve in y direction and there is no obvious fluctuation. The simulation result is reasonable and the expected result is achieved.

5 Finite element analysis of crankshaft

In order to study whether the strength of crankshaft is up to the requirement, Ansys Workbench is used to analyze the crankshaft stress. The following are the specific implementation processes.

-Import model. After saving the crankshaft model drawn in Soildworks as Parasoild format, the static analysis is introduced into the Ansys Workbench.

-Set the crankshaft parameters. The crankshaft material chooses the ductile iron, and its elastic modulus is 173GPa, Poisson's ratio is 0.3, the density is 7.3g/cm^3, the connecting rod material chooses the alloy steel, its elastic modulus is 206GPa, the Poisson's ratio is 0.3, the density is 7.9g/cm^3, the bolt and the piston pin material chooses the carbon steel, and its elastic modulus is 206GPa, Poisson's ratio is 0.24, density is 7.3g/cm^3, piston material chooses cast iron, its elastic modulus is 126GPa, Poisson's ratio is 0.24, density is 7.0g/cm^3

-Dividing the grid. By using the Hexahedral grid method, the cell meshes are divided into 5mm, the number of nodes is 54, 476, and the number of units is 28, 760, as shown in Figure 8.

-Imposing constraints and loads. The left end face of the crankshaft is fully restrained, the housing is formed on the left side of the crankshaft, the drive is added to the right side of the crankshaft, the torque is 15529N·mm, and the acceleration of gravity is added. As shown in Figure 9.

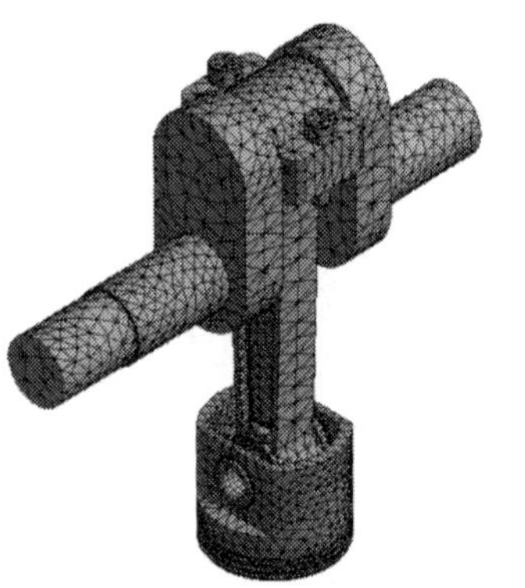

Figure 8. Meshing of the model of the crankshaft, connecting-rod and piston.

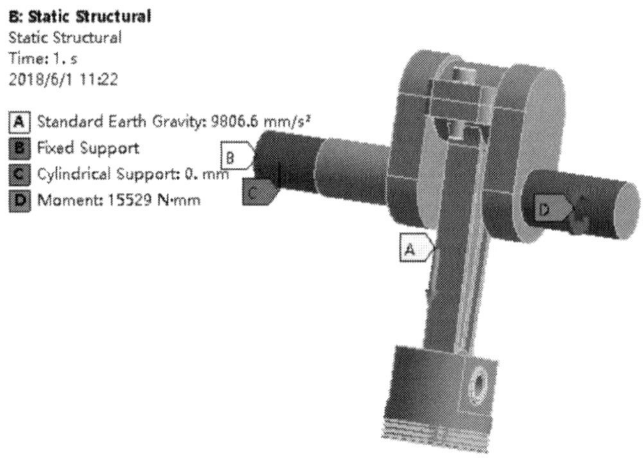

Figure 9. Applys the restraint and load.

-Solving and post processing. The total deformation analysis and the stress cloud of the crankshaft connecting rod piston mechanism are obtained by using the Solve command. As shown in figures 10. and figure 11.

As can be seen from the total displacement diagram, the maximum displacement deformation of the crankshaft is the largest, and the maximum displacement deformation is 9.12×10^{-3}mm, from the equal effect, it can be seen that the stress and deformation at the contact boundary between the crankshaft shaft neck, the bolt and the connecting rod, the piston pin and the connecting rod and the piston are the largest 27.307MPa. Because the material of the crankshaft is carbon-graphite cast iron, its allowable stress value is 320MPa and 320MPa>27.307MPa, so the crankshaft connecting rod piston mechanism satisfies the design requirement.

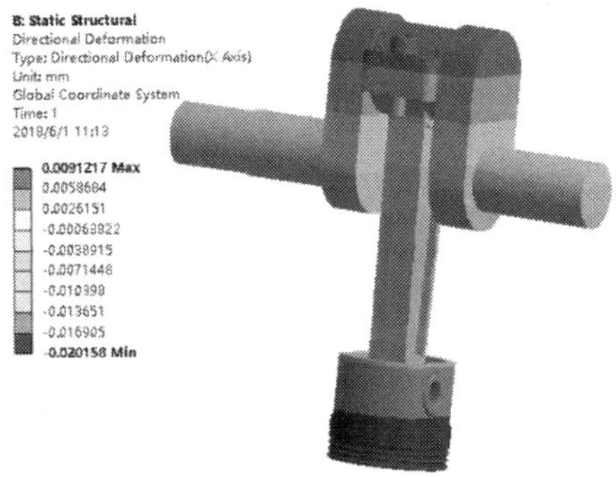

Figure 10. Total deformation.

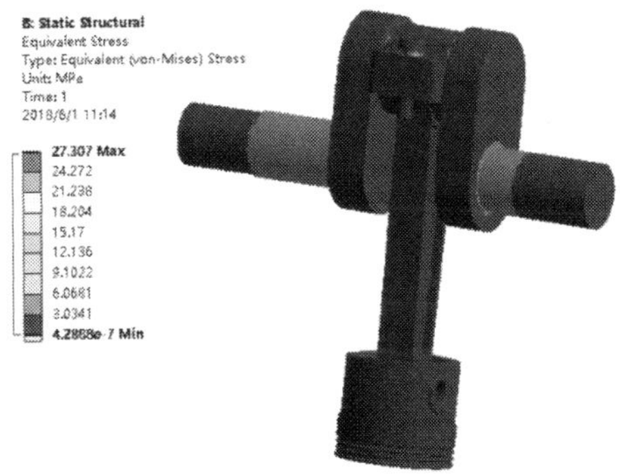

Figure 11. Equivalent stress.

6 Conclusions

The theoretical analysis and calculation of the crank-rod piston mechanism model is provided to provide theoretical basis for the design of ultrasonic composite vibration polishing equipment. The model of crankshaft linkage piston mechanism established in Soildworks software is introduced into Adams software, and the velocity, acceleration and displacement curves of piston are analyzed. The results are similar to the theoretical calculation, when the crankshaft speed is 750r/min, the stroke of the piston is 90mm. In order to study whether the crankshaft strength satisfies the requirement, the stress analysis of the crankshaft connecting rod piston mechanism is carried out by using the Ansys Workbench software, the maximum stress and deformation of crankshaft is 27.307MPa, which is less than the allowable stress of crankshaft 320MPa, so the strength of crankshaft satisfies the requirement. This study provides a new design idea and theoretical basis for similar ultrasonic composite vibrating polishing mechanism.

Acknowledgement

This study is funded by Scientific research project of Key Laboratory of Shaanxi Provincial Department of Education (16JS047).

This study is funded by Science and Technology Research and Development Program of Shaanxi Province(2016GY-013).

References

1. Li Liming, Li Mao, Zhu Yongwei. *Research status and prospect of consolidation abrasive grinding and polishing [J]. Diamond and Abrasives Engineering*, 2009, (05): 17-22.
2. Chen Yu Jun, Tangyu, Miao want to light, Yin Hui. *A review of abrasive jet surface polishing [J]. Surface Technology*, 2015, 44 (11): 119-127.
3. GUO Z N, LEE T C, YUE T M, et al. *The Design of an Ultrasonic Polishing Tool by the Transfer-matrix Method [J]. Materials Processing Technology*, 2000, (102): 122-127.
4. SINGH R, KHAMBA J S. *Ultrasonic Machining of Titanium and Its Alloys: A Review [J]. Materials Processing Technology*, 2006(6): 125–135.
5. ZHAN J M, ZHAO J. *Study of the Contact Force in Free-form-surfaces Compliant EDM Polishing by Robot [J]. Materials Processing Technology*, 2002(129): 186-189.
6. Sun Qi. *Experimental study on ultrasonic-electrochemical polishing of silicon carbide by experiment [D].Harbin: Harbin Institute of Technology*, 2016.
7. Li Ren Chuang. *Experimental study on grinding and polishing of jade surface by ultrasonic Vibration auxiliary device [D].Liaoning: Liaoning University of Science and Technology*, 2017.

Research and Application on the Key Parameters of the Long Compression and Short Pumping Ventilation Dust Removal System

ZHANG She-ji[a], MA Wei and MO Jin-ming

China Coal Technology and Engineering Group Chongqing Research Institute, Chongqing 400037, China

Abstract: For the question of circulation wind emerged in the process of using long compression and short pumping ventilation dust removal system, a new model of air volume matching was proposed, and the empirical formula of the key process parameters such as axial and radial air volume, axial and radial air out distance were derived according to the relevant theories of air jet in the confined space. Then the 20106 return roadway of fully mechanized working face in Wang Jia Ling coal mine was selected to conduct field verification tests. The result showed that the new air volume catching model had good adaptability, and the key process parameters obtained by theoretical calculation and field test were basically identical. Under the condition of optimum technological parameters, the total dust and respiratory dust concentration of the driver's place can be reduced 91.98% and 89.53% respectively.

1 Introduction

Long compression and short pumping ventilation dust removal system was one kind of mixing ventilation methods in the driving face, using dust-clearing fan to trap and purify the dust produced when cutting coal in the head working area of the driving face, on the basis of conventional pressure ventilation. Dust removal system was one of the most effective and mature dust control measures, cooperated with the wall attaching chimney and other dust control technology, dust efficiency can reach above 89% under the condition of optimum process parameters[1~3]. It was strict with the matching of air supplying and drawing volume to avoid windless condition in the overlay segment of air supplying and air drawing tubes, namely "circulation wind" phenomenon. According to experience, air drawing volume was about 80% of the air supplying volume, although dust control measures were used, there were still 20% air cannot enter the dust-clearing fan, and carried dust to pollute the rear working areas, especially serious for the tunneling working face with large amount of air supply and high intensity of dust producing. This paper was aimed at this problem, based on the air distribution with wall attaching chimney, a new model of air volume matching was proposed, and the quantitative relations of axial and radial air supplying volume, air drawing volume were found. Then the empirical formulas of the key process parameters such as axial and radial air out distance were derived according to the relevant

[a] E-mail: mawei0599@163.com

theories of air jet in the confined space, using the concept of controlling wind speed. In the end the research results of the model were verified by field tests, and good dust control effect was achieved.

2 Air volume matching model

The typical flow field of the Long compression and short pumping ventilation dust removal system was shown in figure 1 (a) below [4~5], fresh wind flow from the air supplying tube reached the tunneling working face and blew the dust away, then part of the dust was trapped and purified by the dust-clearing fan, part of the dust spread to pollute the rear working areas with the reason of dust-clearing fan cannot trap all the dust produced when cutting coal. This paper proposed a new air volume matching model on the basis of previous studies [6~7], as shown in the following figure 1 (b). The principle of this model was with the help of air distribution device to achieve axial and radial air supplying in the tunneling face. Then the gas in the front area of tunneling face can be diluted by the axial air supplying, and a fresh wind moving forward from behind can be formed by the radial air supplying, and the dust produced when cutting coal can be airtight in the area front of the diver and can be efficiently tripped and purified with the use of dust-clearing fan. Specific as follows:

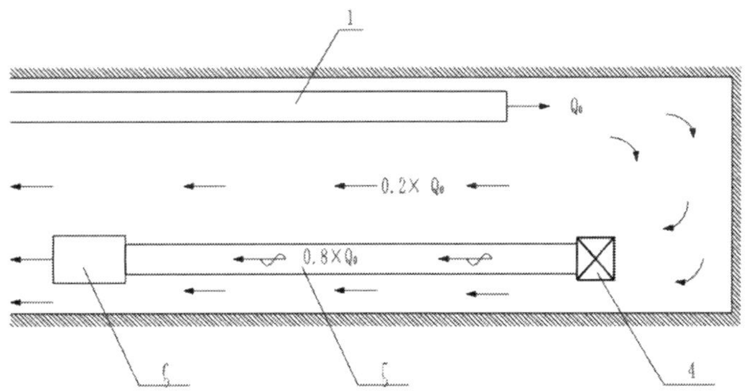

(a)

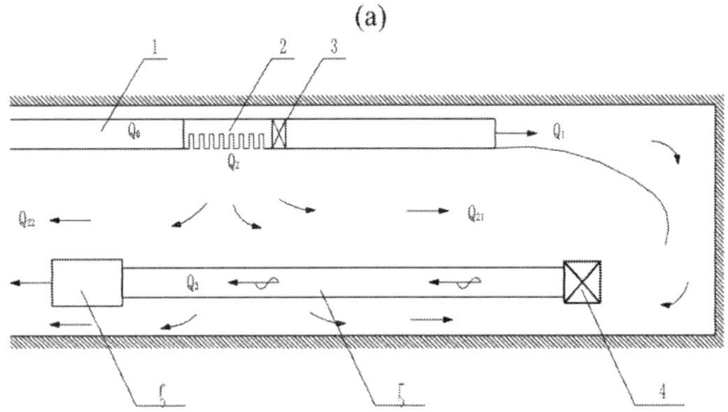

(b)

1—Air supplying tube; 2—Air distribution device; 3—Air regulating valve; 4—Dust collection cover; 5—Air drawing tube; 6—Dust-clearing fan

Figure 1 the flow field for the typical and new air volume matching in fully mechanized tunneling working face with long compression and short pumping ventilation dust removal system

Fresh wind flow from the air supplying tube was changed to axial and radial out air with the using of air distribution device and air regulating valve. Firstly appropriate axial air out distance and velocity (or air outlet type) was selected, and the axial air moving along the top of the roof reached the front area of tunneling face can dilute the gas and dust, and the air velocity should be reduced as possible to avoid wind disturbance and improve the efficiency of dust-cleaning fan in the region where coal was broken down and collapsed. Secondly appropriate radial air out distance, velocity, outlet type and outlet angle were selected, and the tunnel cross-section can be airtight by the radial air to obstruct the dust from spreading to the rear working areas, and the uniform low velocity air flow moving forward from behind can be ensure formed in the area between air distribution device and dust-cleaning fan's entrance to avoid circulation wind phenomenon and make the work safety. Finally, the axial air and forward moving part of the radial air were discharged into the rear of the tunnel through the dust-cleaning fan, while the dust in the wind was also effectively trapped and purified in this process. In addition, part of the radial air was moving from the air distribution device to the rear tunnel to make sure circulation wind phenomenon cannot be appeared in this area, and there was not any large dust source, also won't cause pollution to the return air.

3 Theoretical analyses of the key process parameters

Through the above analysis, the key process parameters in the new model of air volume matching of the Long compression and short pumping ventilation dust removal system in the comprehensive tunneling face included air supplying volume, air drawing volume, axial and radial air volume, axial and radial air out distance and etc. In this model, the axial air and forward moving part of the radial air eventually arrived at the front of the tunneling face, and the former air flow was much in the main section of the latter air flow. So we can assume that there was no interference between the two air flows, and key process parameters of the two flow fields can be calculated respectively.

a) Air supplying and drawing volume

The main purpose of the axial air supplying was diluting the gas in the front area of tunneling face, and the main purpose of the radial air supplying was airtight the tunnel cross-section to obstruct the dust from spreading to the rear working areas. Then the parameters of air supplying volume, air drawing volume, axial and radial air volume can be calculated by the following formulas[8].

$$Q_1=100\times q\times k \tag{1}$$

$$Q_2=Q_{21}+Q_{22} \quad Q_{21}=V_{21}\times S \quad Q_{22}=V_{22}\times S \tag{2}$$

$$Q_0=Q_1+Q_2 \quad Q_3=Q_1+Q_{21} \tag{3}$$

In these formulas:

q — Absolute gas gushing volume of the tunneling face, m^3/s;
k — Irregularity factor of gas gushing;
Q_0— Air supplying volume, m^3/s;
Q_1— Axial air supplying volume, m^3/s;
Q_2— radial air supplying volume, m^3/s;
Q_{21}— Forward moving part of the radial air volume, m^3/s;
Q_{22}— Backward moving part of the radial air volume, m^3/s;
Q_3— Air drawing volume, m^3/s;
V_{21}— Control velocity of the forward moving part of the radial air, m/s;
V_{22}— Control velocity of the backward moving part of the radial air, m/s;
S — Area of the tunnel cross-section, m^2;
100 — Unit air supply coefficient for the absolute gas gushing volume of the tunneling face , the gas concentration of 1% in the return air was used as the conversion factor.

b) Axial air out distance

Axial air flow in the new air volume matching model of Long compression and short pumping ventilation dust removal system belongs to restricted wall-attached jet in the confined space. Then the main segment was selected as the jet segment to reduce the velocity of the wind arrived at the front of the tunneling face, and the simplified structure of the axial air flow field was shown in the following figure 2.

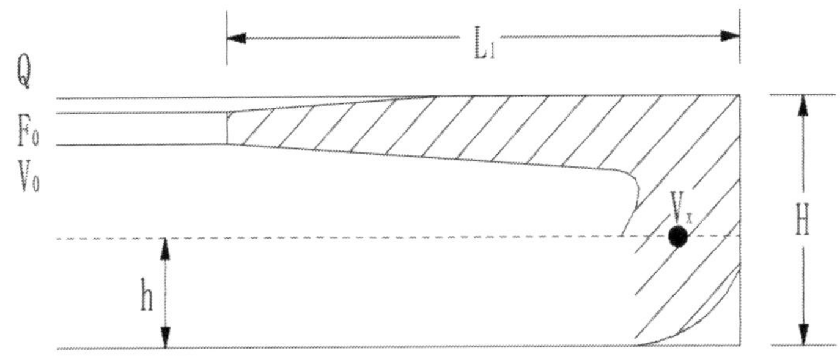

Figure 2 the simplified structure of the axial air flow field

According to the theory of air jet, the axial velocity Vx, which the air flow reached the calculation section of the working area, can be calculated by the following formula[9].

$$\frac{V_x}{V_0}=\frac{Km\sqrt{2F_0}}{x} \tag{4}$$

In this formula:
V_0 — Velocity of the axial air outlet, m/s;
V_x — Velocity of the calculation section in the working area, m/s;
F_0 — Area of the axial outlet, m^2;
K — Correction coefficient for considering the limited air jet;
m — Characteristic parameters related to the outlet type;
x — Total length of the jet stream, which was equal to the calculated cross section from the axial air outlet, , where $x=L_1+(H-h)$, m.
L_1 — Axial air out distance, m;
H — Height of the tunneling face, m;
h — Height of the broken coal, m.

If the air velocity when the wind reached the front of tunneling face can be limited in a certain range, the axial air can well dilute the gas and greatly improve the efficiency of dust-cleaning fan in the region at the same time. Then the axial air out distance can be calculated by the following formula.

$$L_1=\frac{KmQ_1}{V_1}\cdot\sqrt{\frac{2}{F_0}}-(H-h) \tag{5}$$

In this formula:
V_1 — Controlled air flow velocity in the front area in the tunneling face, m/s.

c) Radial air out distance

According to the new model of air volume matching for the Long compression and short pumping ventilation dust removal system in the comprehensive tunneling face, the forward moving part of the radial air flow can be also approximately viewed as restricted wall-attached jet in the confined space, and the simplified structure of the radial air flow field was shown in the following figure 3.

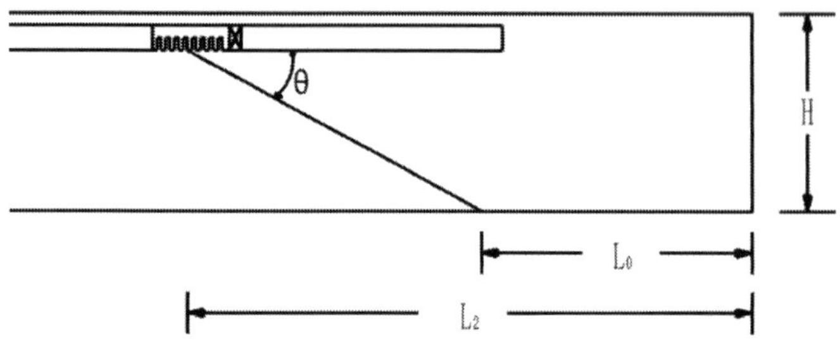

Figure 3 the simplified structure of the radial air flow field

Let us suppose that the distance between the tunnel roof and the radial air outlet can be ignored, and the radius of the air jet should be approximately equal to the height of the tunnel when the restricted wall-attached jet reached the closed cross-section. Then radial air outlet structure characteristics and the size of the tunnel had the following relationship [9]:

$$tg\theta = 3.4\alpha = \frac{3.4\times 0.48}{m} = \frac{H}{L_2 - L_0} \tag{6}$$

In this formula:
θ — Diffusion Angle of the air jet;
α — Turbulence factor, tgθ=3.4α;
m — Characteristic parameters related to the outlet type, m=0.48/α;
L_2 — Radial air out distance, m;
L_0 — Distance between the front area to the closed cross-section, m.

Finally, radial air out distance for the Long compression and short pumping ventilation dust removal system in the comprehensive tunneling face can be calculated by the following formed formula.

$$L_2 = \frac{H\times m}{1.632} + L_0 \tag{7}$$

4 Field verification tests

In order to verify the applicability of the formulas derived from the theory, the 20106 return roadway of fully mechanized working face in Wang Jia Ling coal mine was selected to conduct field verification tests, and the width of the tunneling face was 5.4 m, height was 3.2 m, net area was 17.28 m^2, absolute gas gushing volume was 4.5 x $10^{-3}m^3/$ s, the total air supply volume was $9.625m^3/$ s and the diameter of the air supplying tube was 0.8m.

To ensure the smooth development of field tests, one type of split dust collection covers were deigned according to the comprehensive tunneling machine structure combined with the flow field distribution in the tunneling face firstly, and total area the dust collection covers was 0.38 m^2, arranged in the machine arm roots and its sides respectively. Then one type of air controlling system was also designed according to the situation of the tunneling face, as shown in the following figure 4.

The new system was mainly composed of light weight lifting rail; air formed tube, air regulating valve, air distribution device, negative pressure tube, and tube storage device, etc. It had the characteristics of easy moving and simple adjustment, and can easily adjust the axial and radial air volume, axial and radial air out distance in the stretching process of the tunneling face [10~11].

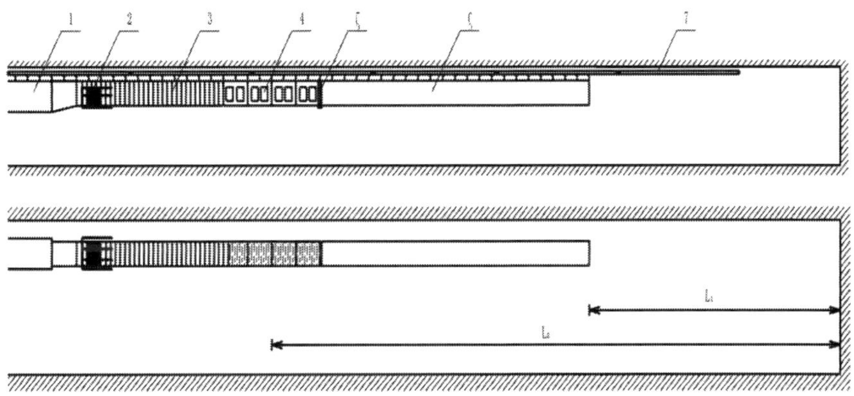

1—Air supplying tube; 2—Tube storage device; 3—Negative pressure tube; 4—Air distribution device; 5—Air regulating valve; 6—Air formed tube; 7—Light weight lifting rail

Figure 4 Component diagram of the new type of air controlling system for the long compression and short pumping ventilation dust removal system in the comprehensive tunneling face

In the process of determining the key parameters, firstly determine the controlled air flow velocity in the front area in the tunneling face V_1=1.0m/s according to the coal mine safety regulation, and selected the Irregularity factor of gas gushing k=3, and used the formula (1) and (5) to calculate Q_1=1.35 m^3/s and L_1=11.5m; Secondly selected V_{21}=0.32m/s and V_{22}=0.16m/s according to the coal mine safety regulation, and used the formula (2) to calculate Q_{21}=5.525 m^3/s, Q_{22}=2.75 m^3/s, Q_2=8.275 m^3/s, and used the formula (3) to calculate Q_0=9.625 m^3/s and Q_3=6.875 m^3/s; Thirdly selected L_0=15m according to the location of the excavator, and used the formula (7) to calculate L_2=20.3m. Finally, the preliminary long compression and short pumping ventilation dust removal system was build and optimized for the field verification tests.

In the process of the field test, axial and radial air volume were adjusted to the calculate value with the using of air distribution device and air regulating valve, axial air out distance was respectively selected to 5 m, 8 m and 10 m, radial air out distance was respectively selected to 15 m, 20 m and 25 m. The optimum technological parameters of the system were determined by continuous testing of the total dust precipitation efficiency of the driver`s place in different air out distances, and the test results was shown in the following figure 5.

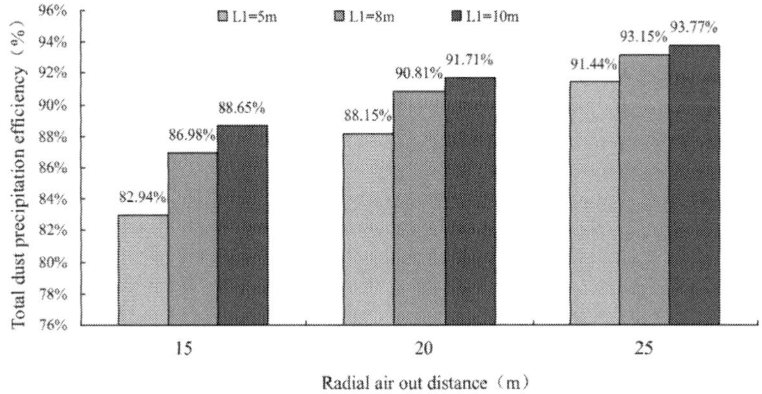

Figure 5 Total dust precipitation efficiency in the position of the driver in different air out distances

Can be seen from the diagram above, under the condition of the same radial air out distance, the total dust precipitation efficiency of the driver`s place was increased with the increase of axial air out distance, which can prove that the farther of axial air out distance, the lower flow velocity when air attaching to front area in the tunneling face, the less dust produced when cutting coal can be blown into the flow, and the more dust can be efficiently tripped and purified with the use of dust-clearing fan. Under the condition of the same axial air out distance, the total dust precipitation efficiency of the driver`s place was increased with the increase of radial air out distance, which can prove that the farther of radial air out distance, the more favorable to form a "air barrier" moving forward from behind, and the dust precipitation efficiency was also higher.

By comparison, the best dust controlling effect appeared under the process parameters with L_1=10m and L_2=25. But considering the machine moved back and forth in the production process, the radial air out distance was finally selected 20m to avoid the air distributing device beyond the outlet of dust-clearing fan, and the dust controlling effect was not very relevant with the L_2=25. Then the best process parameters of this system obtained by field test were L_1=10m and L_2=20m, which was basically identical with the result obtained by theoretical calculation, and under the condition of optimum technological parameters, the total dust and respiratory dust concentration of the driver's place can be reduced 91.98% and 89.53% respectively.

5 Results

(1) The new air volume catching model for the long compression and short pumping ventilation dust removal system had good adaptability. The gas in the front area of tunneling face can be diluted, and the dust can be airtight in the front area of tunneling face with the fresh air flow moving forward from behind, and can be efficiently tripped and purified with the use of dust-clearing fan. In addition, in order to avoid the pollution caused by the fresh radial air flow moving forward, spray and Local sealing measures should be used to control the dust produced at the transfer point.

(2) The best process parameters of this system obtained by field test were basically identical with the result obtained by theoretical calculation, and under the condition of optimum technological parameters, the total dust and respiratory dust concentration of the driver's place can be reduced 91.98% and 89.53% respectively, and can be used to provide reference for the design of long compression and short pumping ventilation dust removal system.

(3) In consideration of the differences of production conditions and air supply conditions in the comprehensive tunneling face, the field test was only for a particular tunneling face, and the empirical formula obtained by theoretical derivation for calculating key process parameters of the long compression and short pumping ventilation dust removal system should be verified by a large number of field experimental data.

References

1 WANG Hong. The 40 years developmental review of the fully mechanized mine roadway heading technology in China, [J]. Coal journal, 2010, 35 (11) : 1815~1820
2 LU Xin-xiao, WANG de-ming, REN Wan-xin et al. Dust production mechanism and dust migration rule in driving face [J], coal mining, 2012, 17 (5) : 19~22.
3 WANG Hui, JIANG Zhongan, HUANG Liting, LIAO Xianxin, WANG Jiran. Experimental research on pressing air-absorption air volumeratio in FPNA ventilation for excavation roadways Journal of Liaoning Technical University (Natural Science) Vol.30 ,No.2 ,Apr ,2011.
4 DU cui-feng, WANG Hui, JIANG Zhong-an, HE Zong-li, HU Guo-yong. Numerical simulations of dust distribution in a fully mechanized excavation face with far-pressing-near-absorption ventilation. Journal of University of Science and Technology Beijing. Vol.32,No 8,Aug,2010.
5 WANG Hai-qiao,SHI Shi-liang,LIURong-hua,LIUHe-qing. Numerical simulation study on ventilation flow field of wall-attached jet in heading face. JOURNALOF CHINA COALSOCIETY Vol.29 No.4, Aug. 2004.
6 WANG Hai-qiao. Study on jet ventilation flow field in heading face. JOURNAL OF CHINA COAL SOCIETY. Vol.24, No.5, Oct.1999.

7 LIU Rong-hua,WANG Hai-qiao,SHI Shi-liang,LIU He-qing. Study on regularity of dust distributing in heading face with forced ventilation. JOURNAL OF CHINA COAL SOCIETY. Vol.27, No.3, June 2002.
8 YANG Sheng-qiang. Theory and technology of dust control [M]. Xuzhou: China mining university press, 2007.
9 ZHAO Rong-yi, FAN Cun-yang, XUE Dian-hua, etc. Air Conditioning (3rd edition) [M]. Beijing: China construction industry press, 1994.
10 MA Wei, LIU Yong, CHEN Fang. Fine dust prevention and control technology of rapid fully-mechanized working face in wetting difficultly seam[J].Coal Science and Technology,2015,43(1):70-73.
11 CHANG Jian-bing, LIU Tao, XU Kui et al. Research on the technology dust prevention and control of the working face [J]. Coal engineering, 2007 (3) : 53-55.

Study of Plantar Pressure Distribution

Chun Kit Ang[1,a], Mahmud Iwan Solihin[1], Weng Jun Chan[1] and Yien Yien Ong[2].

[1]*Department of Mechanical and Mechatronic Engineering, UCSI University, No.1, Jalan Menara Gading, UCSI Heights (Taman Connaught), 56000 Cheras, Kuala Lumpur, Malaysia.*
[2]*Department of Bioscience, Faculty of Applied Sciences (FOAS), Tunku Abdul Rahman University College (TARUC), Jalan Genting Kelang, Setapak, 53300 Kuala Lumpur, Malaysia.*

Abstract. Every region of foot is not equally divided in terms of plantar pressure distribution (PPD) during free standing. This paper is focusing on studying PPD on flat plane and inclined plane and the results obtained from this study may contribute to biomedical researcher in designing orthotic devices. 24 healthy young adults age ranging from 19 to 24 years old and weigh between 50 to 80 kg were invited for experiments purpose. Six regions of both feet were measured which were hallux, medial forefoot, central forefoot, lateral forefoot, lateral midfoot and hindfoot. Remarkable differences were seen in the result as right foot exerted more pressure generally in every region of the foot as to compared with left foot respectively. This is true especially for region such as hallux, medial forefoot and lateral forefoot. On a flat surface, PPD on the hindfoot is the highest. However, at an elevation of 25°, test subjects began to shift their PPD to forefoot regions. While studies of PPD are common, this study provides a new insight for the first time into PPD while standing on different angle of walking plane.

1 Introduction

Most individual spend a large proportion of their lifetimes in the standing position. Walking also known as ambulation is one of the fundamental locomotion for a human being and is the most common activity in our daily life. During standing, our feet and the surface of the ground created a contact. The sole of our feet experienced a force upon landing on the ground. The force of impact on the sole of the foot or the plantar pressure distribution could be irregular and unbalanced. These impacts manipulate changes in pattern of stress applied to the foot, stresses that may causes temporary or permanent strains on the tissues of the foot and therefore induces foot conditions. In a long run, this may cause a foot injury or worsen the disruption of overall health of an individual [1].

Human foot plantar pressure refers to pressure acting upon on the soles of human feet when stepping on surface of ground. In today eras, plantar pressure distributions are profoundly accepted as a reliable biomechanical parameter to study and diagnose various foot disorders. Plantar pressure distributions are affected by various factors such as weight, gender, foot structure and even the way an individual stands or walks. Plantar pressure measurements during standing, walking or other activities provides detailed information for evaluation of diseases or abnormalities, involving the function of the ankle, knee, hip, the back and other patho-mechanics and yield objectives measure to track disease

[a] Corresponding author: ack_kit@hotmail.com

progression [2]. It also provides vital information on mechanical behaviour of the human foot in both static and dynamic conditions.

In the past, a number of studies on foot plantar pressure have been carried out. The study of relationship between plantar foot pressure and supporting surface is called pedobarography. It is often studied for bio-mechanical analysis of human walking or gait analysis and human posture. Pedobarography is extensively practiced and clinically applied in patients with diabetic foot ulceration, a condition which may lead to amputation in extreme cases. Several research groups acknowledge the potential of plantar pressure distributions technology for the diagnosis and analysis of different foot abnormality. In fact, few studies on foot biomechanics have found that plantar pressure distributions are vital to determine the unnatural gait [2].

In most studies, plantar pressure distributions are usually examined during walking. However, studies on plantar pressure distributions while standing on different inclined plane are rare. The knowledge of the changes in plantar pressure distribution of a person standing on an inclined platform may be helpful to further understand the neuropathological mechanism to make physical activities safe and to prevent injury.

2 Experiment Setup

2.1 Sensor

Traditionally, stationary force platforms are used to measure the foot plantar pressure distributions as shown in Figure 1. They are efficient when comes to measuring force applied directly on the platform surfaces. But over the years, a new trend of insole techniques is growing rapidly. The sensor used are typically more flexible and can be placed in flat or curves surfaces. These systems detect the plantar pressure and the shoe that were put on directly and therefore can be custom made accordingly in terms of shoe construction. A general advantage of in-sole technique is that data of multiple steps can be recorded in a single test [3]. Since the position of the foot is mounted on the respective sensor, the data of multiple steps can be averaged easily, thus reducing measurement error, in contrast with platform devices which has only one ground contact that can be generated at a time. Portability plays an important role in collection of data. It encourages a more natural gait movement among test subjects whereas for flat platform, test subjects have to anticipate each and every step so that they can accurately step on the platform itself.

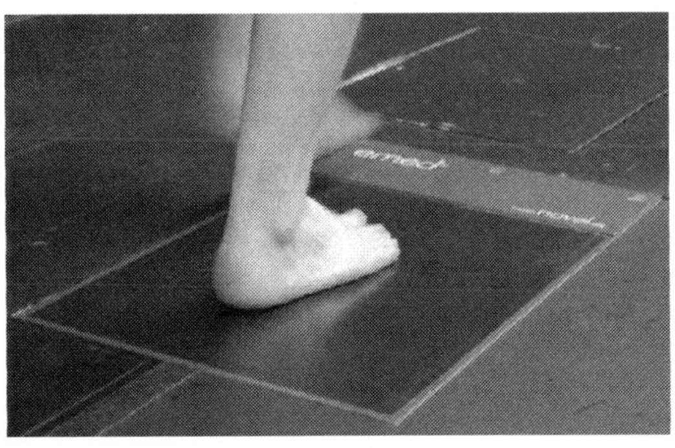

Figure 1. A Platform-Based Foot Plantar Pressure Sensor [4]

Thus, the in-sole sensors that are used in this study are the force sensing resistors (FSRs). FSR is a robust polymer thick film (PFT) components that will decrease in resistance as force applied to the sensing area. The model of the transducer that will be used in this study is the IMS-C20B. The reason this model is chosen compared to its counterpart is because this model may withstand and measure higher force up to 100 kg which facilitate in this study. It has a small sensing area of about 20 mm in diameter. This allow accurate readings to be taken. Apart from that FSR is generally more robust than any other sensor due to its nature to withstand strains and stresses. It is made from ultra-thin filaments, about 0.45 mm thus it is easy to be placed under the sole of the foot and minimizes the error due to its sizes. Besides, FSR has a high repeatable force reading which plays an important role in data collection for this measuring device as test subjects are required to walk on the sensor in repeatable manner. All these factors are to be considered as they can potentially increase the measurement error and device failure.

2.2 Foot region

There have been many studies conducted that are based on the foot of a person. However, PPD is not equally distributed across every region of the foot. In many studies, researchers have their own set of regions of foot of interest. For example, according to Gurney, Kersting and Rosenbaum (2008) the foot was divided into multiple regions which are: Hallux (HA); Second toe (T2); Lateral toes (T3-5); Lateral forefoot (LF); Central forefoot (CF); Medial forefoot (MF); Lateral midfoot (LM); Medial midfoot (MM); Lateral hindfoot (LH); Medial hindfoot (MH) [5] as shown in Figure 2.

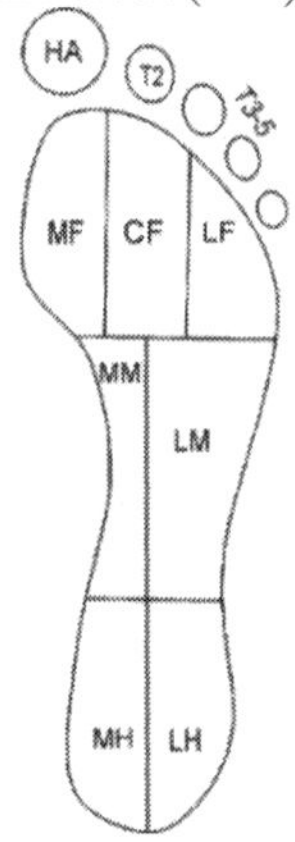

Figure 2. Division of Foot Anatomical Areas of Interest [5]

Similarly, in this study, each foot is divided into 6 regions to enable analysis of every significant region in relation to pressure. The placement of sensors is shown in Figure 3. Region of foot are as following: Hallux (HA); Medial Forefoot (MF); Central Forefoot (CF); Lateral Forefoot (LF); Lateral Midfoot (LM); Hindfoot (HF).

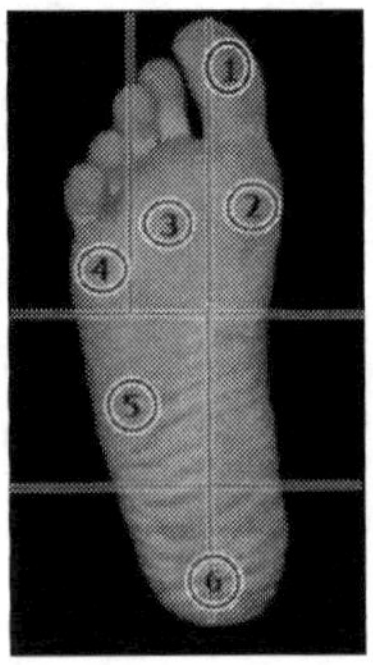

Figure 3. Division of Region of Foot and the Placement of Sensor

2.3 Test subjects

The test subjects in this study consisted of 24 healthy volunteers without any foot disorder, including 14 males and 10 females age ranging from 19 to 24 years old and weighed between 50 to 80 kg. Age, weight and gender were recorded for each test subject to provide information of the sample data taken. Prior to the experiment, test subjects were briefed about the experimental procedure and given some time to familiarize themselves with the procedure. The subjects were instructed to stand on the platform (as shown in Figure 4) at their natural standing position. Test subjects underwent six different tests as there were six inclined planes. Five readings were taken for every type of test to minimizes error.

Figure 4. Adjustable Inclined Platform

2.4 Testing plane

As mentioned earlier, there were six inclined planes with different elevation were tested in this study. The inclined plane was hard and sturdy surfaces to simulate the hard ground upon contact of foot. As shown in Figure 4, these were the following inclined planes: flat plane; 5° inclined plane; 10° inclined plane; 15° inclined plane; 20° inclined plane; 25° inclined plane.

3 Results and Discussion

In this study, PPD for both feet are measured and recorded simultaneously. There are few factors that affect the value such as gender, age, and anatomical structure of the foot. Therefore, there are many ways to analyze the result obtained. The loading of human foot on the ground surfaces is not constant. In Figure 5, the results show an ever changing of pressure on each region of foot for one of the test subjects. The changes are not consistent and unpredictable. It may change to an extend range of 55% of the pressure in a span of 10 seconds. This was observed on every test subjects despite being instructed to stand still. The result is agreeable to the studies conducted by Kei Masani et. al in her work [6]. All the other test subjects have the similar results. In order to reduce the inaccuracy, the test subject was instructed to stand still for 10 seconds and the average value of the readings for each region of the foots was recorded.

The pie chart in Figure 6 shows the average distribution of pressure of both feet of all the test subjects on flat plane. According to the chart, hindfoot exerts the most pressure which is nearly one third of the entire foot pressure combined. The second highest plantar pressure exerted is the medial forefoot which is 19%, followed by central forefoot, 18%. Lateral forefoot exerts 17% of the plantar pressure and Hallux exerts 10%. Lastly, Lateral Midfoot exerts the least pressure of the foot which is 4%. The order from highest PPD to the lowest is as followed: HF > MF > CF > LF > HA > LM.

This validates a previous study [7] in adult by Klimiec in 2016 which found the highest PPD to be at hindfoot region which follow by forefoot region and hallux . Lastly, midfoot region exerted the least pressure among all.

Besides, this study discovers that the loading pressure of the right foot varies with the left foot. Remarkable differences were seen in the result as right foot exerted more pressure generally in every region of the foot as to compare with left foot respectively as shown in Figure 7. This is true especially for region such as hallux, medial forefoot and lateral forefoot.

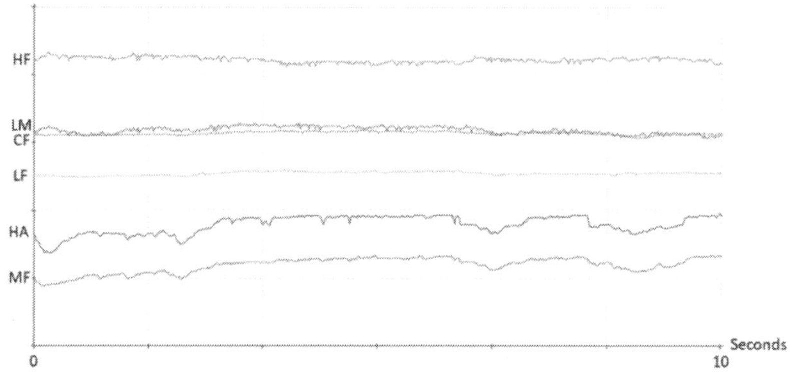

Figure 5. Force on Each Region of Foot

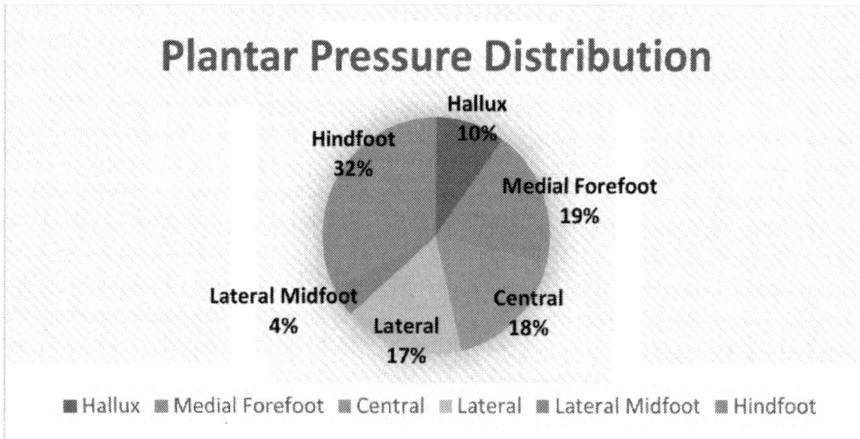

Figure 6. Pie Chart of Plantar Pressure Distribution

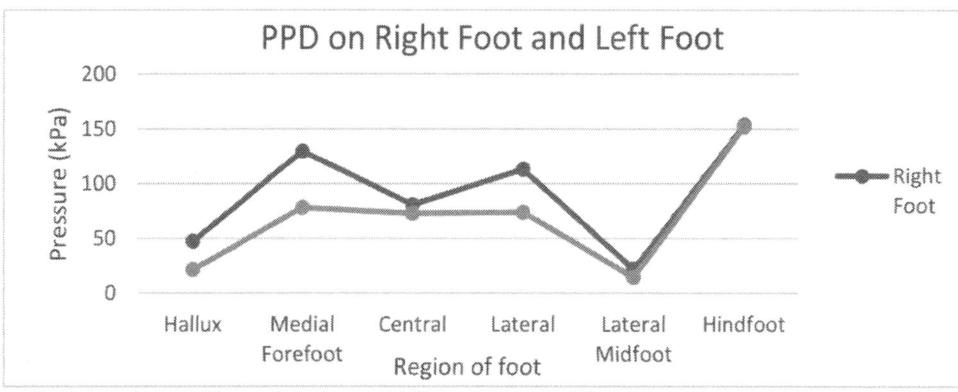

Figure 7. Comparison of Plantar Pressure on Different Regions of Feet

Figure 8 to Figure 11 show the differences of PPD for male and female. Previous studies by Periyasamy in 2011 reported that no significant differences was detected when comparing PPD of men and women of both foot on flat surfaces and mentioned that there were no differences among genders in the midfoot contact area [8]. Similarly, on inclined plane, there are no significant PPD differences between male and female. However, this study discovers that the main difference is the PPD on hindfoot region. The PPD on hindfoot of female is slightly higher than male while standing on inclined plane. This may due to higher contact area of men compared to women.

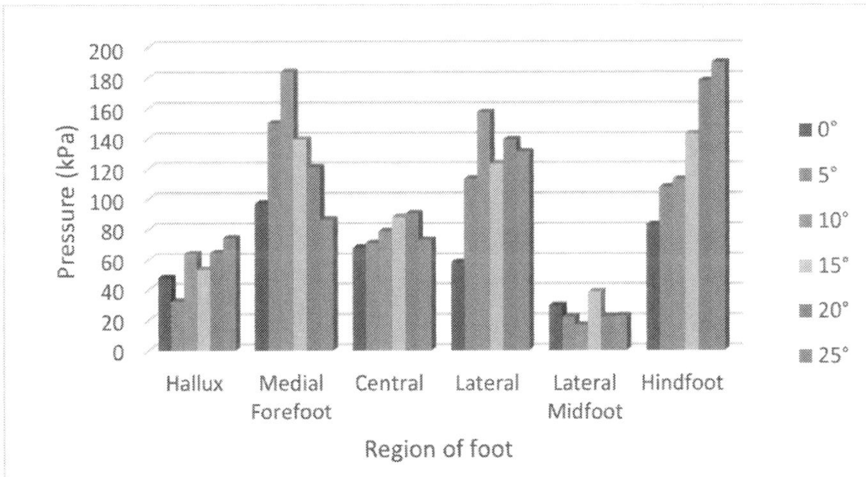

Figure 8. Plantar Pressure of Male Subjects on Right Foot

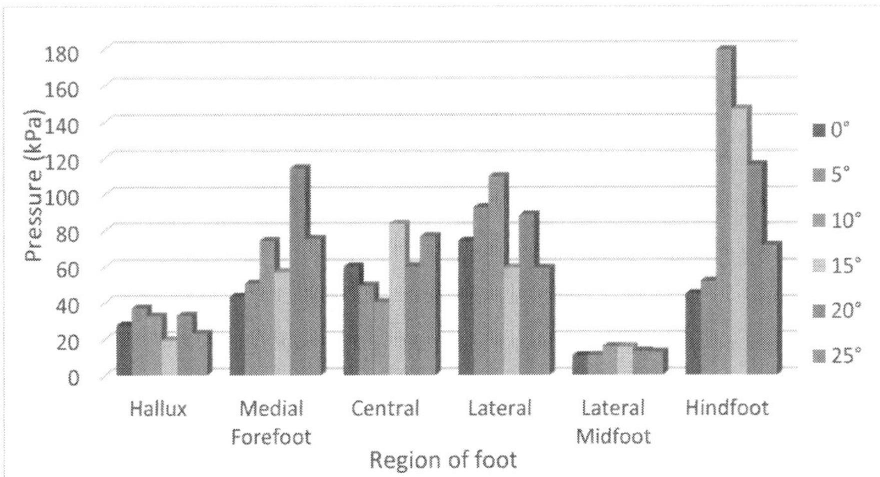

Figure 9. Plantar Pressure of Male Subjects on Left Foot

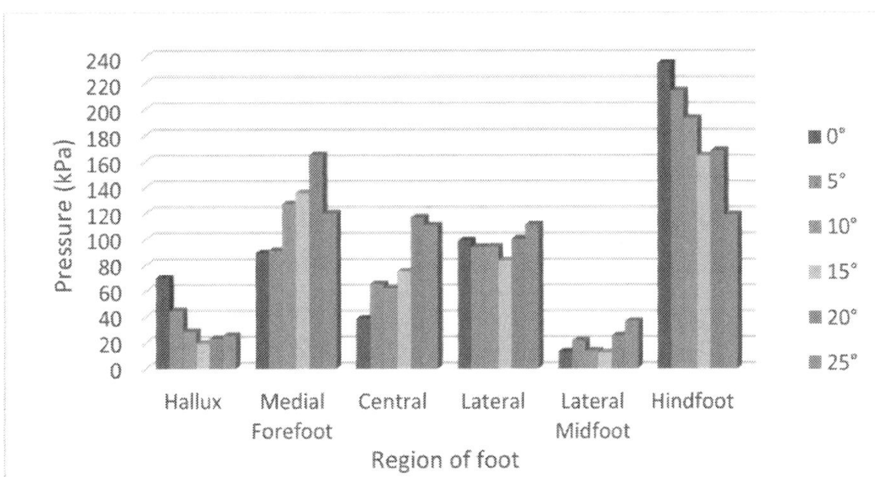

Figure 10. Plantar Pressure of Female Subjects on Right Foot

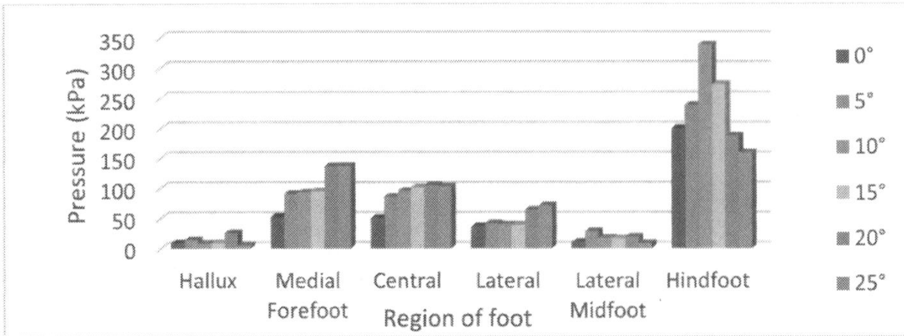

Figure 11. Plantar Pressure of Female Subjects on Left Foot

Furthermore, an astonishing result were shown in Figure 12 as the elevation changes, human feet shifted to compensate the body instability. The shift mainly focused on the medial forefoot, central forefoot, lateral forefoot and the hindfoot. On a flat surface, PPD on the hindfoot is the highest. However, at an elevation of 25°, test subjects began to shift their PPD to the forefoot regions. This is due to human body leaning slightly forward thus changes in pressure distribution to avoid falling backwards. At 30° slope, it is observed that many test subjects failed to balance themselves on the platform. Therefore, the 30° slope test were excluded for this study.

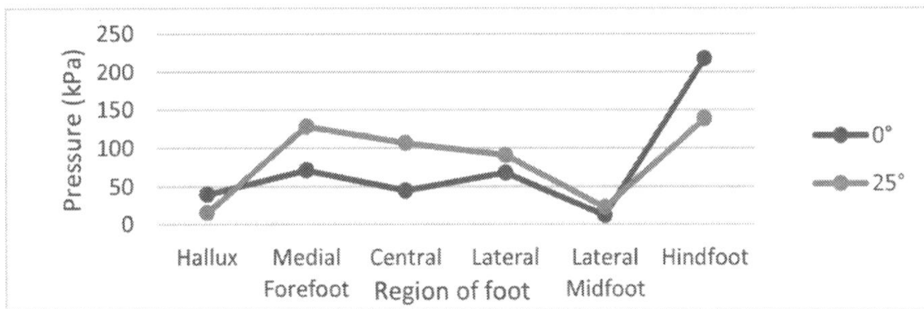

Figure 12. Comparison of 0° and 25° on Both Feet

4 Conclusion

Plantar pressure distribution measurement is extremely useful in analyzing the weight loading of human foot. This study has evaluated that human feet are impossible to stand absolute still as muscles are continuously working to keep the person in balance. PPD on the hindfoot region is the highest while medial midfoot remains the lowest regardless of different inclined plane which less than 25° and gender. Besides that, as the elevation increases, the human foot shifted its PPD to forefoot region to compensate the body instability. While studies of PPD is common, this study provides a new insight for the first time into PPD while standing on different inclined plane.

References

1. Y. T. Chen, K. W. L. and S.Y. Huang, Int J Min Metall Mech Eng (IJMMME) **3**(2) 65–69 (2015).
2. J. A. R. Bautista, *et al.,* Biocybern. Biomed. Eng. **38**(2) 342-361 (2018).
3. D. Rosenbaum and H. P. Becker, Foot and Ankle Surg **3**(1) 1–14 (1997).
4. A. H., Abdul Razak *et al.,* Sensors (Switzerland), **12**(7) 9884–9912 (2012).
5. J. K. Gurney, U. G. Kersting, and D. Rosenbaum, *Gait and Posture* **27**(4) 706–709 (2008).
6. K. Masani, D. G. Sayenko, and A. H. Vette, Gait and Posture **37**(1) 72–77 (2013).
7. E. Klimiec *et al.,* Sens. Actuators, A **247** 504–513 (2016).
8. R. Periyasamy *et al.,* Foot **21**(3)142–148 (2011).

Automatic MHT Method Based on Target Characteristics Applied in the Small Space Targets

Qian ZHAO [1,a], Zhenhong NIU[1], Lian XUE [1], Xin LIU [1], Jianhua LI[1] and Zhifeng LI[1]

[1]*National Key Laboratory of Science and Technology on Test Physics & Numerical Mathematics, China*

Abstract. The Multi-Hypothesis Tracking (MHT) method is one of the most promising approaches to realize automatic tracking of small and dim targets in complex space scene. However, the number of hypotheses generated by this algorithm has an exponentially increasing relationship with the number of trajectories, measurement data, and the number of image frames. In the paper, the automatic detection and tracking method of judging the confidence coefficient of the trajectory obtained by the MHT which is based on the characteristics of dim targets, has been realized in the IRST system. The performance of this method is greatly improved.

1 INTRODUCTION

Automatic tracking technology based on image is one of the important research inspects of Machine Vision Application, and key technology of automatic target detection. It has been widely applied in aerospace systems. Infrared Search and Track (IRST) based on obtaining 2d infrared images, has the advantages of large field of view, small size, and working all day to be used Detecting and tracking space targets by receiving infrared radiation from targets and background. However, it is commonly used in space craft to realize automatic detection of target for Engineering Applications. And image noise, spatial illumination, target motion and other factors have always plagued the realization of target automatic tracking. Difficulties in engineering implementation are reflected in three aspects: 1) The system has a large amount of image data, which makes it difficult for its hardware equipment to complete the real-time detection and tracking of the target. 2) The working environment of the system is complex, the image interference area is large and the SNR is low. 3) Infrared small and dim targets for obtaining images through the IRST have no obvious geometric features [1].

The Multi-Hypothesis Tracking (MHT) method which is first proposed by D. B. Reid [2][3], has been commonly used in target tracking studies based on 2D image. Based on this method, it is to create multiple hypothetical trajectories, calculate the prior probability of each hypothesis, and delete the hypothetical trajectories of low probability. Finally, the correct trajectory can be found based on the measurement point. Although MHT is one of the most promising approaches to solve the problem of tracking multiple targets in complex scenarios, the number of hypotheses generated by this algorithm has an exponentially increasing relationship with the number of trajectories, measurement data, and the number of image frames. Therefore, based on this algorithm, multi-objects detecting and tracking is difficult to realize. Based on the feature of dim targets, the automatic detection and

[a] Corresponding author : bestcherryzhaoqian@163.com

tracking method of judging the confidence coefficient of the trajectory obtained by the MHT is proposed in paper. This method used in the real-time infrared image processing system, has been realized the automatic tracking of dim space target based on infrared image.

2 SYSTEM OVERVIEW

IRST on the spacecraft platform is composed of the IR imaging detector and the real-time infrared image processing system. In complex scenario, the IR imaging detector collects infrared images containing the target, and then transmits it through the connecting cable to the infrared image real-time processing system for processing. The FPGA of infrared image real-time processing system realizes the data stream scheduling and image preprocessing. Meanwhile, DSP accepts the image data from FPGA to complete the complex image processing.

In the space scene, IRST work process is that system scans the certain field according to the prior information, and transmits image data to the infrared image real-time processing system for processing. Firstly, IRST detects and select suspected tracking target. Then the system which is derived from the automatic detection and tracking method, determines the location and size of the target to be detected, and calculates the moving direction and speed information of the target to the spacecraft platform. Finally, the platform controls the detector to track the moving target and keeps target in the field of the detector. During the process, the platform performs target position correction at any time. The working process of IRST is shown in Figure 1.

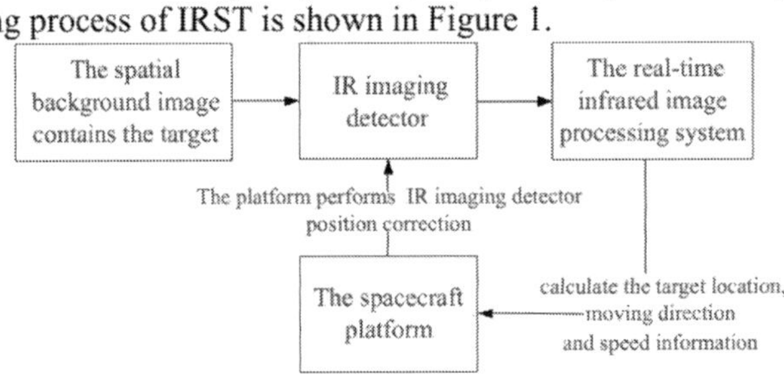

Figure 1. The working process of IRST

3 MRTHOD IMPLEMENTATION

The target detection and tracking methods is divided into Detect-before-track (DBT) and Track-before-detect (TBD). Currently, the TBD method is mainly used to detect the small and dim targets with low SNR. The principle of TBD is not to judge the target, but to track the possible trajectory of the target in the image, and then to complete the threshold decision through a series of calculations [4]. In all methods conforming to the TBD principle, MHT is the classic algorithm to solve the detection problem of small and dim moving targets in complex scenarios.

3.1 The principle of MHT

In the MHT, it is assumed that the maximum range of motion between two adjacent frames is not more than N pixels. From time t, the target at point A may appear at any point B in the square field which is centered at point A and 2N+1 for the side. The target has a total of $(2N+1)^2-1$ possible trajectories from point A to point B. At t+2, the target moves from any point in $(2N+1)^2-1$ at time t+1 to point C. There is a total of $(2(2N+1)-1)^2-1$ possible trajectories. From these points, a new trajectory is formed. This is the root cause of the explosion of the target trajectory as the number of image frames increases. According to the reference, the boundary parameters a, b, λ' and the series estimation parameter K' are used as the threshold parameters in the MHT as the hypothesis trajectory inspection decision [1][2][3]:

$$H_0 : x(i,j,k) = w(i,j,k) \quad \text{There is no target} \tag{1}$$

$$H_1 : x(i,j,k) = s + w(i,j,k) \quad \text{There is a target} \tag{2}$$

If k < K', the decision is as follows:

$$\begin{cases} \sum_{i=1}^{k} x_i \ge a & \text{Judgment for } H_1 \\ \sum_{i=1}^{k} x_i \le b & \text{Judgment for } H_0 \\ b \le \sum_{i=1}^{k} x_i \le a & \text{Wait for the next frame to continue to judge.} \end{cases} \tag{3}$$

If k = K', the decision is as follows:

$$\begin{cases} \sum_{i=1}^{K'} x_i \ge \lambda' & \text{Judgment for } H_1 \\ \sum_{i=1}^{K'} x_i \le \lambda' & \text{Judgment for } H_0 \end{cases} \tag{4}$$

The above four parameters are defined as:

$$a = \ln\left(\frac{1-(1-c_1)\beta'}{(1-c_0)\alpha'}\right) \tag{5}$$

$$b = \ln\left(\frac{(1-c_1)\beta'}{1-(1-c_0)\alpha'}\right) \tag{6}$$

$$K' = (\Phi^{-1}(c_0\alpha') + \Phi^{-1}(c_1\beta'))^2 \left(\frac{\sigma}{s-\mu_0}\right)^2 \tag{7}$$

$$\lambda' = \sqrt{K'}\,(s\Phi^{-1}(c_0\alpha') + \mu_0\Phi^{-1}(c_1\beta'))\left(\frac{\sigma}{s-\mu_0}\right) \tag{8}$$

3.2 Adaptive target optimal track extraction method based on target feature constraint optimization

When tracking targets and multiple interfering targets enter site of the detector at the same time, even if the 2.1 decision condition is used, a large number of candidate trajectories remain, which requires the real-time infrared image processing system to provide large memory and computational overhead. This method relies too much on current data and ignores the use of other image sequence information. At the same time, in the process of track association, there may be several situations such as trajectory bifurcation and trajectory crossing. When multiple measuring points fall into the data association area of a certain trajectory, the trajectory bifurcation occurs;

When a measured data falls within a data association region of multiple trajectories, the trajectory crossing will occur. When a single independent data point appears, a new track may occur. This process is shown in Figure 2.

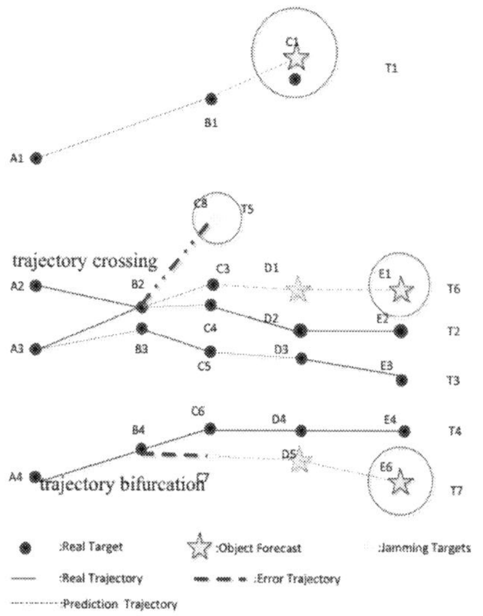

Figure.2. The Process of Track Association

As shown in the figure, multiple error trajectories are obtained through the MHT method, leading to the increase of false alarm rate in the automatic tracking process. Therefore, this paper proposes the adaptive target optimal track extraction method based on target feature constraint optimization. By calculating the confidence of each track, the target motion parameter estimation and trajectory extraction accuracy are improved [5].

For multi-target tracking in IRST, besides the target location information, the target gray information and speed information can be used. At the same time, the speed of movement between the target to be tracked and the target of interference is also different. Therefore, the trajectory judgment criterion of this method is:

(1)If the gray value of the measurement point is close to the gray value of the target, the probability of this measurement point is high.

(2)In the same trajectory, if the measurement point's velocity consistency with the adjacent frame is good, the trajectory is the probability of continuation of the same moving target.

On the contrary, the trace confidence coefficient is low.

After the real-time infrared image processing system accumulates multiple frame information for each currently recorded trajectory, the trace confidence coefficient is automatically evaluated based on the target feature. During the system working, the trajectory of confidence coefficient below the threshold is deleted at any time. According to the reference, the logarithmic form of the trajectory evaluation function m_t is set to be the sum of trace confidence coefficient. The algorithm formula is derived as follows [6]:

$$m_t = m_{t-1} + \Delta m_t \tag{9}$$

The d represents the distance between the predicted value of the target trajectory and the measured value. P_d represents the target detection rate. β_f represents the density of false alarm space. M represents the dimension of measurement space. Pf represents false alarm rate. $P_f=\beta_f \cdot V_k$. V_k represents the volume of the interconnection. Then, the traditional confidence calculation formula is added to the target gray and speed change information as the constraint condition. Δv_t and Δv_m represent the velocity change rate of the trajectory measurement point at t-1 and the velocity change rate of the suspected point at t. The g_t and g_m represent the gray value of the target point at t-1 and the gray value of the suspected point at t.

$$\Delta m_t = \begin{cases} \frac{c_1}{c_1 + |\Delta v_t - \Delta v_s|} \frac{c_2}{c_2 + |g_t - g_s|} \ln\left[\frac{p_d e^{(-d^2/2)}}{\beta_f (2\pi)^{M/2} \sqrt{|S|}}\right], \text{The trajectory has a measure point update.} \\ \ln\left[\frac{1-p_d}{1-p_f}\right], \textit{No update} \end{cases} \tag{10}$$

$$m_1 = \ln\left[\frac{1-\beta_f}{\beta_f}\right] \tag{11}$$

According to (10), measured point and target velocity change rate difference, gray value difference, and distance d between the measurement point and the target prediction value, these factors affect the track confidence. The smaller these three values, the greater the confidence coefficient increment.

4 THE SIMULATION RESULT

The experiment used the IRST system to acquire multiple frames of continuous infrared images. The image contains multiple interfering motion targets and noise points. Finally, we choose 25 frames of continuous images as sample images, in which all the trajectories changed significantly after the 16th frame. Therefore, this paper mainly introduces the experimental results and results of different algorithms in 16 frames to 25 frames.

Calculation steps and results:

Step1: The MHT method is used to determine the non-zero pixel points in the current frame, and the gray value is compared with the high and low thresholds. The pixel points with gray value higher than the high threshold pixel is added to the tree structure and the root node is created, and the result is marked as table 1. The pixel points of the gray value between the high and low threshold are added to the tree structure and the root node is created. The result is marked as 0, and the next frame of image data is waited for further judgment. Discard the pixel gray value less than the low threshold.

Table 1. Examination result

The number of pixels below the low threshold

No.Pic	16th	17th	18th	19th	20th	21th	22th	23th	24th	25th
Number of pixels	32	37	56	54	33	19	15	2	6	0

The number of possible target pixels

No.Pic	16th	17th	18th	19th	20th	21th	22th	23th	24th	25th
Number of pixels	50	56	67	59	57	70	50	41	43	43

Step2: According to the suspected target pixel obtained in step 1, the system automatically performs connected component division and generates a suspected trajectory. Calculated by the MHT method, 169 suspected trajectories are obtained.

Step3: Using the traditional trace confidence coefficient calculation method, each trace confidence coefficient is calculated and the calculation result is shown in Figure.3(b). Using the adaptive target optimal track extraction method based on target feature constraint optimization, the result is shown in Figure.3(c).

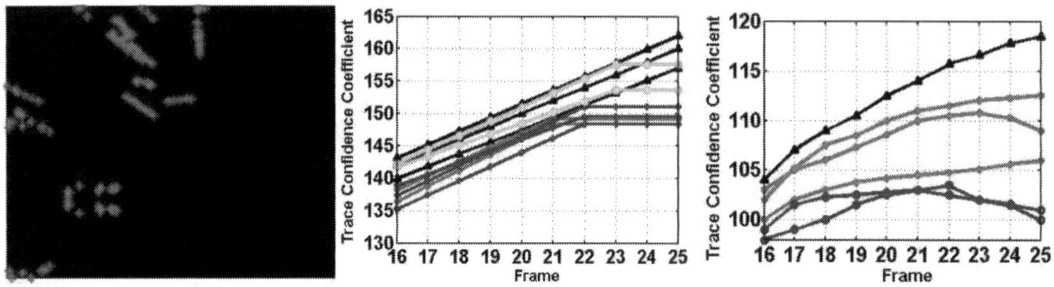

(a) (b) (c)

Figure 3. The result of trace confidence coefficient obtained by two methods

The trajectory obtained using the MHT is shown in Figure. 3(a). Figure. 3(b) is the result of using the traditional trace confidence coefficient calculation method. The difference between true and false trace confidence coefficient is very small. It is difficult to determine the authenticity of the trace. Subsequent data is needed to provide more information to distinguish the true track. According to Figure.3(b), even if the interference target track is associated with the false alarm, its trace confidence coefficient has not changed significantly. According to Figure.3(c), gray value and movement speed of real target are different of other interference target, leading to the obvious change of trace confidence coefficient in continuous frames.

5 THE CONCLUSION

This paper introduces the target feature information into the MHT, and proposes the adaptive target optimal track extraction method based on target feature constraint optimization, which improves trace confidence coefficient calculation method and the trajectory starting condition. The simulation results show that the improved method can effectively distinguish the real target trajectory and the interference trajectory in real time, and realize the automatic tracking of small and dim targets in the spacecraft platform.

ACKNOWLEDGMENTS

This work is supported by the National Natural Science Foundation of China (61671045).

References

1. R.Chai, Acta Photonica Sinica. J. **33**, 1233-1235(2004)
2. D.B.Reid, IEEE Transactions on Automatic Control. J. **24**, 843-854 (1980)
3. Biostein.S, D,Huang.T.S, *Conference on Computer Vision & Pattern Recognition.* 690-695(1988)
4. W.Wang, Z.Li , J.Liu, Y.He, Z.Chen, Procedia Engineering. J. **29**, 1201-1207(2012)
5. L.Liu, Z.Huang, Infrared Physics & Technology. J. **62**, 59-64(2014)
6. Hongren.Zhou, Zhongliang.Jing, Peide.Wang, *Maneuvering Target Tracking. (1991)*

Producing Bio-Oil from Coconut Shell by Fast Pyrolysis Processing

Dewi Selvia Fardhyanti[1,a], Megawati[1], Cepi Kurniawan[1], Retno Ambarwati Sigit Lestari[2] and Bayu Triwibowo1[1]

[1]*Chemical Engineering Department, Faculty of Engineering, Universitas Negeri Semarang, Semarang, Indonesia*
[2]*Chemical Engineering Department, Faculty of Engineering, Universitas 17 Agustus Semarang, Semarang, Indonesia*

Abstract. The utilization of biomass as a source of new and renewable energy is being carried out. One of the technologies to convert biomass as an energy source is pyrolysis which is converting biomass into more valuable products, such as bio-oil. Bio-oil is a liquid which produced by steam condensation process from the pyrolysis of coconut shell. The composition of biomass such as hemicellulose, cellulose and lignin will be oxidized to phenol as the main content of the bio-oil. Production of bio-oil from coconut shell was investigated via fast pyrolysis reactor. Fast pyrolysis was carried out at 500 °C with a heating rate of 10 °C and 1 hour holding time at pyrolysis temperature. The Bio-oil chemical composition was investigated using GC-MS. Percentage value of phenol, 2-methoxy phenol, 3-methoxy 1,2-benzenediol, and 2,6-dimethoxy phenol was 45.42%, 13.37%, 10.09%, and 11.72% respectively.

1 Introduction

Along with the depletion of world oil reserves and environmental issues, various researches on renewable energy are being actively conducted by various parties. Therefore, new technology needed to be able to overcome those issues. One of the technology that can be used is pyrolysis technology where the raw material comes from organic waste (biomass). The pyrolysis technology also has the potential to tackle organic waste problem and convert it into valuable end product.

Pyrolysis is a thermal conversion process for turning biomass into a more valuable product such as bio-oil. Bio-oil is dark-colored liquid fuel, smells like smoke that condensed from pyrolysis vapors product from ingredients that contain lots of lignin, cellulose, hemicellulose and other carbon compounds. Bio-oil consists of carbon, hydrogen, and oxygen with little nitrogen and sulfur. Sulfur and nitrogen content in bio-oil can be eliminated because of its little impact towards bio-oil performance. The largest organic components in bio-oil are lignin, alcohol, organic acids, and carbonyls. These characteristics make bio-oil an environmentally friendly fuel, and potentially have a greater caloric value than other oxygen fuels (such as methanol) and are only slightly lower in caloric value than diesel and other light fuel oils [1].

The biomass used to produce bio-oil can be obtained from agricultural, forest, plantation, industrial and household waste. Tropical countries like Indonesia generally have abundant biomass resources. Biomass raw materials which potentially used as raw materials for making bio-oils are

[a] Corresponding author : dewiselvia@mail.unnes.ac.id

durian leather [2], oak [3], corn cobs [4], powder / sawmills and wheat bark [5], jathropa [6], fruit pulp [7], coconut shell [8].

Coconut shell waste caused many problems in waste management that usually had been left to rot, stacked, and burned. Those handling will have a negative impact on the environment. One alternative that can be taken to overcome those problems are to convert it into valuable products with applicative and cheap technology so that the results are easily applied to the society.

The aim of this research is to convert waste of biomass into new raw materials or products that have an economic value such as Bio-oil (as an alternative energy resources).

This next research needs to be done to extract phenolic compounds as the main component of bio-oil which is corrosive, unstable and so difficult to be upgraded into ideal fuels [1]. Phenolic compounds are useful for disinfectant or floor cleaner. For the rubber industry, phenol is used to increase the quality of the rubber.

2 Materials and Method

2.1 Materials

The coconut shell was obtained from local market at Semarang, Central Java, Indonesia. Materials for supporting the fast pyrolysis process are distilled water. glasswool and N_2 gas.

2.2 Experimental

Coconut shell was washed using distilled water and then cut into small pieces and dried in an oven at 110°C for 1 hour. Dried coconut shell was crushed using hammer mill and separated by size using a sieve shaker.

The 1000 g prepared coconut shell is fed into a fixed bed reactor. The N_2 gas is fed into a vertical reactor (the schematic experimental is shown in Fig.1) at a rate of 200 ml / min from below until it passes through the top of the reactor. The N_2 gas serves as a substitute for air of the reactor, allowing the conditions when the pyrolysis reaction is anaerobic because there is no O_2 gas in the reactor. The steam formed during the pyrolysis process is the bagasse that flows with the N2 gas toward the top of the reactor, then cooled with a water-filled condenser. From the condenser, the vapor containing the mixed gas turns into a liquid with a temperature drop to 50°C. The liquid product from the condenser is collected in a 500 ml sample bottle, while the non-condensable gas, collected in a gas bag.

The schematic experimental of coconut shell reactor pyrolysis is shown in Fig. 1.

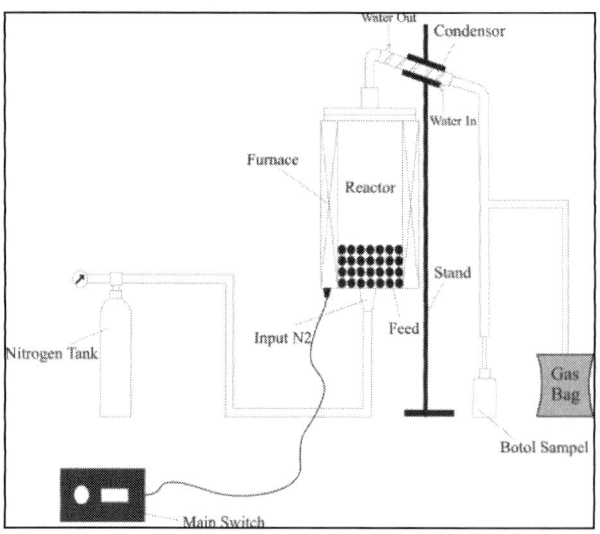

Figure. 1. Fast Pyrolysis Equipment for Producing Bio-Oil from Coconut Shell.

Bio-oil are stored in the refrigerator and used for the next step. The compositions of bio-oil from pyrolysis of coconut shell were analyzed by GC-MS (the principal conditions are shown in Table 1).

Table 1. Conditions of Gas Chromatograph Mass Spectroscopy

GC-MS: Perkin Elmer GC Clarus 680 MS Clarus SQ 8T		
Column:		
Inner diameter	[m]	2.5×10^{-4}
Length	[m]	30
Carrier gas	-	He
Split ratio	[-]	42.2 : 1
Sample volume	m^3	1.10^{-6}
Injection temp.	[K]	583
Column temp.	[K]	333
Pressure column		vacuum
Column flow	mL/min	0.8
MS Scan	[K]	323 - 573
Hold	min	5

3 Results and Discussion

The density of bio-oil obtained in this experiment is 0.961 g/cm^3. Based on the standard range proposed by Mohan et al [9], the average density of bio-oil is 0.94-1.2 g/cm^3. According to Mohan et al [9] the density value is influenced by the molecular weight of the components contained therein. The higher the molecular weight, the density will also be greater. The bio-oil density value that approximates the water density value shows that bio-oil has a considerable water content. The presence of water comes from moisture levels in the biomass feed and indicates that there is still a small amount of oxygen in the reactor allowing the combustion reactions that produce water and carbon dioxide [10]. The condition of the reactor which is free from oxygen needs to be identified accurately.

The bio-oil viscosity obtained in this study was 4,359 cp. The viscosity of bio-oil obtained in this research generally meets the standard range of bio-oil which is 4-7 cp [9]. According to Mohan et al [9], high viscosity mades the bio-oil difficult to flow during the distribution and storage process.

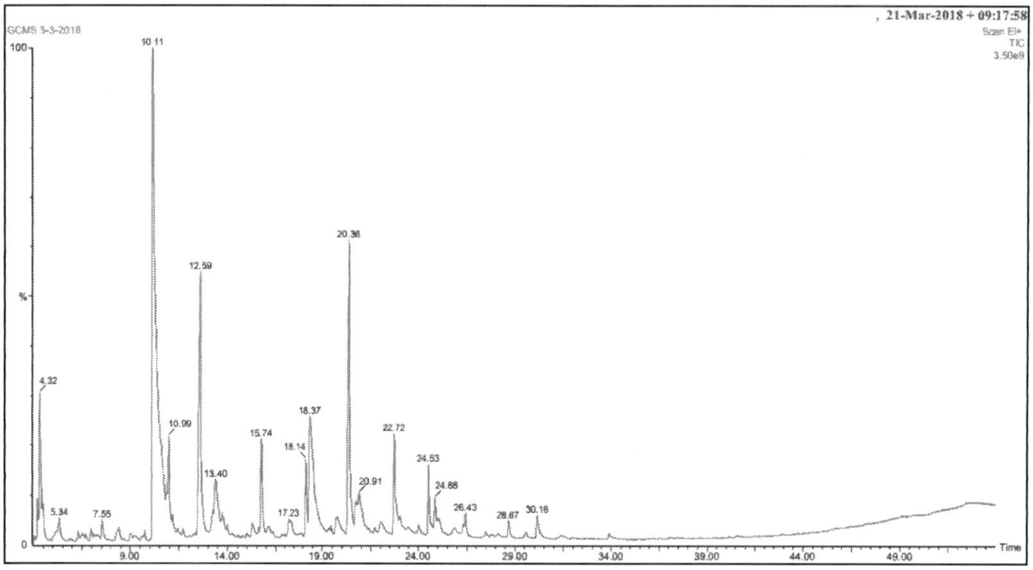

Figure 2. GC-MS Chromatograph of Bio-Oil from Coconut Shell.

The chromatogram results shown in Fig. 2 shows that the composition of the most dominant compounds contained in coconut shell bio-oil is 45.42% of phenol, 13.37% of 2-methoxy-phenol, 10.09% of 3-methoxy-1,2-benzenediol and 11.72% of 2,6-dimethoxy-phenol. The overall bio-oil composition is shown in Table 2.

Bio-oil with phenol content more than 50% would be very good if used as fuel [11]. In this research, the dominant chemical compound of coconut shell bio-oil is phenol. This is because the pyrolysis temperature used is 500°C and most of the lignin in the biomass of the coconut shell has been decomposed.

Table 2. The Compositions of Bio-Oil from Fast Pyrolysis of Coconut Shell

Peak Number	Compounds	Percentage (%)
1	Furfural	3.77%
2	Phenol	45.42%
3	3-methyl-1,2-Cyclopentanedione	4.05%
4	2-methoxy-Phenol	13.37%
5	2-methoxy-4-methyl-Phenol	3.23%
6	4-ethyl-2-methoxy-Phenol	2.19%
7	3-methoxy-1,2-Benzenediol	10.09%
8	2,6-dimethoxy-Phenol	11.72%
9	4-methoxy-3-methoxymethyl-Phenol	3.83%
10	5-tert-Butylpyrogallol	2.31%

4 Conclusions

The bio-oil produced from coconut shell with pyrolysis method has density and viscosity of g/cm^3 and 4,359 cp, respectively.

The main components contained in the bio-oil are phenol (45.42%), 2-methoxy-phenol (13.37%), 3-methoxy-1, 2-benzenediol (10.09%), and 2,6-dimethoxy-phenol (11.72%).

References

1. E. Hambali, S. Mujdalipah, A.H. Tambunan, A.W. Pattiwiri, R. Hendroko, *Teknologi Bioenergi.* (Agromedia Pustaka, Jakarta 2007).
2. S. Ismadji, S., *Prosiding Seminar Nasional Teknik Kimia Soebardjo Brotohardjono I* (Prodi Teknik Kimia UPN Veteran Jawa Timur, 2012).
3. C.J. Ellens, R.C. Brown, Biores. Technol. **103**, pp. 374-380 (2012).
4. I. Demiral, A. Eryazici, S. Sensoz, Biomass Bioenergy, **36**, pp. 43-49 (2012).
5. M. Bertero, G. de la Puente, U. Sedran, Fuel, **95**, pp. 263-271 (2012).
6. A. Haryanti, Norsamsi, P.S.F. Sholiha, N.P. Putri, Konversi, **3,** 2 pp.20-29 (2014).
7. M. Ozbay, E. Apaydin-Varol, B. Uzun, A. Putun, Energy, **33**, pp. 1233-1240 (2008).
8. T. Hidayat, Qomaruddin, *Prosiding SNST ke-6* (Fakultas Teknik Universitas Wahid Hasyim, Semarang 2015).
9. D. Mohan, C.U. Pittman, P.H. Steel, Energy and Fuels, **20**, pp. 848-889 (2006).
10. A.O. Ningrum, *Proses Pembuatan Bio-Oil dari Limbah Kelapa Sawit (Tandan, Cangkang, dan Serat) untuk Bahan Bakar Alternatif dengan Metode Fast Pyrolysis* (Skripsi, Universitas Indonesia, Indonesia, 2011).
11. L. Kusmiati, *Pirolisis Kulit Kayu Pinus (Pinus Merkusii) menjadi Bio-Oil menggunakan Katalis Ni/Lempung* (Skripsi, Universitas Riau, Indonesia, 2015).

Synthesis of Silica Powder from Sugar Cane Bagasse Ash and Its Application as Adsorbent in Adsorptive-distillation of Ethanol-water Solution

Megawati[a], Dewi Selvia Fardhyanti, Radenrara Dewi Artanti Putri, Oki Fianti, Agustin Fitrianingsi Simalango and Afiati Estrina Akhir

Department of Chemical Engineering, Faculty of Engineering, Universitas Negeri Semarang, Indonesia

Abstract. In this study, sugarcane bagasse ash (SCBA) as waste from sugarcane factory was extracted into silica powder. This powder was then used as adsorbent for ethanol purification. Prior to used, the SCBA was washed using HCl solution. The silica extraction was conducted using various NaOH concentrations (0.5; 1; 1.5 and 2 mol/L) as well as extraction times (30, 60 and 90 minutes). After that the mixture was precipitated using HCl solution. The solution was filtrated through a paper filter and its solid particle was dried until its weight was constant. The particle was grounded and sieved using 18 mesh sieves. The silica powder obtained was analysed using FTIR and its result showed that the powder has silica functional groups. The silica yield increases with increasing of concentration of NaOH solution as well as extraction time. The highest silica yield (45.5% w/w) was achieved at 2 mol/L NaOH solution at 90 min. The surface area, pore diameter, and pore volume of silica powder were measured to be 407 m^2/g, 3.81Å, and 2.76 dm^3/g, respectively. From application of silica powder as adsorbent in adsorptive-distillation, ethanol concentration can reach 99.3% w/w which indicates azeotropic point can be passed.

1 Introduction

A sugarcane factory produces sugarcane bagasse ash (SCBA) as by-product. Indonesia's sugarcane production reaches 2.6 tonnes in 2015 [1]. Generally about 280 kg wet bagasse is produced from 1 tonne of cane processed [2]. In addition, sugarcane bagasse used as boiler fuel will result in SCBA as waste (8-10% w/w). SCBA has high silica content (53.1% w/w) [3, 4]. The silica extraction process will produce silica crystal that can be used as hydrophilic adsorbent with surface area between 200-800 m^2/g [5]. The silica powder production process is performed through major steps: raw material preparation, silica extraction, and silica precipitation [6]. The operation condition to obtain the highest silica yield is at 600°C, 3 hours, and using 91.57% HCl solution [7, 8].

At 10% v/v solvent concentration, the silica yielded from rice husk extracted using 1 mol/L KOH and NaOH solutions as solvent were about 10.80 and 22.59% w/w, respectively. Thus, this study was carried out using NaOH solution at various concentrations (0.5; 1; 1.5 and 2 mol/L). Theoretically the solvent concentration increasing increase silica yield as well as HCl solution at pH 7 can be used to

[a] Corresponding author : megawati@mail.unnes.ac.id

precipitate silica after silica extraction [9]. The silica precipitation had been performed using HCl solution at various pH (1, 3, 5, 7) and at pH 7 produced the highest yield (91% w/w) [10].

Accordingly, silica powder can be used as adsorbent in adsorptive-distillation to break azeotropic point of ethanol-water solution for ethanol purification. The polar bond of silica powder works very well to remove water content in ethanol-water solution. Silica powder has 4Å pore diameter so it can be used to adsorb water with 2.8Å pore diameter and entrance ethanol with 4.4Å pore diameter [11]. Adsorptive-distillation for ethanol purification was performed and obtained the highest ethanol concentration using white and blue silica gels as adsorbent at 96.67 and 95.662% v/v, respectively [12]. In this study, the silica powder obtained was applied as adsorbents in adsorptive-distillation of ethanol-water solution.

2 Research methodology

The SCBA used in this research was obtained from sugar company (PT Madubaru, Yogyakarta). Prior to silica extraction, SCBA was heated using furnace for 3 h at 600°C to eliminate carbon content. The ash then was washed with 1 mol/L HCl solution to remove minerals such as extractive mineral and iron composition. The washed products were filtered with filter paper and the filtered solid was stored in a glass bottle. The SCBA before and after washing was analysed using X-Ray Fluorescence (XRF) at Central Laboratory of Mineral and Material Advanced FMIPA, Universitas Negeri Malang.

In silica extraction, 10 g SCBA was extracted using 100 mL NaOH solution to alter the chemical and physical structures. The extraction was conducted in a two-neck flask and heated using a hot plate at 100°C for specific time and stirred using a 200 rpm magnetic stirrer. After that, the mixture was cooled and filtered in order to remove solid residues. The filtrate of sodium silicate was precipitated using 1 mol/L HCl solution until pH 7 and kept for at least 18 hours at room temperature for silica to completely solidified. The slurry then was filtered and washed with distilled water repeatedly until reached neutral pH. The solid was dried using oven at 80°C, grounded and then sieved. In this research, the process variable was NaOH concentrations (0.5; 1; 1.5 and 2 mol/L) as well as extraction times (30, 60, and 90 minutes). The silica powder produced was tested for characteristics of functional groups and surface area (m^2/g), pore diameter (Å), and pore volume (cm^3/g) using Fourier Transform Infrared (FTIR) spectroscopy Perkin Elmer Spectrum Version 10.03. 06 and Brunauer, Emmet, Teller (BET) Quantachrome Nova 1200e.

The adsorptive-distillation of ethanol-water solution using 10 g silica powder produced in this research was conducted in a batch distillation completed with 10 cm adsorbent column. The 100 mL 96% w/w initial ethanol was boiled in three-neck bottle by a heating mantle at 80°C. The vapor flowed to adsorbent in column connected to condenser. The condensate density was analyzed using a picnometer. After that its ethanol content is calculated.

3 Results and discussion

The SCBA from PT Madubaru Yogyakarta was heated in furnace to remove carbon compound [9]. SCBA colour is different after heated in furnace. Before SCBA was heated, it was black and after heating turned to light brown (Figure 1). The heated SCBA was then washed using HCl solution to remove minerals, so that silica composition increases [3, 13]. Both SCBA before and after washed were analyzed by X-Ray Fluorescence (XRF), as in Table 1. The major component in SCBA before washing was Si with 40.2% w/w and after washing became 54% w/w. It obvious that raw material is suitable to be used as source of silica for producing silica powder. The amount of phosphor, chromium, manganese, nickel, copper, barium, and rhenium decreased from about 7.41 to 5.10% w/w. In addition, calcium content can be reduced from 15.7 to 7.12% w/w but potassium and ferric contents cannot be reduced significantly. This fact suggests that SCGA washing using various HCl concentrations should be applied to achieve minimum contents of impurities.

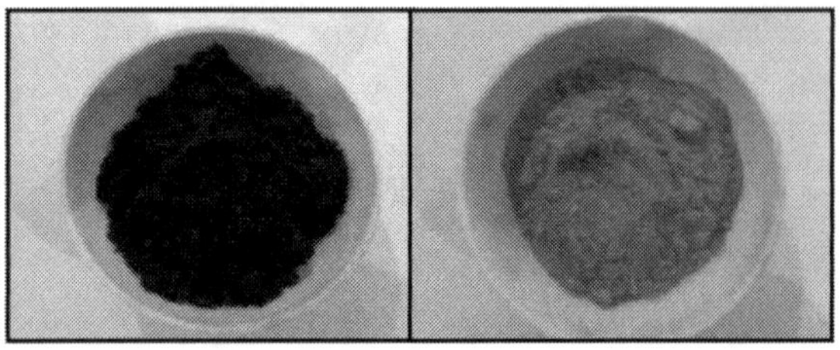

(a) (b)

Figure 1. (a) Before heated in furnace and (b) After heated in furnace.

Table 1. Composition of sugarcane bagasse ash (SCBA).

Composition	**Concentration (% w/w)**	
	Before washed with HCl solution	After washed with HCl solution
Si	40.2	54.0
P	1.40	0.00
K	9.40	8.49
Ca	15.7	7.12
Cr	0.11	0.10
Mn	1.10	0.79
Fe	24.8	23.2
Ni	3.90	3.27
Cu	0.66	0.55
Ba	0.30	0.10
Re	0.40	0.20

The silica extraction process was conducted using NaOH solution and then precipitated using HCl solution, adapting from the literature [14]. These mechanisms can be expressed by Equations 1 and 2 [15]. At the end of these steps, the slurry was filtered and filtered solid was heated in oven. Several photos for these steps show as in Figure 2.

$$2\ NaOH + SiO_2\ (in\ SCGA) \text{ ----> } Na_2SiO_3 + H_2O \quad (1)$$

$$Na2SiO3 + 2\ HCl \text{ ----> } SiO_2\ (pure\ silica\ powder) + H_2O \quad (2)$$

The influence of NaOH concentration to silica yield can be seen in Table 2. The most silica yield (45.5% w/w) was obtained from synthesis with 2 mol/L NaOH solution for 90 min. The production of silica tends to increase in line with NaOH concentration (0.5 to 2 mol/L) at all extraction times tested in this study. This phenomenon is similar to previous research [9]. Figure 3 presents the FTIR spectrum of silica powder, where the presence of three main bands of silica is observed; at 3835.75 and 1638.15 cm^{-1}, the bands correlate with Si-OH, the bands at 1099.21 to 799.62 cm^{-1} is assigned to Si-O-Si, and that at 474.80 cm^{-1} is due to O-Si-O (Table 3). Accordingly, the main bands of silica were O-Si-O, O-H, and Si-O-Si with the peaks around 438, 796, and 1044 cm^{-1}, respectively [13].

However, other research shows that the silica produced had the bands of Si-O-Si, O-Si-O, and Si-OH with wavenumbers of 1077, 464, and 1644 cm^{-1} [16]. Subsequently, the main bands of silica powder produced from this study was similar to previous research.

Figure 2. Silica precipitation (Process condition: washed with 1 mol/L HCl solution, extracted with 1 mol/L NaOH, ratio of SCBA:NaOH = 1:10 w/v, volume = 100 mL; precipitated with 1 mol/L HCl solution).

Table 2. Silica yield of sugarcane bagasse ash (SCBA) extraction using NaOH solution.

NaOH Concentration	Silica Yield (% w/w)		
	30 min	60 min	90 min
0.5	10.6	14.0	14.1
1.0	13.0	24.1	25.1
1.5	25.4	35.7	35.8
2.0	26.2	45.3	45.5

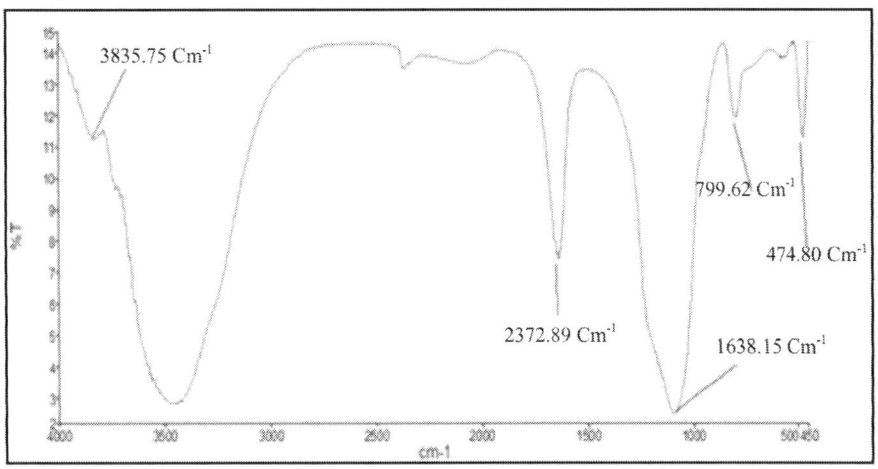

Figure 3. Fourier Transform Infrared (FTIR) of silica powder of sugarcane bagasse ash (SCBA) extraction using NaOH Solution.

Analysis of silica powder using Brunauer, Emmet, Teller (BET) to know powder's characteristics resulted a surface area 407 m^2/g, pore diameter 3.806Å, and pore volume $0.276.10^{-2}$ cm^3/g. Based on previous research, silica crystal has surface area 218 m^2/g, pore diameter 5.56Å, and pore volume $0.32.10^{-2}$ cm^3/g [13]. Moreover, from other research, silica crystal has surface area 265 m^2/g, pore diameter 6.25Å, and pore volume $0.425.10^{-2}$ cm^3/g [17]. The comparison indicates that silica powder produced in this research has physical characteristic as same as with previous research. In text book, surface area of silica crystal is about 200-800 m^2/g [5].

Table 3. The Wavenumber and main bands of silica powder from sugarcane bagasse ssh (SCBA) extraction using NaOH solution.

No	Wavenumber (cm^{-1})	Band
1	3835,75	Si-OH
2	1638,15	Si-OH
3	1099.21	Si-O-Si
4	799.620	Si-O-Si
5	474.800	O-Si-O

In this study, the silica powder produced was applied as adsorbent of water vapor in adsorptive-distillation of ethanol-water solution in order to increase the ethanol concentration. The adsorption capacity of silica powder will increase if the silica was heated prior to use. Before silica was heated, the capacity was just 0.02%, but after heating, the capacity increased up to 3.2%. Subsequently, ethanol concentration increased from 96% v/v to 99,3% v/v. This condition shows that azeotropic point of ethanol-water solution is passed. Hence, continuous distillation to obtain absolute concentration can be conducted using simple distillation.

4 Conclusions

Silica powder was successfully extracted from sugarcane bagasse ash conducted by a heating process followed extraction with NaOH solution and precipitation with HCl solution. XRF analysis showed SCBA composition data indicating that Si content was about 54% w/w after heating and washing. The production of silica tends to increase in line with NaOH concentration (0.5 to 2 mol/L). The most silica yield (45.5% w/w) was obtained from synthesis process using 2 mol/L NaOH solution for 90 min. Based on BET analysis, the silica powder has characteristics such as surface area 407 m^2/g, pore diameter 3.806Å, and pore volume $0.276.10^{-2}$ cm^3/g. Furthermore, the application of silica powder as adsorbent showed significant increase of ethanol concentration, from 96 to 99.3% v/v.

References

1. Badan Pusat Statistika, *Statistik Tebu Indonesia* (Badan Pusat Statistika, Jakarta, 2015)
2. S. Yokoyama, *Buku Panduan Untuk Produksi dan Pemanfaatan Biomassa* (The Japan Institut of Energy, Tokyo, 2008)
3. S. Norsuraya, H. Fazlena, R. Norhasyimi, *Proceedia Eng.* E **148** (2016)
4. S. Rovani, J.J. Santos, P. Corio, D.A. Fungaro, ACS Omega E **3** (2018)
5. K. Othmer, *Encyclopedia of Chemical Technology* (Wiley-Interscience, New York, 1997)
6. V. Hariharan, G. Sivakumar, I. J. of Chem. Tech. Res. E **5**, 3 (2013)
7. V. B. Carmona, R. M. Oliveira, W. T. L. Silva, Ind. Crops and Products E **42**, 1 (2013)
8. P. Worathanakul, W. Payubnob, A. Muangpet, I. J. Chem. Molecul. Nuclear, Mat. and Metallurgical Eng. E **3**, 8 (2009)
9. M. Soleh, *Ekstraksi Silika dari Sekam Padi dengan Metode Pelarutan dan Pengendapan Silika serta Analisis EDX dan FTIR*. Thesis (Institut Pertanian Bogor, Bogor, 2014)
10. U. Kalapathy, A. Proctor, J. Shultz, Bioresour. Technol. E **73** (2000)

11. H. J. Huang, S. Ramaswamy, U. W. Tschimer, B. V. Ramarao, *Separation and Purification Technology* (Woodhead Publishing, 2008)
12. Megawati, R. A. Jannah, I. Rahayuningtyas, *AIP Conference Proceeding* E **1788** (2017)
13. R. A. Bakar, R. Yahya, S. N. Gan, Proceedia Chem. E **19** (2016)
14. P. A. Handayani, E. Nurjanah, W. D. P. Rengga, Jurnal Bahan Alam Terbarukan E **4**, 2 (2014)
15. B. S. Todkar. O. A. Deorukhkar, S. M. Deshmukh, I. J. of Eng. Res. and Dev. E **12**, 3 (2016)
16. A. Tadjarodi, M. Haghverdi, V. Mohammadi, Mat. Res. Bulletin E **47** (2012)
17. R. H. Alves, T. V da S. Reis, S. Rovani, D. A. Fungaro, J. of Chem. E **1-9** (2017)

Numerical Investigation on Physical Foaming and Decay Process Using Multicomponent Thermal Lattice Boltzmann Model

Fu-Min Liu[1,a] and An-Lin Wang[1]

[1]School of Mechanical Engineering, Tongji University,4800 Caoan Highway, Shanghai, China

Abstract. Foam materials produced by physical process are common in the field of material engineering such as foamed asphalt. Considering heat and phase transfer, the complicated nonlinear phenomenon of foaming and decay process is difficult to be described numerically. In this paper, a multicomponent thermal lattice Boltzmann model for simulation of physical foaming and decay process is proposed. The model combines a thermal single-component phase change model with a multicomponent model, and the two models are verified separately. The physical foaming and decay process is solved numerically based on the proposed model. Simulation results show the influence of the distribution and content of the phase-change component on the expansion rate and decay time during physical foaming and decay process.

1 Introduction

Physical foaming and decay process caused by bubbles generation and crushing of one component (the phase-change component) in multicomponent fluids, such as foamed asphalt foaming process [1] is widespread in material engineering and other field. The macro dynamic behaviour of such foam-flow is very complex because of the interaction between different fluids and the process of mass and heat transfer. To investigate the foam-flow, multicomponent multiphase fluids accompanied by phase change with heat transfer need to be described numerically. However, conventional computational fluid dynamics (CFD) methods face difficulties in predicting such complex flow.

Different from conventional CFD methods, based on microscopic nature and mesoscopic characteristics, the lattice Boltzmann method (LBM) performs well in numerical simulations of many kinds of complex flow. Many thermal lattice Boltzmann models have been developed in the last few years [2-4]. However, those models were only used to study the phase change in single component multiphase fluids. In this work, we focus on thermal multicomponent multiphase fluids, of which one component is accompanied by the phase change process. The single-component phase change model presented by Gong and Cheng [5] is combined with a multicomponent model to simulate the phase change process of one component in multicomponent fluids with heat transfer, which can be used to investigate the foaming and decay process of foam-flow. The rest of the paper is organized as follows. Section 2 reviews the presented thermal phase change model and multicomponent model, and describes the combination of the two models. In Section 3, the phase change model and the multicomponent model are validated separately. In Section 4, foaming and decay process is simulated

[a] Corresponding author : 1410288@tongji.edu.cn

and the influence of the distribution and content of the phase-change component on the two important indicators (expansion rate and decay time) for studying foam-flow is discussed. Conclusions are made in Section 5.

2 Model description

2.1 Thermal single-component phase change model

In LBM, the fluid is described by the evolution of density distribution functions. The evolution equation of the density distribution function with the Bhatnagar-Gross-Krook collision operator [6] is written as:

$$f_\alpha(\mathbf{r}+\mathbf{e}_\alpha\delta_t,t+\delta_t)-f_\alpha(\mathbf{r},t)=-\frac{1}{\tau}[f_\alpha(\mathbf{r},t)-f_\alpha^{eq}(\mathbf{r},t)]+\Delta f_\alpha(\mathbf{r},t)\qquad \alpha=0,1\cdots N \tag{1}$$

where $f_\alpha(\mathbf{r},t)$ is the density distribution function along α direction at the lattice site $\mathbf{r}$ and time t, δ_t is the time step, $\mathbf{e}_\alpha$ is the discrete velocities, τ is the dimensionless relaxation time, $\Delta f_\alpha(\mathbf{r},t)$ is the body force term, N is the number of the discrete velocities(In this work, we use two dimensions with nine directions (D2Q9) lattice configuration), $f_\alpha^{eq}(\mathbf{r},t)$ is its corresponding equilibrium distribution function, which has the following form:

$$f_\alpha^{eq}=\rho w_\alpha[1+\frac{\mathbf{e}_\alpha\cdot\mathbf{u}}{c_s^2}+\frac{(\mathbf{e}_\alpha\cdot\mathbf{u})^2}{2c_s^4}-\frac{\mathbf{u}^2}{2c_s^2}] \tag{2}$$

where c_s is the lattice sound speed which is equal to $1/\sqrt{3}$, w_α is the weight coefficient, and $w_0=4/9, w_{1-4}=1/9, w_{5-8}=1/36$. ρ and $\mathbf{u}$ are the density and the macroscopic velocity, respectively. They are defined as:

$$\rho=\sum_\alpha f_\alpha,\quad \mathbf{u}=\frac{1}{\rho}\sum_\alpha f_\alpha\mathbf{e}_\alpha \tag{3}$$

The exact-difference-method force scheme [7] is adopted and $\Delta f(\mathbf{r},t)$ is given by:

$$\Delta f_\alpha(\mathbf{r},t)=f_\alpha^{eq}(\rho(\mathbf{r},t),\mathbf{u}+\Delta\mathbf{u})-f_\alpha^{eq}(\rho(\mathbf{r},t),\mathbf{u}) \tag{4}$$

where $\Delta\mathbf{u}=\mathbf{F}\delta_t/\rho$ is the velocity change caused by body force during time step δ_t, $\mathbf{F}$ is the body force including the interparticle interaction force $\mathbf{F}_{\text{int}}$, the gravitational force $\mathbf{F}_g$ and so on. $\mathbf{F}_{\text{int}}$ and $\mathbf{F}_g$ can be obtained as follows:

$$\mathbf{F}_{\text{int}}(\mathbf{r})=-c_0\psi(\mathbf{r})g\nabla\psi(\mathbf{r}),\quad \mathbf{F}_g(\mathbf{r})=G\cdot(\rho(\mathbf{r})-\rho_{ave}) \tag{5}$$

where c_0 = 6.0 [8] and g is a Green's function, which depends on the strength of the interaction force between particles, and represents attractive (repulsive) force when $g < 0$ ($g > 0$). G is the acceleration of gravity and ρ_{ave} is the average density of the whole computation domain. $\psi(\mathbf{r})$ is the effective mass at lattice site $\mathbf{r}$, and it is a function of ρ , which can be given by:

$$\psi(\mathbf{r})=\sqrt{\frac{2(P(\mathbf{r})-c_s^2\rho(\mathbf{r}))}{c_0 g}} \tag{6}$$

where $P(\mathbf{r})$ is the local pressure. Peng–Robinson (P–R) equation of state (EOS) is selected to calculate local pressure and introduce the influence of temperature field on density distribution. It is given by:

$$P=\frac{\rho RT}{1-b\rho}-\frac{a\rho^2\varepsilon(T)}{1+2b\rho-b^2\rho^2} \tag{7}$$

where $\varepsilon(T)$=[1+(0.374 64+1.542 26ω-0.269 92ω^2)(1-$\sqrt{T/T_c}$)]^2 , ω is the acentric factor, and ω = 0.344 for water. a = 0.457 24$R^2T_c^2/P_c$ and b = 0.0778RT_c/P_c in Equation (7), T_c and P_c represent the critical temperature and the critical pressure. We set a = 2/49, b = 2/21, and R = 1 in the following simulations [8]. Then T_c = 0.0729 can be obtained.

By averaging the moment before and after the collision, the actual physical velocity **U** is given by:

$$\rho \mathbf{U} = \sum_\alpha f_\alpha \mathbf{e}_\alpha + \delta_t / 2\mathbf{F} \tag{8}$$

To solve energy equation, another distribution function is introduced. The evolution equation of the temperature distribution function is written as:

$$h_\alpha(\mathbf{r}+\mathbf{e}_\alpha\delta_t, t+\delta_t) - h_\alpha(\mathbf{r},t) = -1/\tau_T [h_\alpha(\mathbf{r},t) - h_\alpha^{eq}(\mathbf{r},t)] + \delta_t \omega_\alpha \phi \qquad \alpha = 0,1\cdots N \tag{9}$$

where τ_T is the relaxation time for temperature with τ_T=0.5+3$\chi\delta_t / \delta_x^2$, in which χ=$\lambda / \rho c_v$ is the thermal diffusivity, λ is the thermal conductivity and c_υ is the specific heat capacity. $h_\alpha^{eq}(\mathbf{r},t)$ is the equilibrium temperature distribution function which has the following form:

$$h_\alpha^{eq} = T w_\alpha [1 + \frac{\mathbf{e}_\alpha \cdot \mathbf{U}}{c_s^2} + \frac{(\mathbf{e}_\alpha \cdot \mathbf{U})^2}{2c_s^4} - \frac{\mathbf{U}^2}{2c_s^2}] \tag{10}$$

T is the macroscopic temperature which is obtained by $T = \sum_\alpha h_\alpha$, φ in Equation (9) is the source term describing the energy change due to phase change. Gong and Cheng [5] derived the source term from the thermodynamic relation of entropy. And it can be written as:

$$\phi = T[1 - \frac{1}{\rho c_v}(\frac{\partial P}{\partial T})_\rho]\nabla \cdot \mathbf{U} \tag{11}$$

2.2 Multicomponent model

The single-component pseudopotential model can be easily extended to a multicomponent model by treating each component using a separate density distribution function. Each set of density distribution functions can be evolved separately following the Equation (1). In order to separate different components, the interparticle force between different components is introduced into the multicomponent model. Summing over all the components and interacting sites, the interparticle force between different components acting on the particles of the σth component at site **r**, $\mathbf{F}_{\text{intmc}}$, has the following form like Equation (5):

$$\mathbf{F}_{\sigma,\text{int mc}}(\mathbf{r}) = -c_0 \psi_\sigma(\mathbf{r}) \sum_{\bar{\sigma}} g_{\sigma\bar{\sigma}}(\mathbf{r},\mathbf{r}') \nabla \psi_{\bar{\sigma}}(\mathbf{r}') \tag{12}$$

And the macroscopic density of multicomponent fluids is defined as:

$$\rho = \sum_\sigma \sum_\alpha f_{\sigma,\alpha} \tag{13}$$

The macroscopic velocity and the actual physical velocity can be obtained as follows:

$$\mathbf{u} = \sum_\sigma \frac{\rho_\sigma \mathbf{u}_\sigma}{\tau_\sigma} \Big/ \sum_\sigma \frac{\rho_\sigma}{\tau_\sigma}, \quad \mathbf{U} = \frac{1}{\rho}\left(\sum_\sigma \rho_\sigma \mathbf{u}_\sigma + \frac{1}{2} \sum_\sigma \mathbf{F}_\sigma \right) \tag{14}$$

where $\mathbf{F}_\sigma$ is the total body force of the σth component.

2.3 Combined model

To simulate foaming and decay process, the two models mentioned above are combined. For simplicity, we consider fluids consisting of two components. We set water as the first component and the effective mass ψ_1 is calculated by Equation (6) with P-R EOS. As the temperature rises, a liquid-vapour phase change process of the first component will appear. The second component is set to be a kind of component without phase change in our simulation, and the effective mass ψ_2 is given by $\psi_2 = \rho_2$. The EOS of the multicomponent fluids [9] is calculated by:

$$P = (\rho_1 + \rho_2)c_s^2 + c_0/2(g_{11}\psi_1\psi_1 + g_{12}\psi_1\psi_2 + g_{21}\psi_2\psi_1 + g_{22}\psi_2\psi_2) \tag{15}$$

where $g_{12} = g_{21}$ according to the Newton's 3rd Law. And substituting it into Equation (11), the source term in the evolution equation of the temperature can be written as:

$$\phi \approx \mathrm{T}[1 - \frac{1}{\rho c_v}(\frac{\rho_1 R}{1 - b\rho_1} + \frac{g_{12}\rho_1\rho_2 R}{g_{11}(1 - b\rho_1)\psi_1})]\nabla \cdot \mathbf{U} \tag{16}$$

It should also be mentioned that density of the first component is different compared with which that in the single-component thermal phase change model at the same temperature. That is caused by the interparticle force between different components. In this work, we set $\rho_2 = 1.0$, $g_{11} = 0.1$(or -0.1), $g_{12} = 0.005$, and $g_{22} = 0$. For the single-component flow, the specific heat capacity (c_v) is considered constant. However, in a multicomponent system, c_v of the mixture flow is no longer a constant. It is related to the distribution of each component. We define it as a parameter, which changes along with the mass fraction of each component. c_v can be expressed as:

$$c_v = 1/\rho\left(\sum_\sigma \rho_\sigma c_{v\sigma}\right) \tag{17}$$

3 Model verification

As the combined model is proposed by integrating the thermal phase change model and the multicomponent model, we verify two models separately. The phase change model is validated through the simulation of bubble growth and departure from a horizontal surface. Meanwhile we verify the multicomponent model for surface tension by binary static droplet simulations. It should be noted that variables in this paper are in lattice units, which can be converted into real physical properties based on the principle of reduced properties.

3.1 Thermal phase change model validation

Bubble growth and departure from a heated horizontal isothermal surface is simulated to study the relationship between departure diameter of a bubble and gravity acceleration. And the results are compared with analytical solutions and the simulation results in Gong and Cheng's [5] work.

A 150 × 450 lattice structure is chosen to simulate the bubble growth and departure from a heated horizontal isothermal surface. Initially, the computation domain is occupied by the liquid phase. At the centre of the bottom wall, a high temperature core is specified to ensure that nucleation starts there. And simulation conditions are set as Gong and Cheng's [5] work. As time goes by, a bubble will grow and depart from the bottom wall. The process of bubble growth and departure under the condition of $G = 0.000\ 04$ is shown in Figure 1.

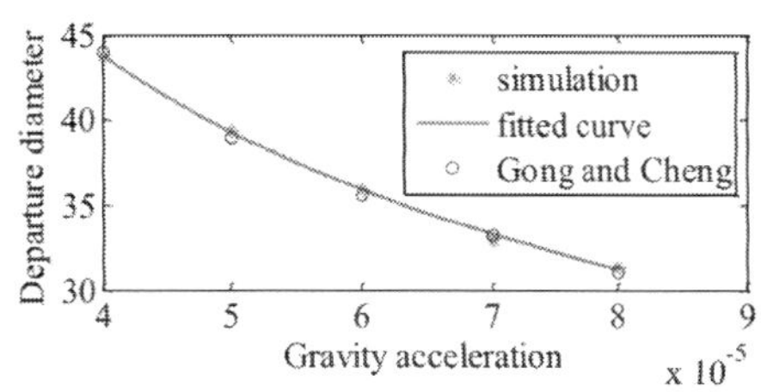

Figure 1. The process of bubble growth and departure. **Figure 2.** Gravity effect on bubble departure diameter.

The relationship between departure diameter of the bubble and gravity acceleration [10] is given as follows:

$$D \sim |G|^{-0.5} \tag{18}$$

To verify this relationship, a series of simulations are performed under different values of gravity acceleration. The results are compared with the reference and presented in Figure 2. We can see that at the simulation results of bubble departure diameter D are similar with the reference results in Gong and Cheng's [5] work, and can be fitted into regression formula $D = 0.3061|G|^{-0.4902}$. The exponent of G is -0.4902, which agrees well with the analytical result given by Equation (18). And simulation results showed above verify the correctness of the thermal phase change model.

3.2 Multicomponent model validation

A series of binary static-droplet simulations are performed to verify the properties of surface tension and show the computational stability of the model under isothermal conditions.

We simulate a static liquid droplet of the first component with different radii r in a 200 × 200 gravity-free lattices field full of the second component at a constant temperature $T = 0.82T_c$. The initial centre of the droplet is located at (100, 100). Periodic boundary conditions are used in this simulation. Surface tension force can be calculated by the Laplace law which can be expressed as:

$$\Delta P = P_{in} - P_{out} = \kappa/r \tag{19}$$

where k is a constant, P_{in} and P_{out} are the pressure inside and outside of droplet, respectively. A series of simulations are conducted by changing r. Simulation and linear fitting results are shown in Figure 3. As shown, our simulation results are satisfied with the Laplace's law.

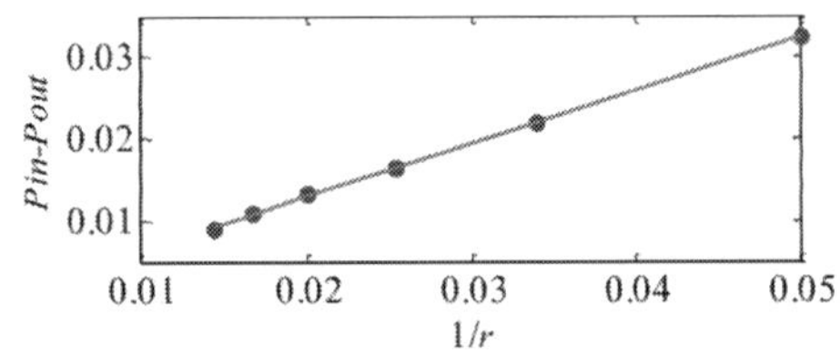

Figure 3. Relationship between $P_{in}-P_{out}$ and $1/r$ for binary static droplet system.

4 Physical foaming and decay process simulation

Physical foaming and decay process caused by bubbles generation and crushing of one component in multicomponent fluids is simulated in a 200 × 150 field. The field of Y > 100 is filled with the high-temperature gaseous first component initially. And liquid droplets of the first component are distributed in the high-temperature second component which fills the rest of the field. The initial temperature of the first component liquid droplets is $0.82T_c$ (the initial density of the first component is the corresponding saturation liquid density at $0.82T_c$), while the initial temperature of the second component and the gaseous first component is $2.07T_c$. Constant pressure is kept at the top boundary. Constant temperature wall boundary conditions are applied at horizontal and bottom boundaries (the values of temperature is 2.0 T_c). The specific heat capacity of both components is set to 1.0. Physical foaming and decay process is shown in Figure 4.

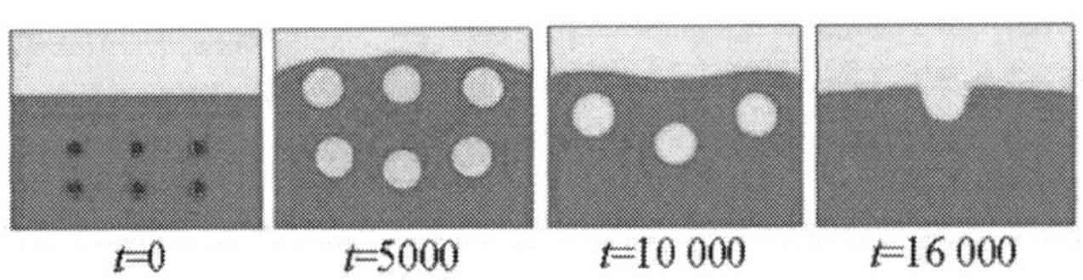

Figure 4. Physical foaming and decay process.

Expansion rate and decay time are two important indicators for studying foam-flow. The liquid level of the second component, which can be used to investigate the expansion rate and decay time of foaming and decay process, is measured during bubble generation and crushing. The influence of distribution and content of the first component on the expansion rate and decay time is investigated using the control variable method. Figure 5(a) shows the average liquid level of the second component during foaming and decay process under different first component distributions (the initial number of the droplets is 2×3, 2×4, 2×5 separately at the same content of the first component). We can see that droplets distribution does not affect the expansion rate, but the more uniform the distribution, the slower the decay is. And Figure 5(b) shows the average liquid level of the second component during foaming and decay process under different first component contents (the initial radius of droplets is 4, 5, 7 separately at the same number of the droplets). We can see that as the content of the first component increases, the expansion rate increases and decay becomes faster. And step decay occurs happens when bubbles collapse as shown in Figure 5. It is because the bubbles generated in this simulation are large and the number of the bubbles is small, just as Jenkins mentioned [11].

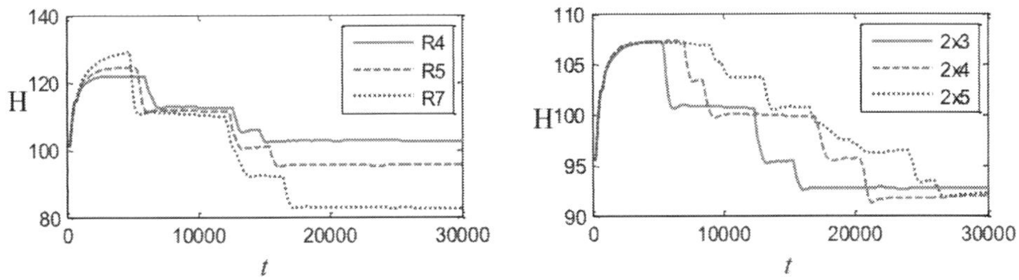

5(a) 5(b)

Figure 5. Average liquid level of the second component during foaming and decay process. (5(a)-under different first component distributions; 5(b)-under different first component contents)

5 Conclusion

In this paper, we propose a multicomponent lattice Boltzmann model for simulation of physical foaming and decay process caused by bubbles generation and crushing of one component (the phase-change component) in multicomponent fluids. The model combines a single-component thermal phase change model and a multicomponent model. Both models are verified. The combined model is used to simulate physical foaming and decay process. It is found that the distribution of the phase-change component does not affect the expansion rate, but the more uniform the distribution, the slower the decay is. And simulation results also show that as the content of the phase-change component increases, the expansion rate increases and decay becomes faster.

Acknowledgments

This work is supported by the International Exchange Program for Graduate Students, Tongji University (No. 201801223).

References

1. B. W. Hailesilassie, M. Hugener, M. N. Partl, Constr. Build. Mater. **85**, 65 (2015)
2. Y. Feng, W. Q. Tao, Numer. Heat Tr. B-Fund. **66**, 544 (2014)
3. N. Frapolli, S. S. Chikatamarla, I. V. Karlin, Phys. Rev. E **90**, 043306 (2014)
4. Q. Li, Q. J. Kang, M. M. Francois, Y. L. He, K. H. Luo, Int. J. Heat Mass Tran. **85**, 787 (2015)
5. S. Gong, P. Cheng, Int. J. Heat Mass Tran. **55**, 4923 (2012)
6. P. L. Bhatnagar, E. P. Gross, M. Krook, Phys. Rev. **94**, 511 (1954)
7. A. L. Kupershtokh, D. A. Medvedev, D. I. Karpov, Comput. Math. Appl. **58**, 965 (2009)
8. P. Yuan, L. Schaefer, Phys. Fluids **18**, 042101 (2006)
9. X. Shan, G. Doolen, J. Stat. Phys. **81**, 379 (1995)
10. W. Fritz, Phys. Z. **36**, 379 (1935)
11. Jenkins, K. J. (2000) Mix design considerations for cold and half-warm bituminous mixes with emphasis of foamed bitumen (Doctoral dissertation, Stellenbosch: Stellenbosch University)

Swelling Properties of Propranolol HCl-loaded Polyvinyl Alcohol-*graft*-lactic Acid Hydrogel Films

Jirapornchai Suksaeree[1, a] and Chomnapas Chuchote[2, b]

[1] *Department of Pharmaceutical Chemistry, College of Pharmacy, Rangsit University, Pathum Thani, Thailand*
[2] *Department of Pharmacognosy, College of Pharmacy, Rangsit University, Pathum Thani, Thailand*

Abstract. This paper presented the swelling properties of propranolol HCl-loaded polyvinyl alcohol-*graft*-lactic acid (PPH-loaded PVA-g-LA) hydrogel films. The swelling properties including swelling-deswelling measurement and swelling behaviour in different environmental stimuli fluids such as water, various pH, and various ionic strength were determined. It was found that the swelling properties of the PPH-loaded PVA-*g*-LA hydrogel films depend on the LA amount addition. They had a good swelling and deswelling in water and acetone. The PPH-loaded PVA-*g*-LA hydrogel films showed the high swelling in the medium pH 7. Thus, the prepared PPH-loaded PVA-*g*-LA hydrogel films had a good swelling property that could be used to controlled drug release in pharmaceutical product.

1 Introduction

The adhesive hydrogel films are polymer networks holding a significant amount of water whilst maintaining their structural integrity without dissolution whereas such thin films are adhering on human skin. They are ideal biocompatible materials that can be used as adhesives for several technologies and applications. They will be providing the controlled drug release from their formulation ; as well as, the active drugs can easily diffuse and through-out from the adhesive hydrogel films. The adhesive hydrogel films can be prepared by two method, chemical and physical hydrogel films that are replaced the hydrogen bond as cross-linked networks by intramolecular and prevented by physical interactions between different more than two polymer chains by intermolecular, respectively [1-5].

Polyvinyl alcohol (PVA) is hydrolyzed from polyvinyl acetate and produced a high percentage of repeat units which is mainly composed of C–C bonds with pendant hydroxyl groups. It is a water-soluble polymer, low toxicity, and biodegradable. It can be modified by chemical reaction and blending such as grafting with starch [6], blending with methylated cornstarch-based film [7], blending with modified xylan [8], and crosslink with maleic acid [9]. In addition, modified PVA hydrogel can be desired as a hydrophilic polymer that can accordingly swell or shrink in water or environmental stimuli fluids such as pH, temperature, ionic strength, etc. PVA hydrogels have become especially attractive to the various fields such as pharmaceutical, biomedical, and tissue engineering including ophthalmic contact lenses, drug delivery, and tendon repair [10]. Lactic acid (LA) is a wide

[a] Corresponding author: jirapornchai.s@rsu.ac.th
[b] Corresponding author: chomnapas@gmail.com

availability and relatively low cost that commonly used in biomaterial applications. The LA can be grafted with hydroxyl groups of PVA by esterification method may decrease the hydrophilicity, and intra- and intermolecular hydrogen bonds of the PVA. The reactions between alcohols of PVA and carboxylics of lactic acid can take place with catalysts at various temperatures. For example, the grafting on backbone of PVA with a copolymer of lactide and glycolide by ring-opening melt polymerization using stannous octoate as catalyst is reported by Breitenbach and Kissel [11]. Their physicochemical properties, such as melting point, glass transition point, degree of crystallinity, molecular composition and architecture, could be systematically adjusted as new class of biodegradable polymers to the considerable potential as drug delivery. Moreover, the PVA-*graft*-LA (PVA-*g*-LA) is fully soluble in cold water and improving flexibility and elasticity as well as tear resistance when it is synthesized by melt polymerization at 100°C under nitrogen atmosphere and continuously stirred [12], using stannous chloride [13], stannous 2-ethyl hexanoate [14], and manganese acetate [15] as a catalyst. However, the previous reports of PVA-*g*-LA reaction have been carried out under organic solvents, initiator, and catalysts that are environmental concerns and high cost and react under nitrogen atmosphere. Therefore, the present research focused on a green route synthesis has not yet been reported for the grafting reaction of LA onto PVA in a water-based system. The green route synthesis is prepared by eco-friendly curing method which is a one term of polymer chemistry and polymer process engineering without chemical agents.

Our previous publication reported the thermal properties including differential scanning calorimetry and thermogravimetric analysis, the confirmation of main functional groups by Fourier transform infrared spectroscopy, the swelling-deswelling behavior, and the moisture measurement per dry mass of modified polyvinyl alcohol-*graft*-lactic acid (PVA-*g*-LA) hydrogel films that synthesized by eco-friendly synthesis method [16]. Presently, we studied the preparation of the propranolol HCl (PPH)-loaded PVA-*g*-LA hydrogel film. The hydrogel films were synthesized in aqueous solutions without initiator and catalyst. The different amounts of PVA and LA were 1:0.5, 1:1, 1:1.5, and 1:2, and different curing times were 120, 150, and 180 min at 100°C. The swelling behaviors were characterized by swelling-deswelling measurement and swelling behavior in different environmental stimuli fluids such as water, various pH, and various ionic strength.

2 Experimental Works

The compositions of PPH-loaded PVA-*g*-LA hydrogel films are also shown in Table 1. Firstly, PVA (Mw = 195,000 g/mol) was dissolved in ultrapure water by heating it on a water bath at 70±2°C and being continuously stirred until clearly PVA solution. Subsequently, it was cooled down to room temperature. 80% L-(+)-LA solution was slowly added into clearly PVA solution with the help of a mechanical stirrer for 60 min to form a homogeneously gel-like mixture solution. The PPH powder was mixed in the mixture. After that, the gel-like mixture solution was cured at high temperature in an oven.

The swelling-deswelling behaviors, area 2 cm × 2 cm of PPH-loaded PVA-*g*-LA hydrogel films were immersed in excess distilled water at room temperature and then swollen samples were weighted (W_s). Then, swollen PPH-loaded PVA-g-LA hydrogel films were transferred into acetone at room temperature and were weighted at time interval (W_n) until equilibrium point. The reswelling and deswelling behaviors were measured again in distilled water and acetone, respectively at room temperature. The swelling or reswelling was calculated in terms of the relative gel volume (V_{rel}) using the following equation (1) [17].

$$V_{rel} = (W_n / W_s)^3 \quad (1)$$

Table 1. Ingredient composition and preparation conditions of PPH-loaded PVA-*g*-LA

Code	PVA (g)	LA (g)	PPH (g)	Water (g)	Temperature (°C)	Time (min)
PL1						120
PL2		2.5		87.5	100	150
PL3						180
PL4						120
PL5		5.0		85.0	100	150
PL6	5.0		5.0			180
PL7						120
PL8		7.5		82.5	100	150
PL9						180
PL10						120
PL11		10.0		80.0	100	150
PL12						180

The 1.0-1.5 g of PPH-loaded PVA-*g*-LA hydrogel films were placed on aluminum pan and then heat at 120°C. The percentage of moisture content was measured by moisture analyzer (model: MAC 50/NH, Poland) and calculated according to equation (2).

$$\text{Moisture content (\%)} = (W_{initial} - W_{dry}) / W_{initial} \quad (2)$$

Swelling measurement of PPH-loaded PVA-*g*-LA hydrogel films were investigated the effect of pH and different cations. Various pH preparations were adjusted to 2, 4, 7, and 10 using 1 mol/L sodium hydroxide and 1 mol/L hydrogen chloride aqueous solution in distilled water. The 10 mmol/L chloride salt of sodium chloride or potassium chloride were prepared as swelling media. Then, the samples were immersed in swelling media at time intervals. The equilibrium swollen samples were weighted and calculated according to equation (3). The all data were reported.

$$\text{Ratio of water absorption amount} = (W_2 - W_1) / W_1 \quad (3)$$

where W_1 was dried sample and W_2 was swollen sample.

3 Results and discussion

The swelling-deswelling behaviors of the PPH-loaded PVA-*g*-LA hydrogel films were determined in distilled water and acetone that are shown in Figure 1. The deswelling behavior of swollen PPH-loaded PVA-*g*-LA hydrogel films in acetone as well as the reswelling behavior of the PPH-loaded PVA-*g*-LA hydrogel films in water were investigated. The curing time and curing temperature condition significantly affected to the ratio of water absorption and the ratio of water retention due to the higher strength interaction of ester bond between the hydroxyl groups of PVA and the carboxyl groups of LA in their structure [6, 10].

The moisture content in the PPH-loaded PVA-*g*-LA hydrogel films was measured by moisture analyzer and calculated as a percentage of moisture content that are shown in Figure 2. When increased the amount of the LA, it was found that the moisture content in the PPH-loaded PVA-*g*-LA hydrogel films increased, but at the maximum amount of the LA mixing found the decreasing of the moisture content in the PPH-loaded PVA-*g*-LA hydrogel films because the higher strength interaction between of ester bond between the hydroxyl groups of PVA and the carboxyl groups of LA might be obtained.

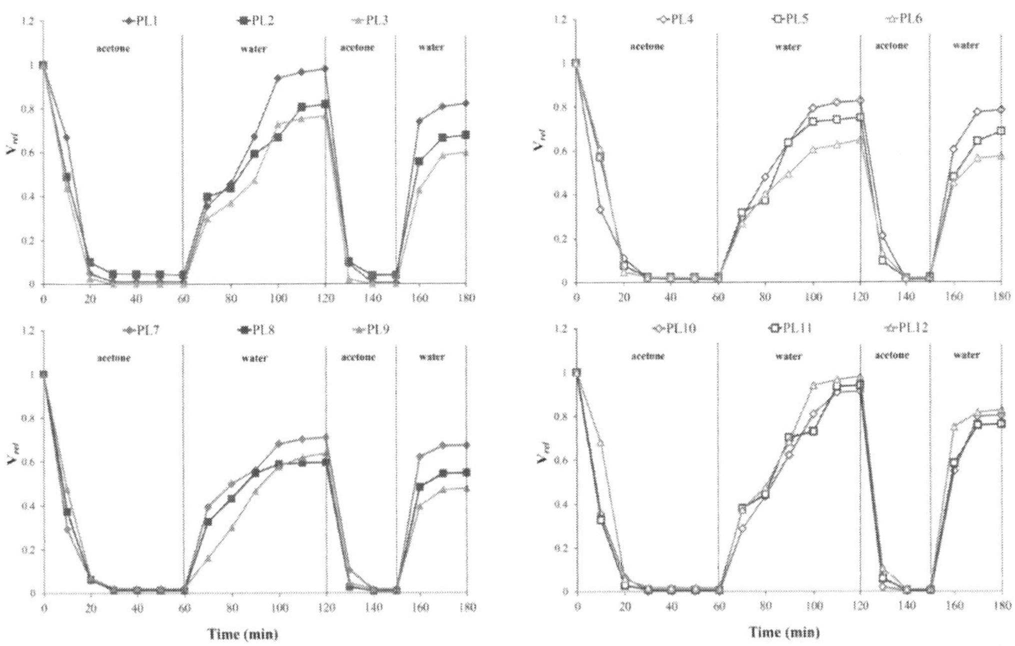

Figure 1. The swelling-deswelling behaviors of PPH-loaded PVA-*g*-LA hydrogel films.

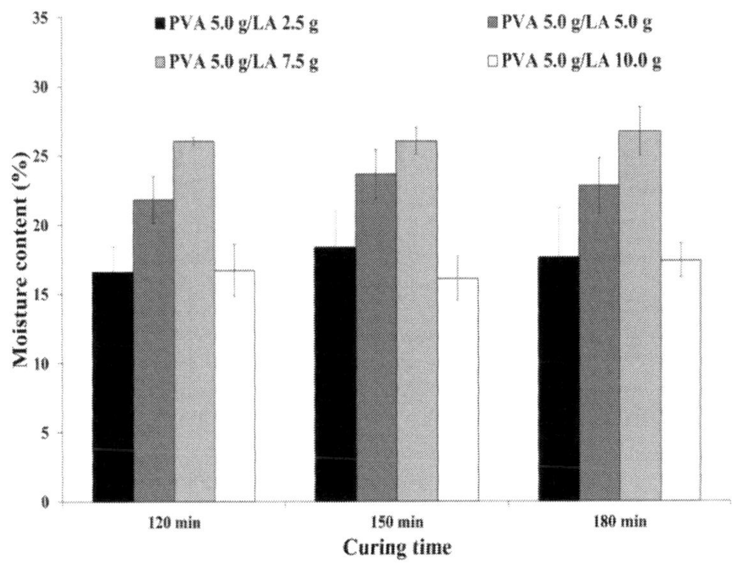

Figure 2. The moisture content of PPH-loaded PVA-*g*-LA hydrogel films.

The PPH-loaded PVA-*g*-LA hydrogel films were swollen in water at room temperature that had a value of the ratio of the water absorption at 2.20±0.53. It was clear that the latter floats in water, whereas the PPH-loaded PVA-*g*-LA hydrogel films was measured that are shown in Figure 3 and 4. To better understand the swelling behaviors of the PPH-loaded PVA-*g*-LA hydrogel films in presence of different cations, the chloride salts of Na^+ and K^+ with the concentration of 10 mmol/L were employed as swelling media. The results of the ratio of the water absorption in saline solutions was clearly decreased that comparing with distilled water as swelling media. This is due to the theoretical equilibrium swelling capacity and the initial swelling rate constant in saline solution are much lower than in distilled water that is reported [18]. Because of the additional cations causing a nonperfect anion–anion electrostatic repulsion, leading to a decreased osmotic pressure or ionic pressure difference between the hydrogel network and the external solution [19]. The pH value is one of the most important parameters affecting the absorption process [20], thus this work studied the influence of pH values ranged from 2 to 10 on water absorption of PPH-loaded PVA-*g*-LA hydrogel films. It was found that the swelling behaviors were depended on pH valus. When the pH value was changing in the range of 2 to 7, the swelling capacity of PPH-loaded PVA-*g*-LA hydrogel films increased, but decreased at pH 10. Because of the low pH value had the higher protonation degree of carboxyl groups. The carboxyl groups become ionized and the electrostatic repulsion between the molecular chains was predominated at high pH value, leading the network more expanding [21]. But the higher pH value at 10, the decreased of swelling capacity could be explained by the theoretical equilibrium swelling capacity and the initial swelling rate constant in swelling media are much lower than in distilled water that related to the above description. The theoretical equilibrium swelling capacity and the initial swelling rate constant increased as pH increasing from 2 to 6, but decreased at pH more than 8 [18].

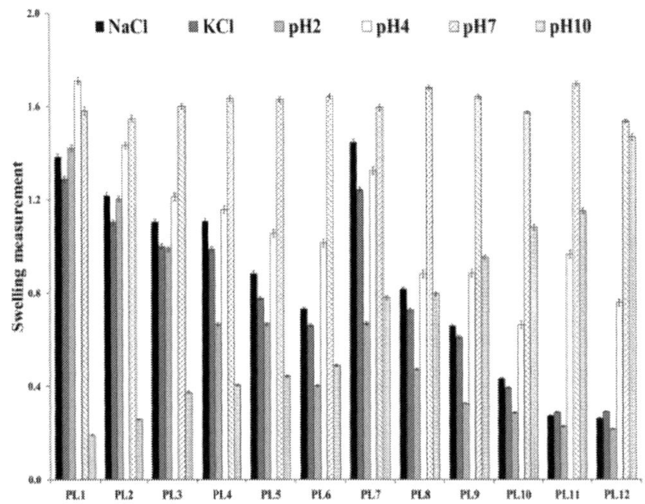

Figure 3. The swelling measurement of PPH-loaded PVA-*g*-LA hydrogel films in various aqueous solutions.

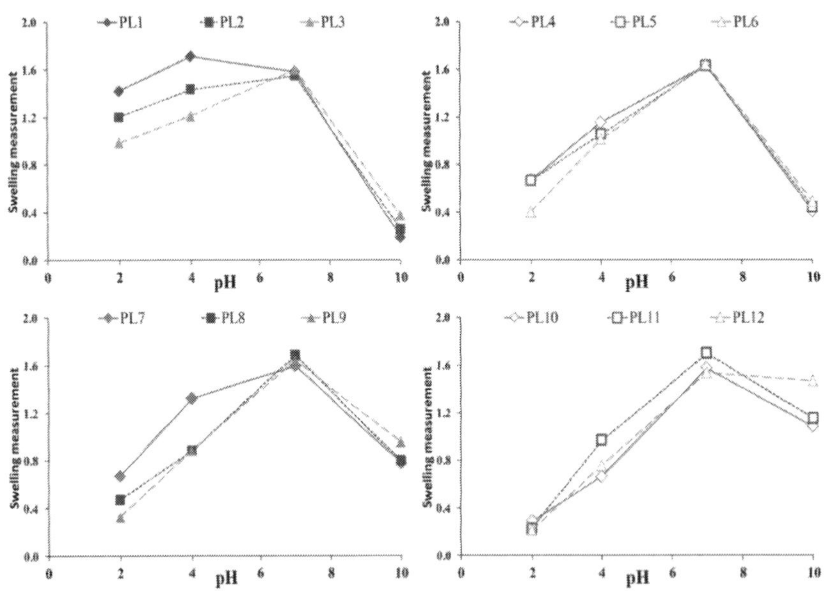

Figure 4. The swelling measurement of PPH-loaded PVA-*g*-LA hydrogel films in various pH conditions.

4 Conclusions

The swelling properties of PPH-loaded PVA-g-LA hydrogel films were investigated and measured. It was found that the different environmental stimuli fluids such as water, various pH, and various ionic strength and the amount of the LA were affected on swelling properties of PPH-loaded PVA-g-LA hydrogel films. The PPH-loaded PVA-g-LA hydrogel films showed the high swelling behavior in water, but rapidly deswelling behavior in acetone. The PPH-loaded PVA-g-LA hydrogel films had a high swelling in the medium pH 7. Thus, the prepared PPH-loaded PVA-g-LA hydrogel films had a good swelling property that might be used to controlled release of the drug in pharmaceutical product.

Acknowledgements

The authors would like to acknowledge the Faculty of Pharmacy and the Research Institute of Rangsit University for financial supports.

References

1. E.M. Ahmed, J. Adv. Res. **6**, 105–21 (2015)
2. T. Cerchiara, B. Luppi, F. Bigucci, I. Orienti, V. Zecchi, J. Pharm. Pharmacol. **54**, 1453-9 (2002)
3. W.E. Hennink, C.F. van Nostrum, Adv. Drug Deliv. Rev. **54**, 13-36 (2002)
4. J.M. Rosiak, F. Yoshii, Nucl. Instrum. Meth. B. **151**, 56-64 (1999)
5. L.L. Smith, US 6719699 B2, (2004)
6. C. Xiao, M. Yang, Carbohydr. Polym. **64**, 37-40 (2006)
7. Z. Guohua, L. Ya, F. Cuilan, Z. Min, Z. Caiqiong, C. Zongdao, Polym. Degrad. Stabil. **91**, 703-11 (2006)
8. S. Tanodekaew, S. Channasanon, P. Uppanan, J. Appl. Polym. Sci. **100**, 1914-8 (2006)
9. S.-A. Riyajan, S. Chaiponban, K. Tanbumrung, Chem. Eng. J. **153**, 199-205 (2009)
10. C.M. Hassan, N.A. Peppas, *Structure and applications of poly(vinyl alcohol) hydrogels produced by conventional crosslinking or by freezing/thawing methods*. (Berlin Heidelberg, Springer, 2000)
11. A. Breitenbach, T. Kissel, Polymer. **39**, 3261-71 (1998)
12. S.J. Carlotti, O. Giani-Beaune, F. Schué, J. Appl. Polym. Sci. **80**, 142-7 (2001)
13. J. Ding, S.-C. Chen, X.-L. Wang, Y.-Z. Wang, Ind. Eng. Chem. Res. **48**, 788-93 (2008)
14. J.M. Onyari, S.J. Huang, J. Appl. Polym. Sci. **113**, 2053-61 (2009)
15. N. Tudorachi, A. Chiriac, J. Polym. Environ. **19**, 546-58 (2011)
16. J. Suksaeree, C. Luprasong, C. Monton, L. Charoenchai, W. Pichayakorn, J. Therm. Anal. Calorim. **120**, 929-36 (2015)
17. D. Ceylan, M.M. Ozmen, O. Okay, J. Appl. Polym. Sci. **99**, 319-25 (2006)
18. J. Liu, Q. Li, Y. Su, Q. Yue, B. Gao, Carbohydr. Polym. **107**, 232-40 (2014)
19. Y. Bao, J. Ma, N. Li, Carbohydr. Polym. **84**, 76-82 (2011)
20. Y. Zheng, Y. Liu, A. Wang, Chem. Eng. J. **171**, 1201-8 (2011)
21. X. Li, S. Xu, J. Wang, X. Chen, S. Feng, Carbohydr. Polym. **75**, 688-93 (2009)

Formulation and Characterization of Topical Anti-acne Spot Gel Containing Herbal Extracts

Jirapornchai Suksaeree[1, a] and Chomnapas Chuchote[2, b]

[1] *Department of Pharmaceutical Chemistry, College of Pharmacy, Rangsit University, Pathum Thani, Thailand*
[2] *Department of Pharmacognosy, College of Pharmacy, Rangsit University, Pathum Thani, Thailand*

Abstract. This research work aimed to prepare the topical anti-acne spot gel incorporating the crude polyherbal extract that is a traditional Thai herbal recipe, namely "Ya-Sa-Marn-Phlae". The topical anti-acne spot gels were prepared by add the crude polyherbal extract in each microemulsion or self-microemulsifying system and subsequently mixed in the blank anti-acne spot gel. The pH, viscosity, stability, contamination test, and skin irritation test were evaluated. The topical anti-acne spot gel formulations had a yellowish in color that was visually checked by researcher and photographed by digital camera. They had a pH near to a human skin and good viscosity which ease to use and safe to contact to the skin. They had a good stability and no contamination of *Staphylococcus aureus* and *Escherichia coli*. The topical anti-acne spot gels exhibited no oedema formation, and very slight erythema formation. Thus, the topical anti-acne spot gels were safe to apply on the skin without irritation.

1 Introduction

In the recent years, herbal formulations from Thai herbal compounds are the good demand resulting in its rapid growth in Thailand. They are more acceptable and interesting in the view that they are less side effects and safe than the synthetic products. The herbal products are continuously developed and has been reported by incorporated the Thai herbal compounds such as curcumin [1], quercetin [2], paclitaxel [3], tea catechins [4], capsaicin [5, 6], *Zingiber cassumunar* [7, 8] etc. The main objectives of herbal products are delivery of the active compound with constant release rate over the period of treatment and channel for entity of the active compound to the site of action [9, 10].

This present work used the traditional Thai herbal recipe, namely "Ya-Sa-Marn-Phlae", to formulate the topical anti-acne spot gel formulation. This traditional Thai herbal recipe composed of *Curcuma longa L., Areca catechu L., Oryza sativa L.,* and *Garcinia mangostana L.* in ratio of 1:1:1:1, respectively found as a treatment of wounds and skin infections that described in elsewhere [11, 12]. The crude polyherbal extract is prepared and tested by Dr.Chonlatid Sontimuang, Faculty of Traditional Thai Medicine, Prince of Songkla University.

The objective of this study was the preparation of the topical anti-acne spot gel containing the polyherbal extract. The polyherbal extract was entrapped in each microemulsion (ME) or self-microemulsifying (SME) system before mixing in the topical anti-acne spot gel. However, the suitable

[a] Corresponding author : jirapornchai.s@rsu.ac.th
[b] Corresponding author : chomnapas@gmail.com

ME and SME systems are observed and optimized via our group which reported in elsewhere [13, 14]. The only one of the optimized system for each ME and SME was selected to prepare the topical anti-acne spot gel. The topical anti-acne spot gels were determined and evaluated the pH, viscosity, stability, contamination test, and skin irritation test.

2 Experimental Works

2.1 Raw materials and chemicals

Caprylic acid, Tween® 20, Tween®80, propylene glycol, carbomer, Triethanolamine, etc. were purchased from P.C. Drug Center Co., Ltd, Thailand. The polyherbal extract is prepared by Dr.Chonlatid Sontimuang. Briefly, the polyherbal formulation composed of *C. longa* L., *A. catechu* L., *O. sativa* L., and *G. mangostana* L. in ratio of 1:1:1:1, respectively. The mixture herb was extracted in 95% ethanol and water at room temperature for 7 days. Then, the solution of polyherbal extract was evaporated the ethanol by rotary evaporator and freeze-dried to remove the water. Finally, the crude polyherbal extract was sterilized at -20°C.

2.2 Preparation of ME and SME

The composition of the ME and SME base showed a clear and thermodynamically stable system are chosen according to preliminary trials explained of the pseudoternary phase diagrams in a previous work [13, 14]. ME system composted of the 20 %w/w of caprylic acid as an oil phase, 70 %w/w of mixture of Tween®80 and absolute ethanol (8:2) as a surfactant, and 10 %w/w of water phase. SME system composed of 10 %w/w of caprylic acid as an oil phase, 60 %w/w of mixture of Tween®80 and absolute ethanol (8:2) as a surfactant, and 30 %w/w of propylene glycol as a co-surfactant phase without water phase. Then, the 1 %w/w Thai herbal extract was dissolved in optimized ME or SME system. The systems were prepared by vortex-mixed for 30 second. The monophasic formulations were formed spontaneously at room temperature.

2.3 Preparation of topical anti-acne spot gel formulation containing polyherbal extract

The clear and thermodynamically stable system that had a good clear emulsion in appearance was mixed in base formulation of topical anti-acne spot gel that are presented in Table 1. All gradients were easily mixed in beaker.

Table 1. The composition of blank and topical anti-acne spot gel formulations

Formulas	Polyherbal extract (%w/w)	ME (%w/w)	SME (%w/w)	Carbomer (%w/w)	Propylene glycol (%w/w)	Paraben concentration (%w/w)	Triethanolamine (%w/w)
B1	-	10	-	0.9	20	1	0.45
B2	-	-	10	0.9	20	1	0.45
F1	1	10	-	0.9	20	1	0.45
F2	1	-	10	0.9	20	1	0.45

2.4 Characterization

The Nanoplus 3 (Particulate system, USA) was used to determine the particle size, size distribution, and zeta potential at 25±2°C. The pH and viscosity of topical anti-acne spot gel containing polyherbal extract was measured by a SevenEasy S-20 pH meter (Mettler Toledo, Switzerland) and a Brookfield viscometer (Brookfield engineering laboratories Inc, USA) at 25 ± 2°C with the S-96 spindle at 100 rpm, respectively. Each topical anti-acne spot gel containing polyherbal extract was measured in triplicate. The

2.5 Stability test

The formulation was kept at 40±2°C and 75%RH for 0, 1, 2, and 3 months (Method I). Freeze – thaw stability test, the formulation was kept in refrigerator at 4±2°C for 24 hours and subsequently moved to 40±2°C for 24 hours and repeated all step for 6 cycles (Method II).

2.6 Contamination test

The formulation was diluted to 10^{-1}, 10^{-2}, and 10^{-3} µg/mL of concentration with 0.9% sodium chloride solution, and then speeded onto Petri-dish that contained each manital salt agar (*Staphylococcus aureus*) or MacConkey agar (*Escherichia coli*) contamination test. They were incubated at 37±2°C for 24 hours.

2.7 Skin irritation test

The topical anti-acne spot gel formulation containing polyherbal extract was applied onto the skin of healthy adult albino rabbits; the New Zealand white hybrid strain obtained from the Department of Animal Science, Faculty of Agriculture, Kasetsart University, Thailand. It was evaluated by Thailand Institute of Scientific and Technology Research following the OECD guidelines for testing of chemicals (TG 404) with the Animal Ethics Committee approval number TS-59001 [15]. The rabbit's skin was assessed on each site at 1, 24, 48, and 72 hours after removal of the patches. The skin irritation was graded according to the scale originally proposed by Draize test skin.

3 Results and discussion

The crude polyherbal extract is prepared and tested by Dr.Chonlatid Sontimuang. The crude polyherbal extract is the traditional Thai herbal recipe, namely "Ya-Sa-Marn-Phlae" is found that the antibacterial, anti-inflammatory, and antioxidant activities and low toxicity that described in elsewhere [11, 12] that will be used to develop a topical formula based on traditional knowledge. The Ya-Sa-Marn-Phlae recipe composed of the 4 Thai herbs that are *C. longa L., A. catechu L., O. sativa L.*, and *G. mangostana L.* in ratio of 1:1:1:1, respectively.

From the previous work reported the preparation of the pseudoternary phase diagrams for ME and SME system that showed a clear and thermodynamically stable system [13, 14]. The best systems were 20 %w/w of caprylic acid as an oil phase, 70 %w/w of mixture of Tween®80 and absolute ethanol (8:2) as a surfactant, and 10 20 %w/w of water phase for ME system, and 10 %w/w of caprylic acid as an oil phase, 60 %w/w of mixture of Tween®80 and absolute ethanol (8:2) as a surfactant, and 30 %w/w of propylene glycol as a co-surfactant phase without water phase for SME system. The crude polyherbal extract was added in concentration of 1 %w/w that showed a clear and thermodynamically stable system. The zeta potential, droplet size, and size distribution of optimized ME and SME system were determined that are shown in Table 2. Both ME and SME showed that the good stability and the droplet size of the dispersed phase was less than 200 nm with narrow distribution. The transparency of the both systems was observed which would not show aggregation effect due to charge of the droplets [16].

Table 2. The zeta potential, droplet size, and size distribution of optimized ME and SME system

Formulas	Zeta potential (mV)	Droplet size (nm)	PI
ME base	-7.19±1.23	192.30±1.97	0.22
Crude polyherbal extract – loaded ME	-11.26±2.42	177.00±0.26	0.21
SME base	-8.22±0.59	148.47±0.93	0.30
Crude polyherbal extract – loaded SME	-7.89±0.65	133.03±1.70	0.23

Figure 1 shows the appearance of blank topical anti-acne spot gels (left) and topical anti-acne spot gels (right). The appearance of the blank topical anti-acne spot gels was a white in color while the intensity of its color increased with the addition the polyherbal extract, which was a yellowish in color.

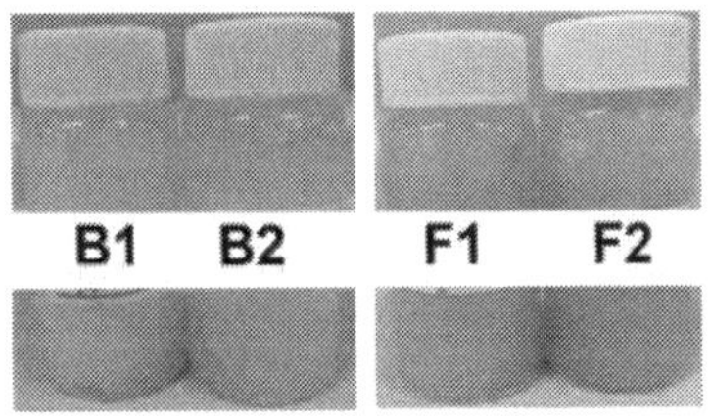

Figure 1. The appearance of blank topical gels (left) and topical anti-acne spot gels (right)

The pH and viscosity value of topical anti-acne spot gels are shown in Table 3. After crude polyherbal extract was added, it was found that not changed from those blank gel formulations. Thus, the crude polyherbal extract not affected on these values. The pH value of fresh preparation of topical anti-acne spot gel formulations were 6.01±0.01 and 5.85±0.02 for F1 and F2 formula, while the pH values were found that varied from 5.60 – 5.80 after storage at 40±2°C for 3 months, and 5.79 to 5.66 after freeze – thaw test. However, this pH value was safe to use for skin applications that human skin surface pH is on average between 5.00 – 6.00. Similarly, the viscosity after stability test of topical anti-acne spot gels slightly decreased, when compared to those fresh topical anti-acne spot gels. This might be the temperature effects which slightly decreased the movement of polymer chain. The appearance of topical anti-acne spot gel formulation after stability testing is shown in Figure 2. It was found that the intensity of its color was clearly observed to dark yellow after stability test. Results showed that the topical anti-acne spot gels stood stable at stability testing condition, both kept at 40±2°C with 75%RH in stability chamber for 3 months and freeze – thaw test for 6 cycles, with a little change.

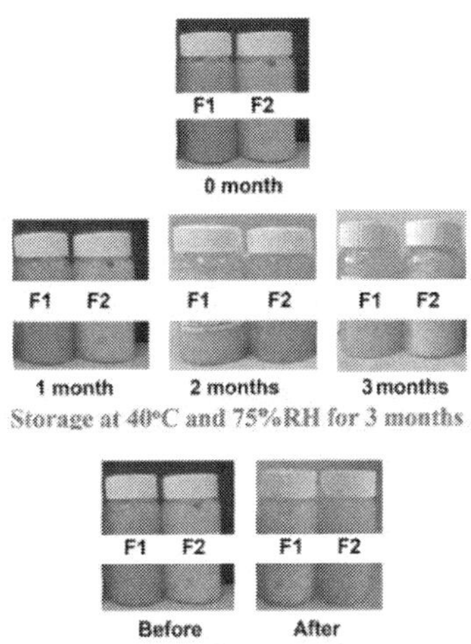

Figure 2 The appearance of topical anti-acne spot gel formulations after stability testing

Table 3. pH and viscosity of blank and topical anti-acne spot gel formulations

			pH value		
Formulas	0 month	1 month	2 months	3 months	Freeze – thaw stability
B1	6.03±0.02	6.00±0.01	6.01±0.01	5.93±0.02	5.90±0.04
B2	5.90±0.02	5.82±0.03	5.80±0.02	5.77±0.03	5.70±0.03
F1	6.01±0.01	5.88±0.01	5.81±0.01	5.80±0.02	5.79±0.02
F2	5.85±0.02	5.72±0.02	5.64±0.01	5.60±0.02	5.66±0.04
			Viscosity value ($\times10^3$ cps)		
B1	340.89±30.71	306.98±20.01	288.96±20.82	228.82±26.80	338.10±31.82
B2	366.88±25.07	340.70±29.02	300.19±20.77	280.72±28.91	256.66±26.93
F1	334.33±25.90	297.59±18.00	245.39±21.08	221.20±22.49	302.75±24.30
F2	347.36±22.09	322.15±21.12	304.15±28.88	269.47±26.54	248.38±22.67

The contamination test of topical anti-acne spot gels was observed by using Manital Salt agar and MacConkey agar for *S. Aureus* and *E. coli* testing, respectively at 37±2°C for 24 hours which presents in Figure 3. The contamination was not found in topical anti-acne spot gels at different diluted concentration

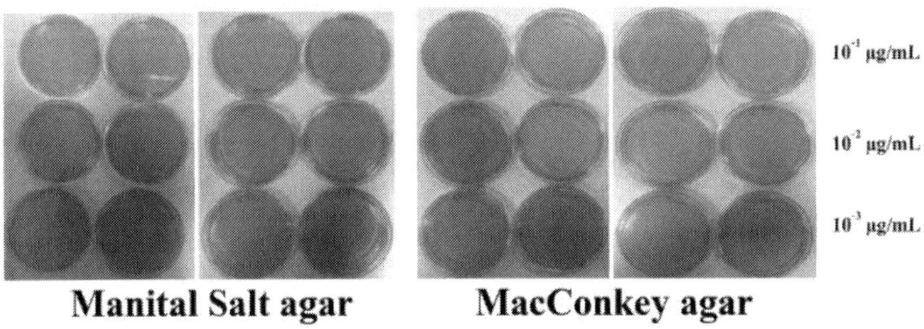

Figure 3 The antimicrobial sensitivity testing for *S. aureus* (Manital Salt agar) and *E. Coli* (MacConkey agar)

The topical anti-acne spot gels were further tested *in vivo* by acute dermal irritation test in New Zealand white rabbits. The scores of dermal reactions: erythema and eschar formation, and oedema formation, were used as an index of skin damage caused by these topical anti-acne spot gels. When the distilled water was treated as a control area on rabbit's test, it was found that no sign of any erythema and eschar formation, and oedema formation. Unlucky, when the topical anti-acne spot gels were treated on tested rabbit's skin, one of all tested rabbits exhibited a very slight erythema formation (score = 1) on the tested skin at 1 hour after removal of the gels and any residual, and this effect not exhibited at 24 hours, while the other rabbit showed no erythema (scale 0) at any time of the investigations. After that, they were not found the erythema and eschar formation until 72 hours. All tested rabbits showed that no oedema formation (score = 0). The effect of high content of surfactant was used to describe for this damage on the rabbit's skin. The degree of skin irritation potential of test substances that had various number of factors such as species of animals used. The results of skin irritation may be changed if the different animal models have been employed. Generally, the laboratory animals are more susceptible to chemicals than humans [17]. However, the skin irritation on human skin testing will be tested to provide practical information on potential skin damages of these topical anti-acne spot gels.

4 Conclusions

The topical anti-acne spot gels were initially prepared by our research group by add the crude polyherbal extract in each ME or SME system and subsequently mixed in blank anti-acne spot gel.

The topical anti-acne spot gels had a yellowish in color with good stability and no contamination. The pH and viscosity were found that safe to use on the skin. The topical anti-acne spot gels exhibited no oedema formation, and very slight erythema formation at 1 hour and was not presented until 72 hours with no eschar formation. Thus, the topical anti-acne spot gel formulations were safe to apply on the skin without irritation.

Acknowledgements

The authors wish to thank KI Tull for English language assistance and Dr.Chonlatid Sontimuang, Faculty of Traditional Thai Medicine, Prince of Songkla University for prepared the polyherbal extract. This work was supported by the Faculty of Pharmacy and the Research Institute of Rangsit University.

References

1. K. Maiti, K. Mukherjee, A. Gantait, B.P. Saha, P.K. Mukherjee, Int. J. Pharm. **330**, 155-63 (2007)
2. A. Priprem, J. Watanatorn, S. Sutthiparinyanont, W. Phachonpai, S. Muchimapura, Nanomed. Nanotechnol. Biol. Med. **4**, 70-8
3. S. Rane, B. Prabhakar, Indian J. Pharm. Sci. . **75**, 420-6 (2013)
4. J.-Y. Fang, T.-L. Hwang, Y.-L. Huang, C.-L. Fang, Int. J. Pharm. **310**, 131-8 (2006)
5. P.R. Desai, S. Marepally, A.R. Patel, C. Voshavar, A. Chaudhuri, M. Singh, J. Controlled Release. **170**, 51-63 (2013)
6. Y. Zhu, W. Peng, J. Zhang, M. Wang, C.K. Firempong, C. Feng, et al., J. Funct. Foods. **8**, 358-66 (2014)
7. J. Suksaeree, L. Charoenchai, F. Madaka, C. Monton, A. Sakunpak, T. Charoonratana, et al., Asian J. Pharm. Sci. **10**, 341–9 (2015)
8. J. Suksaeree, C. Monton, F. Madaka, T. Chusut, W. Saingam, W. Pichayakorn, et al., AAPS PharmSciTech. **16**, 171-81 (2015)
9. S. Parasuraman, G.S. Thing, S.A. Dhanaraj, Pharmacogn. Rev. **8**, 73-80 (2014)
10. Ajazuddin, S. Saraf, Fitoterapia. **81**, 680-9 (2010)
11. S. Chusri, S. Tongrod, J. Chokpaisarn, S. Limsuwan, S.P. Voravuthikunchai, BioMed Research International. **2014**, 6 (2014)
12. S. Chusri, S. Settharaksa, J. Chokpaisarn, S. Limsuwan, S.P. Voravuthikunchai, J. Altern. Complement. Med. **19**, 671-6 (2013)
13. P. Maneewattanapinyo, K. Kantapak, P. Sansiri, N. Chatpitukpong, C. Sontimuang, J. Suksaeree, Thai J. Pharm. Sci. **40**, 53-6 (2016)
14. P. Maneewattanapinyo, K. Kantapak, P. Sansiri, N. Chatpitukpong, C. Sontimuang, J. Suksaeree, Thai J. Pharm. Sci. **40**, 57-60 (2016)
15. Organization for Economic Co-operation and Development (OECD), *OECD Guidelines for Testing of Chemicals, Section 4: Health Effects, Test No. 404: Acute Dermal Irritation/Corrosion*. (Paris Cedex 162002)
16. S.-x. Cui, S.-f. Nie, L. Li, C.-g. Wang, W.-s. Pan, J.-p. Sun, Drug Dev. Ind. Pharm. **35**, 603-11 (2009)
17. S. Songkro, W. Pichayakorn, S. Sungkarak, J. Wungsintaweekul, J. drug deliv. sci. tech. **21**, 401-10 (2011)

Polymer Composite Manufacturing by FDM 3D Printing Technology

Katarzyna Bryll[1,a], Elżbieta Piesowicz[2], Paweł Szymański[3], Wojciech Ślączka[4] and Marek Pijanowski[1]

[1]Institute of Basic Technical Sciences, Faculty of Marine Engineering, Maritime University of Szczecin, ul. Waly Chrobrego 1-2, 70-500 Szczecin, Poland
[2] Institute of Materials Science and Engineering, Faculty of Mechanical Engineering and Mechatronics, West Pomeranian University of Technology, Al. Piastów 19, 70-310 Szczecin, Poland.
[3] Institute of Materials Technology, Faculty of of Machine Construction and Management, Poznan University of Technology ul. Piotrowo 3, 60-965 Poznan,
[4] Maritime Risk Analisis Center, , Faculty of Navigation, Maritime University of Szczecin ul. Waly Chrobrego 1-2, 70-500 Szczecin, Poland

Abstract. 3D printing technology was developed nearly 30 years ago. One of its characteristics is that instead of removing materials, 3D printing creates 3D elements directly from CAD models, adding one layer of material on another. This offers a beneficial capability of making complex elements in terms of shape and materials, impossible to be manufactured by traditional methods. Owing to intensive research in recent years, considerable progress has been achieved in the development and commercialisation of new innovative processes of 3D printing by fused deposition modeling (FDM), including printing of composite materials. The study outlines the main methods of creating polymer composite structures using FDM technology.

1 Introduction

The idea of 3D printing has been known for decades [1]. Additive manufacturing is today used for making tools and rapid prototypes at the design and construction stage in aviation, automobile and military industries, architecture, medicine, goldsmithing, food and footwear industries, where in a relatively short time a real model can be made to be presented, apart from printed documentation, to a customer or sponsor. The development of 3D printing technology has enabled making prototypes as well as spare parts or complete devices with attractive properties, usually of small size. 3D printers in medicine are used for printing prosthetics, implants, and bioprints using cells (wound healing) [1-5].

Extended applicability and wider dissemination of 3D printing, as well as higher performance requirements necessitate seeking multimaterial solutions and use of state-of-the-art composites [2-9].

Composite properties depend on many factors, i.e. properties of component phases, their relative quantities and geometry (particle size, shape, orientation and distribution of the reinforcement phase in the matrix) as well as the strength of interfacial boundaries or production methods [10].

This study reviews the state of the art in FDM printing of composite materials.

[a] Corresponding author : k.bryll@am.szczecin.pl

2 Fused deposition modeling (FDM)

FDM technology also known as fused filament fabrication (FFF) is currently the most widespread way to rapid production of items utilizing additive manufacturing [1-5]. This method is a well-known technology patented in 1989 by Scott Crump, a co-founder of the Stratasys company [1].

3D printing by FDM consists in depositing a filament of thermoplastic material. A portion of filament from a reel passes through a hot head, with a temperature higher than the melting point of the filament, then is extruded in the XY plane creating a layer of solid material on the build plate.

Creating a model can be done by depositing a layer contour, then filling the inside with plasticized material by zigzag movement of the head. After printing one layer the head moves along the Z axis initiating the build-up of the next layer. Using this technique, we can create complex shapes with a minimum of preparatory processes. The production process begins with creating a model in the CAD program, then the model is incorporated into the program enabling control of process parameters such as head movement, feed rate, layer thickness, infill, head and table temperatures, slicing, support application, etc. Such a program generates a G-code, which uploaded in a 3D printer enables making a real model. The model removed from the printer needs finish machining, for instance to remove the supports and imperfections [2-9].

The diagram of 3D printing procedure by FDM method is shown in Fig. 1.

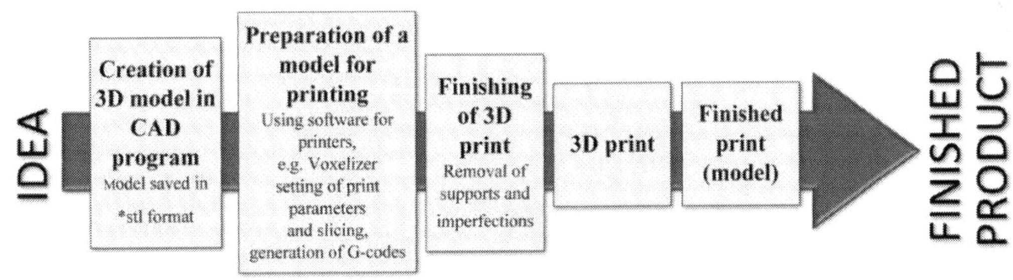

Figure. 1. Procedure of FDM printing

The most common drawbacks associated with FDM technology include [2-9]:

- stepped layers - visible paths of the deposited material due to a certain distance between edges of subsequent layers. This effect can be minimised by printing lower layers as well as reducing the diameter of the extruder (which will extend the process and raise the cost).

- overhang and bridging - printing of each model begins from the first layer applied onto a build plate. A subsequent layer is printed onto the surface of the previous one. If an item has elements set at an angle relative to the vertical axis (e.g. at 45°) the filament may not have a support and collapse. In such cases, supports should be generated at the model preparation stage to prevent damage. A bridge, that is parts of the model supported on its two ends, is a particular, troublesome case of overhang [9]. Therefore, in the case of long overhangs, auxiliary supports have to be designed, then removed during the final machining of the print. Alternatively, we can try to divide a model and place it on the build platform so that the bridges can be set upright.

- stringing (oozing) - occurs if during the extruder movement between two points some of the plasticized filament leaks down due to gravitation. In order to remove this effect, appropriate retraction should be set, a process of shifting the filament back into the nozzle.

- 'warping' of the model occurs when the edges or corners of the model rise above the table, deforming the model. The process takes place due to material shrinking and uneven temperature distribution across the model. We can partly counteract the process by changing temperature settings of the table, or putting a kapton tape on the table surface.

- hygroscopicity (term commonly used in works [2-7], although these authors propose to call this irregularity 'occluded or precipitated porosity'), is a property of polymer materials, leading to deteriorated print quality. Under the influence of nozzle temperature, moisture in the material leads to the creation of air bubbles in the model structure. This process can be minimised if filament is stored in appropriate conditions or dried before printing.

- 'structural inhomogeneity', a term proposed by the authors, means inequality, discontinuity, undulation, heterogeneity of structure particle size or insufficient density of the object, a characteristic of the manufacturing process. Structural inhomogeneity is connected with selective deposition of the filament layer by layer on the surface, by temperature differences, or porosity. This can be minimised by reducing the length between the spits of the filament.
- print damage by the printer head due to insufficient accuracy of nozzle movement.

3. Methods of printing polymer composites using FDM technology

Based on the review of publications [2-24] it is proposed to divide methods of printing composite materials or multimaterial structures by type of the extruder (printing head) Fig 2.

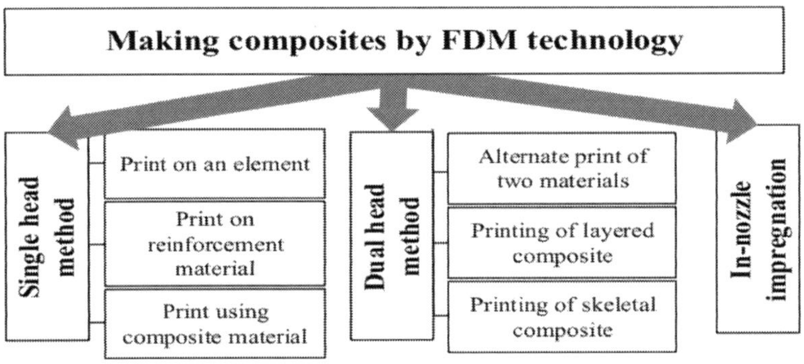

Figure 2. Composite manufacturing by FDM technology

3.1. Creating multimaterial structures by printing on an element

An interesting method of making multimaterial structures is printing a layer of polymer material on the surface of a ceramic, metal, wooden or polymer material. Structural elements obtained in this way may be composed of a few materials, in which the built-in element occupies much of the entire volume [5-6].

The biggest limitation of this method is the need to divide the model into parts separated by parallel planes, identifying subsequent printing stages. Besides, the printing head has to be controlled and stopped so that additional elements can be placed. The technological problem in this method is resuming printing without breaking the continuity of the material [5-6].

3.2. Creating a composite structure by single nozzle printing on the reinforcement material

Manufacturing a composite structure using a single head by printing on the reinforcement material (Fig 3a), is a method similar to creating multimaterial structures. Composite materials are obtained in a multi-stage process, where the number of stages depends on the number of reinforcement layers in a finished element. Each printing stage is designed by specifying a division plate, which enables stopping the printing and depositing the reinforcement phase on the print. Layered composites made by this method can be reinforced with continuous fibre, cloth, or mat [5].

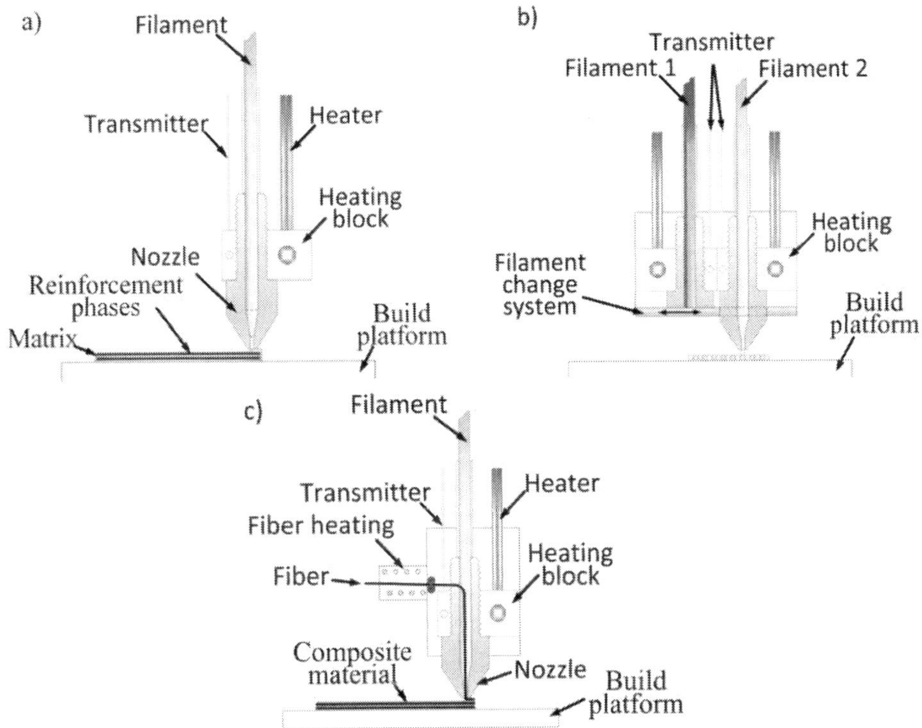

Figure. 3 Manufacturing and composite structure by:
a) single head printing on a reinforcement material
b) dual materials printing
c) in-nozzle impregnation

Disadvantages of this type of printing include restrictions associated with the process and with the thickness of individual layers of the reinforcement material in the composite. If a bunch of reinforcing fibres is too thick, the reinforcement phase may get deformed because fibers stuck to a jet of plasticized material may get shifted. The layer of reinforcement commonly found in the literature [2-4] is approximately 0.5 mm thick.

3.3. Creating a composite structure with a single nozzle using a specific filament

Manufacturing a composite structure with a single nozzle uses a different filament from that in traditional 3D printing technology; which besides polymeric material contains an additive: rubber microspheres, particles of glass or carbon fibre, wood flour etc.

The biggest limitation to this method is the availability of the material for printing. A prepared filament can be purchased, but currently the market offers a limited quantity of such products [4-6, 10-13].

Composite filament could also be produced as a result of extruding a monofilament of defined diameter. However, it is a process requiring specialist equipment and know-how. For instance, to obtain the required filament diameter (1.75 mm or 3 mm), it should be extruded at a specific speed. The greater the speed, the smaller is the filament diameter. There are also limitations to the particle size and volumetric fraction of reinforcement in the filament. Excessive particle sizes and/or volume fraction may lead to blocking the flow of 3D printer nozzle, which will result in defective print [4-6, 10-13].

3.4. Creating a composite structure by alternate dual material printing

A diagram illustrating the making of a composite structure by means of alternate printing using two materials is given in Figure 3b.

The process of selective printing with two materials is an advanced 3D printing method, getting increasingly more common. It can be done using: a head consisting of two nozzles or newer solutions of one nozzle with a filament change system.

The method consists in alternate deposition of different materials in one printing process. The advantage of this method is the possibility to control the properties of a finished item at the design stage, because we can design and then make any composite structure (Fig. 4 a-b).

The greatest limitations of that manufacturing method stem from dual material printing imperfections, e.g. contamination due to the impossibility of transition from one filament to the other, fragment after fragment, as well as problems in proper configuring of the G-code, and the requirement of using materials with similar thermal characteristics. These problems are being constantly solved by introducing new systems of filament change and software updates by printer manufacturers.

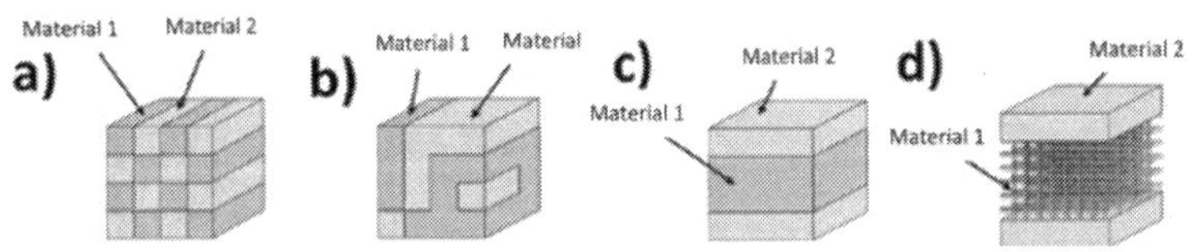

Figure. 4 Exemple print structures of dual materials composite
a-b) varied distribution of materials c) sandwich structured composite - layered, d) sandwich structured composite - skeletal.

3.5. Creating a layered composite structure using two materials

Creating a layered composite structure is the form of printing using two materials. An essential difference between these methods is the manner of material deposition, which in this case is layered, first material no 1, then material no 2, no 1 again etc. A schematic structure obtained by this method is shown in Fig. 4c.

3.6. Creating a skeletal composite structure using two materials

Printing a skeletal composite is similar to printing a layered composite, the biggest difference being that one of the materials is not solid, but a skeletal structure (Fig. 4d).

To produce such structures we can use algorithms reducing infills, known from printing prototypes technology, which have helped to save time and materials in the process of printing. The most common skeletal structures include hexagonal cells, square cells or honeycomb; other structures individually designed can be used as well, given that the FDM method has a limitation concerning overhangs known as bridges.

3.7. Creating a layered composite structure using a special nozzle allowing introduction of the continuous fiber

Figure 3c schematically shows the making of continuous composites by in-nozzle impregnation.

Filament being the matrix material and continuous reinforcing fibers are supplied separately to the printer nozzle. In order to increase fibre impregnation with thermoplastic material, before its insertion in the nozzle the reinforcement phase is heated. Fibers are automatically supplied to the head through filament motion. Filament is plasticized inside the printer head and connected with the reinforcement phase. Subsequent stages of the printing process are similar to traditional 3D printing.

4. Summary

3D printing enables making single pieces of complex parts based on a CAD model without the use of expensive tools or moulds. The development of this technology results in prints in different areas of life. Of a few 3D printing systems, the one based on FDM modelling is particularly widespread thanks to the simplicity and capabilities of the method. Other advantages include prices of the printers dropping, better software and a broadening scope of materials used, including composites, improving mechanical properties of 3D prints.

The literature on the subject [2-22] contains numerous publications addressing 3D printing of composite materials. Various types of reinforcement can be used, for instance carbon structures [14,16-17] reinforcing plates [11] and short fibres: carbon [14], polymer [15] and glass [18].

However, the majority of research refers to the preparation and printing of composites from filament. Other methods using 3D printing with two heads are mentioned in few publications [7, 18-19]. These composites usually show worse mechanical properties compared to composites produced by conventional methods, because composites reinforced with short fibres or particles are mechanically worse than those strengthened with continuous fibres [20-22]. That is why 3D printing using continuous fibres is an interesting solution [23-24]. 3D printed composites are designed to make elements of high electric conductivity [13,17], low coefficient of thermal expansion and high thermal conductivity [22].

Despite increased interest in printing composite materials and many publications on the developments of this technology, there are still numerous issues requiring further analysis. Essential technical problems to be solved are associated with intermittent printing or filament changes. It is also necessary to seek new material connections and adapt them for printing, provided that such solutions are cost-effective and environmentally acceptable.

This requirement may be satisfied by single polymer composites, in which both the matrix and reinforcement are made of the same or similar polymeric material, where components can vary in molecular weight, density or degree of branching [25].

Issues of 3D FDM printing of single polymer polyester composites will be discussed by these authors in subsequent publications.

Acknowledgments

The research presentedin this article was carried out under the Grant of the Ministry of Science and Higher Education of Poland nr 1/S/IESO/17 performed at the Maritime University of Szczecin, Poland.

References

1. Patent US 5121329 A: Apparatus and method for creating three-dimensional objects; 9 Cze 1992
2. Fuda Ning, Weilong Cong, Jingjing Qiu, Junhua Wei, Shiren Wang: Additive manufacturing of carbon fiber reinforced thermoplastic composites using fused deposition modeling; Composites Part B: Engineering;Volume 80, 2015, Pages 369-378
3. Nannan GuoMing C. Leu: Additive manufacturing: technology, applications and research needs; Frontiers of Mechanical Engineering; 2013, Volume 8, Issue 3, pp 215–243
4. Kumar S. & Kruth J.-P. Composites by rapid prototyping technology. Materials & Design 31, 850–856 (2010).
5. Dudek P.: FDM 3D printing technology in manufacturing composite elements. Archives of Metallurgy and Materials, 58(2013),nr.4,1415-1418
6. David Bak, "Rapid prototyping or rapid production? 3D printing processes move industry towards the latter", Assembly Automation, 2003, Vol. 23 Issue: 4, pp.340-345
7. Sung-Hoon AhnEmail authorKyung-Tae LeeHyung-Jung KimRenzhe WuJi-Soo KimSung-Hyuk Song Smart soft composite: An integrated 3D soft morphing structure using bend-twist coupling

of anisotropic materials International Journal of Precision Engineering and Manufacturing; 2012, Volume 13, Issue 4, pp 631–634
8. Bakarich S. E., Gorkin R. 3rd, in het Panhuis M. & Spinks G. M. Three-dimensional printing fiber reinforced hydrogel composites. ACS applied materials & interfaces 6, 15998–16006 (2014). [PubMed]
9. Suwanprateeb J., Sanngam R., Suvannapruk W. & Panyathanmaporn T. Mechanical and in vitro performance of apatite-wollastonite glass ceramic reinforced hydroxyapatite composite fabricated by 3D-printing. J Mater Sci Mater Med 20, 1281–1289 (2009). [PubMed]
10. Chybowski L., Gawdzińska K., On the Possibilities of Applying the AHP Method to a Multi-criteria Component Importance Analysis of Complex Technical Objects. Advances in Intelligent Systems and Computing, Volume 445, Springer International Publishing 2016, pp. 701-710 [ISSN 2194-5357, ISBN 978-3-31306-1, online ISBN 978-3-319-31307-8, DOI 10.1007/978-3-319-31307-8_71]
11. Dimas L. S. & Buehler M. J. Modeling and additive manufacturing of bio-inspired composites with tunable fracture mechanical properties. Soft Matter 10, 4436–4442 (2014). [PubMed]
12. Compton B. G. & Lewis J. A. 3D-printing of lightweight cellular composites. Adv Mater 26, 5930–5935 (2014). [PubMed]
13. Leigh S. J., Bradley R. J., Purssell C. P., Billson D. R. & Hutchins D. A. A simple, low-cost conductive composite material for 3D printing of electronic sensors. PLoS One 7, e49365 (2012). [PMC free article] [PubMed]
14. Tekinalp H. L. et al. Highly oriented carbon fiber–polymer composites via additive manufacturing. Compos Sci Technol 105, 144–150 (2014).
15. Gray R. W. IV, Baird D. G. & Bohn J. H. Thermoplastic composites reinforced with long fiber thermotropic liquid crystalline polymers for fused deposition modeling. Polym Compos 19, 383–394 (1998).
16. Shofner M. L., Lozano K., Rodriguez-Macias F. J. & Barrera E. V. Nanofiber-reinforced polymers prepared by fused deposition modeling. Journal of Applied Polymer Science 89, 3081–3090 (2003).
17. Pidcock G. C. & in het Panhuis M. Extrusion Printing of Flexible Electrically Conducting Carbon Nanotube Networks. Advanced Functional Materials 22, 4790–4800 (2012).
18. Zhong W., Li F., Zhang Z., Song L. & Li Z. Short fiber reinforced composites for fused deposition modeling. Materials Science and Engineering A 301, 125–130 (2001).
19. Scheithauer U., Bergner A., Schwarzer E., Richter H.-J. & Moritz T. Studies on thermoplastic 3D printing of steel–zirconia composites. Journal of Materials Research 29, 1931–1940 (2014).
20. Hsueh C.-H. Young's modulus of unidirectional discontinous-fibre composites. Compos Sci Technol 60, 2671–2680 (2000).
21. Hine P. J., Lusti H. R. & Gusev A. A. Numerical simulation of the effects of volume fraction, aspect ratio and fibre length distribution on the elastic and thermoelastic properties of short fibre composites. Compos Sci Technol 62, 1445–1453 (2002).
22. Love L. J. et al. The importance of carbon fiber to polymer additive manufacturing. Journal of Materials Research 29, 1893–1898 (2014).
23. Ryosuke Matsuzaki, Masahito Ueda, Masaki Namiki, Tae-Kun Jeong, Hirosuke Asahara, Keisuke Horiguchi, Taishi Nakamura, Akira Todoroki, and Yoshiyasu Hirano:Three-dimensional printing of continuous-fiber composites by in-nozzle impregnation; Sci Rep. 2016; 6: 23058.
24. Frank Van Der Klift, Yoichiro Koga, Akira Todoroki, Masahito Ueda, Yoshiyasu Hirano, Ryosuke Matsuzaki:3D Printing of Continuous Carbon Fibre Reinforced Thermo-Plastic (CFRTP) Tensile Test Specimens; Open Journal of Composite Materials, 2016, 6, 18-27
25. K. Bryll, K. Gawdzińska, M. Nabiałek, P. Pawłowska; The effect of degradation in aqueous media on viscosity average molecular weight of single polymer polyester composites; Revista de Chimie 2017; 68(9): 2034-2038.

Innovative Multilayer Polymer-glass Composites Reinforced with Metal-ceramic Foam

Katarzyna Gawdzińska[1, a], Janusz Grabian[1], Dorota Nagolska[2] and Barbara Kwiecińska[3]

[1]Maritime University of Szczecin, Faculty of Marine Engineering, ul. Willowa 2-4, 71-650 Szczecin, Poland
[2] Poznan University of Technology, Faculty of Mechanical Engineering and Management, ul. Piotrowo 3, 60-965 Poznan, Poland
[3] Maritime University of Szczecin, Faculty of Navigation, ul. Willowa 2-4, 71-650 Szczecin, Poland

Abstract. When selecting the materials used in transport, in addition to the set of strength properties, which form the basis of the structural calculations, the possibility of weight reduction and the limitation of fire risks and maintenance activities should also be taken into account. The article provides a short description of both metal and metal-ceramic foams, as well as studies on multilayer sandwich composite materials. They consist of polyester-glass laminates and composite-ceramic foams. These materials were subjected to a technological three-point bending test. Such materials can replace the polyester-glass laminates that are commonly used in maritime transport and other industries.

1. Introduction

Composite materials constitute the most promising and quickly developing group of modern construction materials [1-5], which, due to desirable usability properties, are very popular among representatives in the fields of science, technology, industry and business. The guiding principle that has lead to the construction of composites is the desire to obtain a material with specific, expected properties, which largely depend on the technology used. In order to obtain the desired result, two or more materials with different properties are usually combined. The bi-component or multi-component nature of the composite, allows the beneficial characteristics of the components to be used, while reducing the adverse ones. Composite polymeric materials are widely utilized, due to their low density and high mechanical strength, as well as their high resistance to many acids and bases. The use of composites of various materials, participation and type of consolidation phase is significantly reducing the mass of all types of structures and at the same time is reinforcing them [5-6]. A popular way of strengthening materials is to reinforce them with the use of sandwich materials in the form of mats and fabrics. Materials that have this type of reinforcement are called layered composites, laminates or sandwiches (in the latter, they combine ceramics, polymers and metals). Foams are another group of materials. The forecasts developed in the early 1990s, regarding the predictions of technologies and applications of new materials, indicated that more and more often in mechanical engineering, especially in equipment and transport systems, cellular materials produced from metals [7-8] will be used, which are commonly known as: "porous metals" "metal foams" or "metal sponges". The dynamics of the trends related to the anticipated introduction of these materials for machine

[a] Corresponding author : k.gawdzinska@am.szczecin.pl

construction have been shown in Figure 1. These forecasts proved to be accurate and at the beginning of the second decade of the 21st century the wider use of these materials is being observed, especially those made of light metals. Unlike monolithic materials, foams are characterized by the occurrence of designed discontinuities in the material that have a repetitive or varied shape, whose size and arrangement in the entire volume of the product is in an orderly or disordered manner, which results in significant differences in properties, changing the reactions of this material to external factors, e.g. temperature or load.

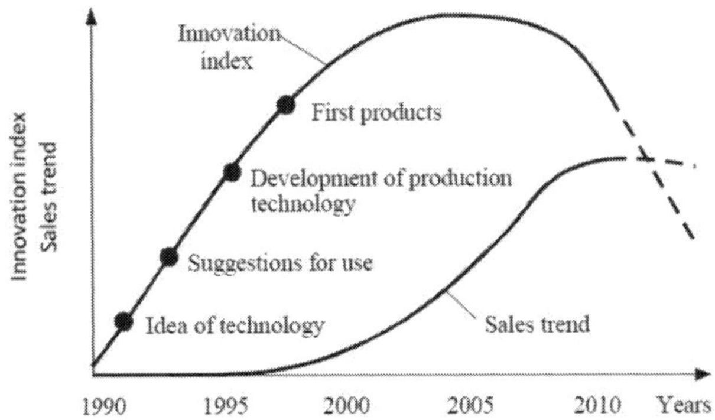

Figure 1. Forecast for the use of metal cellular materials in machine construction [1]

Metal foams (Fig. 2) can practically be made from all kinds of metals (except mercury). The most common foams are made of pure aluminium, as well as: magnesium, titanium, zinc, nickel, copper and iron or their alloys [6].

Figure 2. Metal foam (aluminium)

Metal foams have many unique properties that make them widely used in machine construction. Their low density means that they are an ideal filling material for layered structures that need high stiffness. Their low thermal conductivity makes it possible for them to be used as an insulating material, and the ability to damp vibrations suggests their use as damping liners. On the other hand, their tendency to deform considerably under the influence of loads can be used in systems that absorb impact energy or limit the effects of an explosion, as well as in the production of packaging. Cellular materials may exhibit both isotropy and anisotropy in their properties. According to the definition given in the literature [1, 3, 6], metal foams are:

-porous materials, the construction of which can be described as a geometrically disordered arrangement of pores in a metal matrix,

-a metal material containing a considerable number of gas-filled pores in its volume.

Metal foams with open pores (Fig. 3a) are used in heat exchangers, energy absorbing elements, as filters in the chemical industry, and catalyst components for internal combustion engines. They also have biological applications as silencers in flow installations, flame arresters, carriers of elements and as chemical compounds acting as catalysts for chemical processes, coming into contact with gaseous or liquid media, which are often chemically aggressive, and also have high temperatures. They are made of materials such as copper, nickel, iron, titanium, tantalum and their alloys [7-8].

a) b)

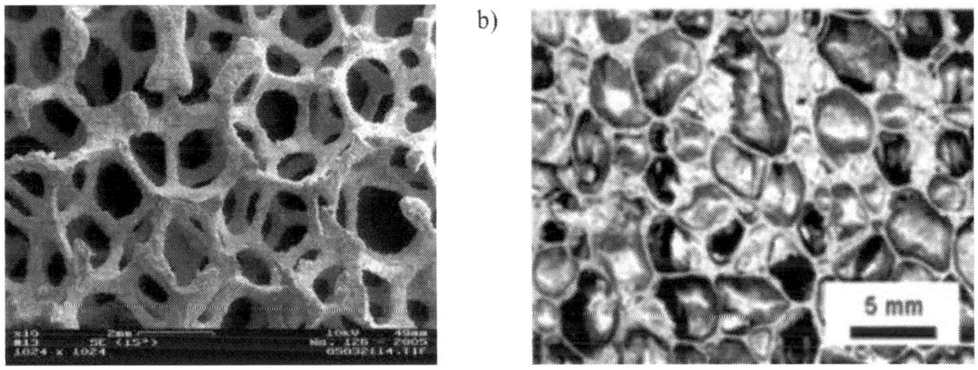

Figure 3. Example of foam structure: a) open pores [1], b) closed pores

Metal foams with closed pores (Fig. 3b) are used in mass-produced land transport, mainly in passenger cars, (in the body parts absorbing impact energy) increasing their stiffness as well as improving comfort of use by reducing vibrations and noise. This type of foam is also used in aircraft construction, general construction and elements of road infrastructure. Foams with closed pores are most often made of light metals, e.g. from pure aluminium, often with the addition of calcium.
The production of metal foams by the appropriate processing of pure metals or their alloys gives a certain, rather limited, area of change in the parameters which characterize these foams. The expansion of the functional range of the foams, as well as the possibilities of controlling them, can be provided by multiphase materials, these include composite materials. Metal-ceramic composites, with a wide range of composition, can be the starting material for the production of foams with special features specific to the shipbuilding area. Composite metal-ceramic foams, developed in accordance with the principles of material engineering, expand the possibilities of design. This paper has described multi-layer sandwich composite materials, in which the material of the spacer was composite metal-ceramic foam placed between polyester-glass composites in the form of a laminate. Plates made of this material were subjected to a three-point bending technological test.

2. Research material

The introduction of a spacer with special properties, between the two layers of the polymer-glass composite, is justified due to the way the material will work – under compressive stress as a result of local pressure and also resulting from the bending state, which may take place in the operating conditions of the vessel and in possible collisions. An innovative material that can be used in this case is aluminium foam with a specific gravity of 0.3-0.5 g/cm^3, which is made of aluminium foamed with a foam concentrate [8-9]. The foam can also be produced from an aluminium-ceramic composite by blowing gas into the liquid metal, and its internal structure can be controlled to some extent [9]. Such composite foams (AlSi11 matrix, SiC reinforcement, argon foaming gas) are manufactured at the Maritime University of Szczecin. Such foams have a specific resistance to compression, as shown in Figure 4.

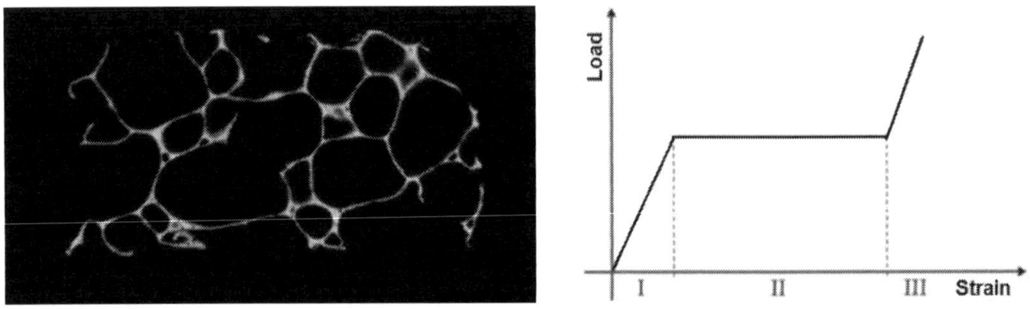

Figure 4. Aluminium-ceramic foam as a material for the spacer and its strength characteristics

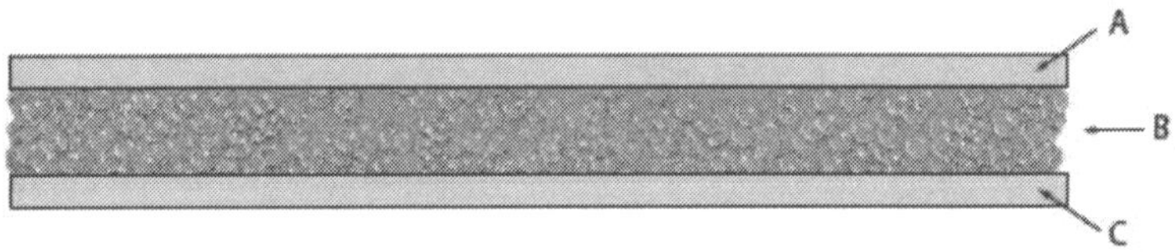

Figure 5. The proposed three-layer material for preliminary testing: A – polyester-glass composite, B – aluminium foam, C – polyester-glass composite

The preliminary assessment of the suitability of the proposed material (Fig. 5) was based on a technological three-point bending test [9-12]. For research purposes related to the technological three-point bending test, test panels with dimensions of 430 × 200 mm were made as shown in Figure 6. Polymer-glass composite laminates were made using Estromal 17 LM01 polyester resin together with METOX 50 hardener (in proportions of 1.5 percent by weight of hardener per unit volume) and twelve infiltrated layers of glass fiber fabric with a weight of 350 g/m^2. Multilayer composites were made using AlSi11/SiC foam layers 25 mm thick, laid and glued between two layers of polymer laminate. After 48 hours, 320 × 60 mm samples were cut from the sample boards.

Figure 6. Test plates with dimensions of 430 × 200 mm

3. Aims and test conditions

The purpose of the preliminary tests was to determine the possible suitability of the multi-layer material presented above, as a replacement for the widely used polymer-glass composite. The satisfactory results of the technological three-point bending test, based on standard normalized bending tests of metals, polymers, plywood, etc., which determine sample dimensions and the geometry of the support system and load element, have been accepted as the basic criterion. To determine the behavior of the laminar material under load, resulting in deformation and its destruction, the load diagram shown in Figure 7 was adopted and carried out with the use of a testing machine using the tool from its equipment.

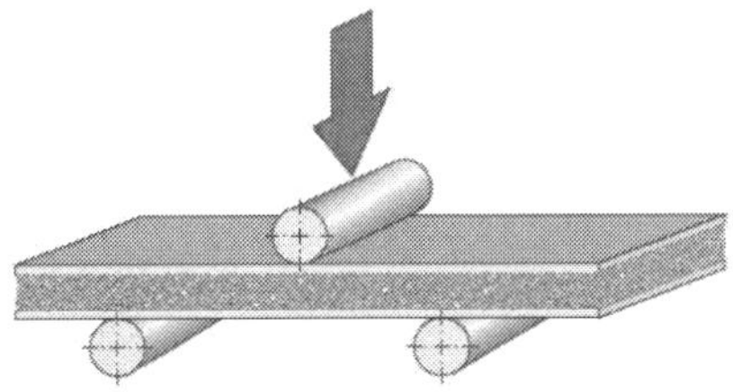

Figure 7. Scheme of the station for the adopted technological three-point bending test

A technological three-point bending test was carried out with the use of a testing machine and the appropriate instrumentation, examining the samples cut from test panels made of laminate-composite glass fiber – polyester resin as a reference level to the classical material, and new multilayer materials using fibreglass composite – polyester resin and aluminium-ceramic foams. The results of the technological test, carried out for samples with dimensions of 200 × 60 mm, which were laid on the support rollers which were at distance of 150 mm apart (for all samples), are shown in Figure 8.

Figure 9 shows the appearance of the samples after the bending test. It was possible to observe the various mechanisms of destruction of the multilayer materials, such as loosening of the outer layer, cracking in the axis of symmetry, as well as the foam spacer having been crushed in the area where the action of the deforming force had been applied.

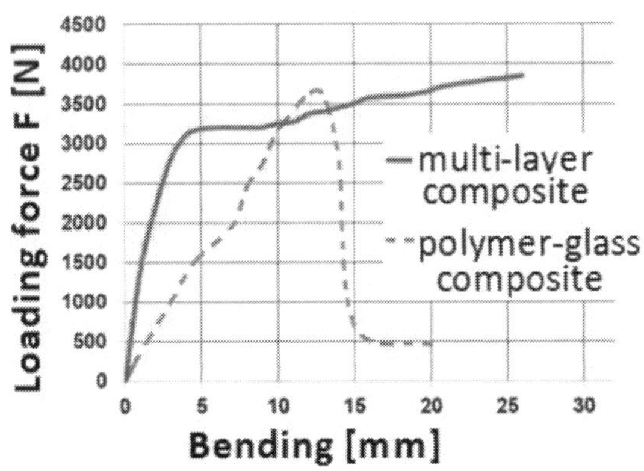

Figure 8. The combination of three-point bending curves of samples cut from different sample boards.

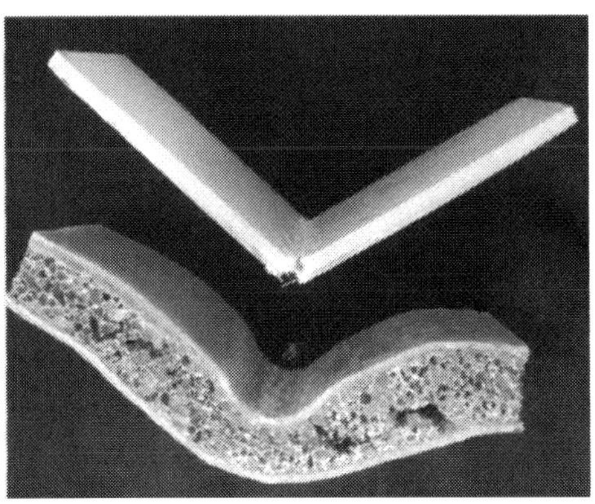

Figure 9. View of the destroyed samples

4. Examples of the use of multilayer composites

The properties of foamed metals, and especially the properties of the proposed multi-layer sandwich composite materials, make them a potential construction material for the transport industry, machine tool construction, the electrotechnical industry (applications in the construction of batteries, fuel cells and supercapacitors) as well as in the manufacture of chemical apparatus. Examples of the use of composite metal foams - metal in machine construction are shown in Figures 10-11.

Figure 10. Acoustic cover of a large machine tool (milling machine) [13]

Figure 11. The use of aluminum foam in a car body structure [1]

5. Conclusion

Replacing widely used multi-layer composites with metal composites and polyester-glass composite will contribute to reducing the mass of the finished product. Adding a metal-ceramic foam spacer will increase the bending strength and thus also significantly increase their safety, e.g. in the case of ship collisions, due to its energy absorbing properties. This means that the materials in question can be widely used in the automotive and aerospace industries [14-15].

The mechanical, acoustic, chemical, fire resistant, thermal and electrical properties of composite metal foams and polyester-glass composites, as well as their considerable resistance to environmental conditions and the ever reducing costs of their production, make them competitive in some areas compared to the traditional materials used in various fields of mechanical engineering. They enable the use of complex construction solutions while reducing energy and fuel consumption, favourably affecting environmental protection, as well as improving the safety and comfort of the use of technical devices.

Acknowledgments

The research presented in this article was carried out under the Grant of the Ministry of Science and Higher Education of Poland nr 1/S/IESO/17: "Increasing operational effectiveness of complex technical systems by systematic development and implementation of innovations using novel materials and modifying the object's structure" performed at the Maritime University of Szczecin, Poland.

References

1. J. Grabian, *Composite metal foam in the shipbuilding*, (2012) (in Polish);
2. W. Ignalewski, *Wpływ nowoczesnych technologii na rozwój statków śródlądowych w Europie*, (2017);
3. K. Gawdzińska, L. Chybowski, A. Bejger, S. Krile, *Determination of technological parameters of saturated composites based on sic by means of a model liquid*, Metalurgija, **Vol. 55 No. 4** November 2016, pp. 659-662, (2016);
4. F. Ashby, A. Evans, A. Flech, *Metal Foams a Design Guide*, (2000);
5. C. Fiebig, M. E. Steffen, S. Caba, M. Koch, *Hybrid Composites of Plastic and Aluminium Foam*, Euro Hybrid Materials and Structures 2016, p. 170–176, (2016);
6. K. Gawdzińska, L. Chybowski, W. Przetakiewicz, *Proper matrix-reinforcement bonding in cast metal matrix composites as a factor of their good quality,* Archives of Civil and Mechanical Engineering **16(3)**, p. 553-563, (2015);
7. G. Kaptay, Interfracial criteria for stabilization of liquid foams by solid particles. Colloids and Surfaces A: Physicochemical and Engineering Aspects **Vol. 230 (1–3),** 2004, 67–80;
8. PN-EN ISO 7438:2006 *Metale – Próba zginania. Statyczna próba zginania techynologicznego*;
9. K. Gawdzinska, K. Bryll, D. Nagolska, *Influence of Heat Treatment on Abrasive Wear Resistance of Silumin Matrix Composite Castings*, Arch. Metall. Mater., **Vol. 61 (2016), No 1**, p. 177–182, (2016);
10. J. Grabian, W. Ślączka, P. Pawłowska, W. Kostrzewa, *The role of innovative composite materials in the safe and efficient operation of floating marine structures*, Scientific Journals of the Maritime University of Szczecin, **Vol. 52 (124)**, pp. 23–29, (2017);
11. R. Neugebauer, T. Hipke, J. Hohlfeld, R. Thümmler, *Highly Damped Machine Tools with Metal Foam*, International Conference "Advanced Metallic Materials", Smolenica, Slovakia, (5–7 November 2003);
12. D. Pulikowski, F. Lackner, C. Scheuerlein, D. Meinel, F. Savary, D. Tommasini, M. Pajor, *Testing Mechanical Behavior of Nb3Sn Rutherford Cable During Coil Winding* **Vol. 27/4** (2017);
13. Aluminium foam, www.alueurope.eu/talat/lecture/1410.pdf.
14. M. Garcia-Avila, M. Portanova, A. Rabiei, *Ballistic performance of composite metal foams*. Composite Structures **Vol. 125**, 2015, 202–211;
15. T. Klepka, *Construction of axial-symmetric polymeric extrudates of complex forms,* Polimery **Vol. 53/5.** 2008, p. 390-395.

An Investigation of Tool Performance in Interrupted Turning of Inconel 718

Tadeusz Chwalczuk[1,a], Martyna Wiciak[1], Agata Felusiak[1] and Piotr Kieruj[1]
[1]*Poznan University of Technology, Piotrowo 3 St., 60-965 Poznan, Poland*

Abstract. The paper presents the result of an investigation of cemented carbides cutting tools performance in interrupted turning of Inconel 718. The discussion is based on case study revealed from industry experience. The spacing created for operational reasons cause interruption in turning process what has crucial effect on tool life and wear. Furthermore cutting parameters and tool geometry have significant influence on creating burr at the edge of spacing which leads dimensional deviation. The research based on two stages, preliminary and modelling, was introduced. In the first an effect of cutting speed, feed, depth of cut, rake angle and tool major cutting edge angle on wear and burr size was investigated. The second step was a regression based modelling of major parameter influence on various process aspects such as values of forces, tool wear and burr size. It was found that the consequential effect on plastic deformation of edge and tool performance have depth of cut. It also has been proven that higher cutting speed, lower values of angles provides better quality of machined element. The paper gives recommendations for turning parameters selection to acceptable burr forms, tool life and process performance.

1 Introduction

Interrupted turning is process that might be challenging from the technological point of view. Additionally when the material such as Inconel 718 is being considered the process is even less effective. Cutting tool performance is being drastically reduced via severe variable loading. The inserts' wear is mainly connected with cracks and thermal fatigue. What is more Ni-based materials' tendencies for creation of build ups on tool and machined surface, low thermal conductivity, high and hardness strength in elevated temperature leads to difficulties in process optimization [1-2].

The crucial aspect on interrupted turning performance has the temperature in cutting zone. Jiang et al. [3] proved the Salomon's hypothesis for interrupted cutting. In their work the relationship between temperature and cutting speed was analysed. The temperature of tool and workpiece decreased at high cutting speeds. It is caused by increased feed rate of removed material. The main heat flux do not interfere with tool and machined surface but mainly it is consumed by removed chip. Nevertheless the temperature generated on cutting insert has great significances on tool performance. What is also important the cutting materials with high thermal conductivity allows to reduces cutting temperatures near the cutting edge [3]. Olviera et al. [4] investigated the ceramic and PCBN cutting tool materials performance when turning under continuous and interrupted conditions of hardened steel. They found

[a] Corresponding author: tadeusz.chwalczuk@put.poznan.pl

that increasing temperature in cutting region during continuous turning cause reduction in the tool's hardness. Such phenome do not occurs in interrupted cutting. In their work it was also stated that at lower temperatures, thermal softening is hindered and the lower hardness and thermal conductivity of Al-based ceramics with SiC whiskers is less important then in PCBN. What is most surprising experiment proved that where the tool life is longer for interrupted cutting then in continuous. Probably in turning of hardened steel the thermal influence has greater impact of tool performance then mechanical aspects of interrupted cutting.

Not only tools material have influence on cutting tool performance but also cutting tool micro and macro geometry [5]. In results presented in [6] it was stated that different cutting edge preparation is a reason of higher material deformation which leads to higher process temperature. Habrat [7] and Rybicki at al. [1] investigated a tool geometry impact on aerospace alloys machinability. It was found that chip-braking is decisive to process performance.

Cutting vibrations during milling processes is specific example of interrupted cutting. Many basic research in the field of machining mechanics is reflected in milling. During this process, key factors such as static and dynamic cutting tools displacements significantly affects the machinability indicators such as roughness and tool performance [8].

Interrupted turning is connected with the necessity of applying advanced methods of cutting parameters optimization and tool condition monitoring during the process [3, 9-11]. One of the more promising methods of diagnostics is the method of measuring dynamic response of tool and machined element [8, 11]. Ratava at al. [12] proposed a complex algorithm for detecting tool displacement based on values depth of cut, cutting speed and feed rate. Application of artificial neural networks improves the decision making process and give abilities to even better automatization [13].

The research results and analysis presented in this paper concern the assessment of an impact of technological parameters and the geometrical of tool on a technological effects of interrupted turning. This topic is strongly connected with industrial practice. In contrast to the previously mentioned papers, in this work Ni-based superalloy machinability is evaluated. Inconel 718 has tendencies to creating build-ups on tool and machined surface what is also connected with plastic deformation of components' edge.

2 Experiment

The investigation have been carried out on Inconel 718 pipe, hardness aprox. 35 HRC (Figure 1a). Tool holder were chosen with dimension according to ISO: PSSNR 2020-12 and PSRNR 2020-12. Cutting inserts were used: SNMG 120408 RP and SNMG 120408 MS by Kennametal.

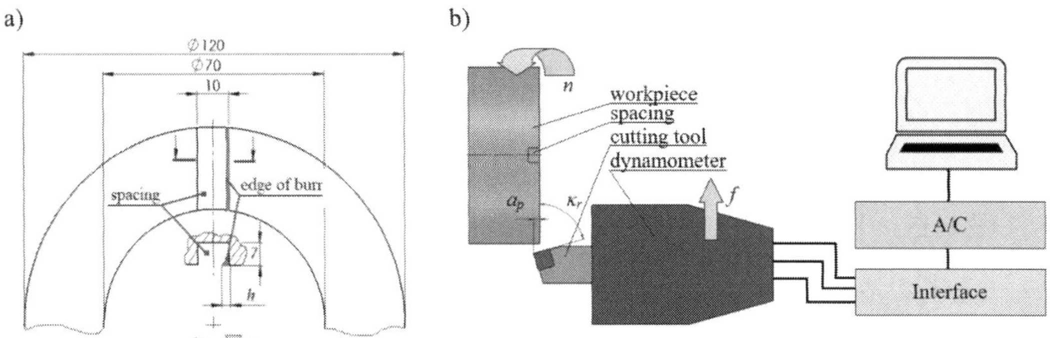

Figure 1. a) Workpiece dimensions; h – high of the burr; b) The set-up of measurement equipment.

Cutting insert material: cemented carbides with AlTiN coating (Kennametal's code – KC5510). Such selection of tool holders and inserts allows to fulfil research program shown in Table 1. The force dynameter was connected through amplifiers and band-pass filters to a data acquisition computer. The signals were analysed and procced with the application of dedicated software (Figure 1b).

Research presented in this paper is based on factional plan $2^{(k-p)}$ for five variables: feed f, depth of cut a_p, cutting speed v_c, orthogonal rake angle γ_0 and tool major cutting edge angle κ_r. The applied program based on 16 random combinations of variables presented in Table 1 was repeated twice for each combination – that gives 32 tries. Design of experiment and its analysis were performed with Statistica 13.3 software by Statsoft.

The study adopted two technological criteria related to the presented research program. The first was the wear indicator VB_c measured on the flank face. The second was the height of the burr produced inside the technological gap measured in the plane consist the vector of cutting speed on the exit edge of processed material.

Table 1. Cutting parameters applied in the research.

Level of research program and variables values	f [mm/rev]	a_p [mm]	v_c [m/min]	γ_0 [deg]	κ_r [deg]
-1	0,13	0,10	30	5	45
1	0,28	0,50	88	12	75

In the second stage of research the new program was implemented involving only one variable that have the most significant influence on analysed parameters. The linear correlation was estimated. The research program is based on mathematical regression where investigated parameters has four levels and tests were repeated three times.

3 Results and discussion

The graphs presented in Figure 2 illustrates the results of Pareto's analysis. Taking into concern the analyzed input variables from the statistic point of view, the most important is technological parameters is the depth of cut. The values of a_p affect both the value of wear after machining and the height of the burr. Other input factors with the assumption of no interaction between them are statistically insignificant.

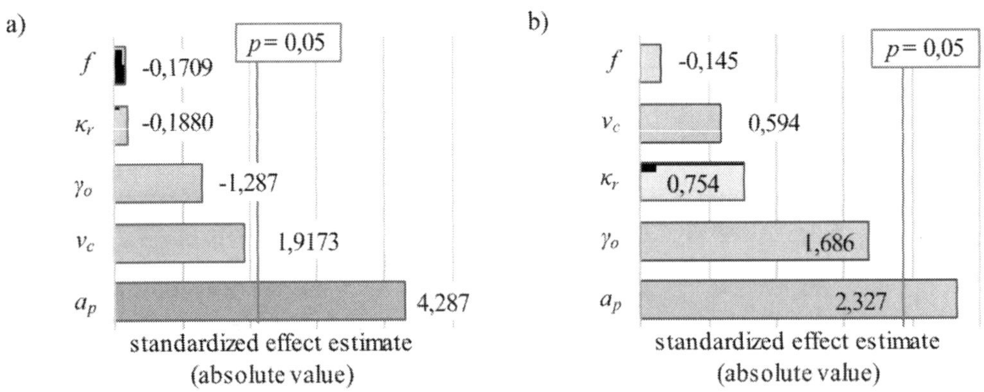

Figure 2. Pareto effects charts for input variables on a) wear indicator VB_c and b) burr high h.

Figure 3 presents the impact of the analyzed parameters on the VB_c wear after the same cutting time. The largest estimated wear value is achieved for test where a_p = 0,5 mm. The difference in VB_c in the studied range of a_p reaches over 1,2 mm. Despite the lack of statistical influence of the cutting speed and rake angle in the quantitative mean analysis, the differences between the boundary values reach 0.45 mm and 0.25 mm, respectively. Such differences are significant for technological reasons. The low effect of feed f on the VB_c value might be found surprising. Both the change in a_p and f is related to the change of the cross-sectional area of the cutting layer and thus to the value of the cutting force. So that the affect of those parameters should be similar. However, it must be that the change of the a_p increases the share of the rake face during cutting and consequently also the chip breaker on this surface.

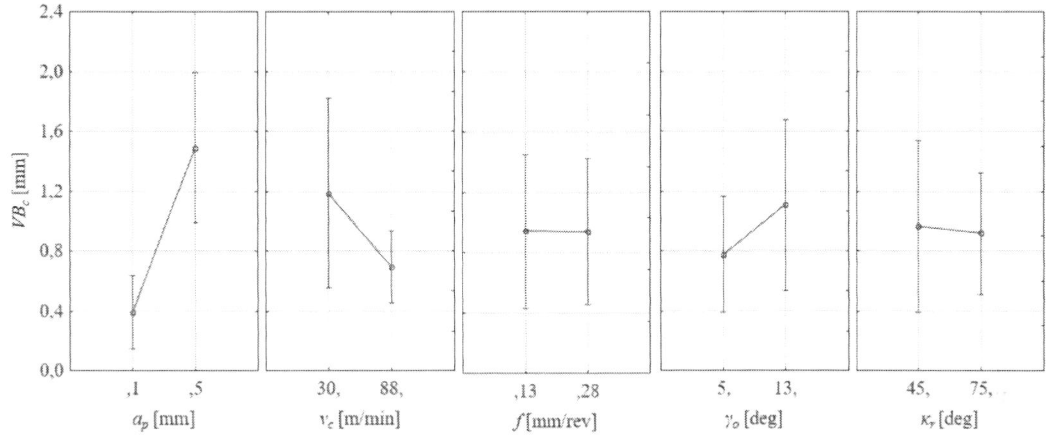

Figure 3. Input parameters (tool geometry and cutting parameters values) influence on a wear VB_c.

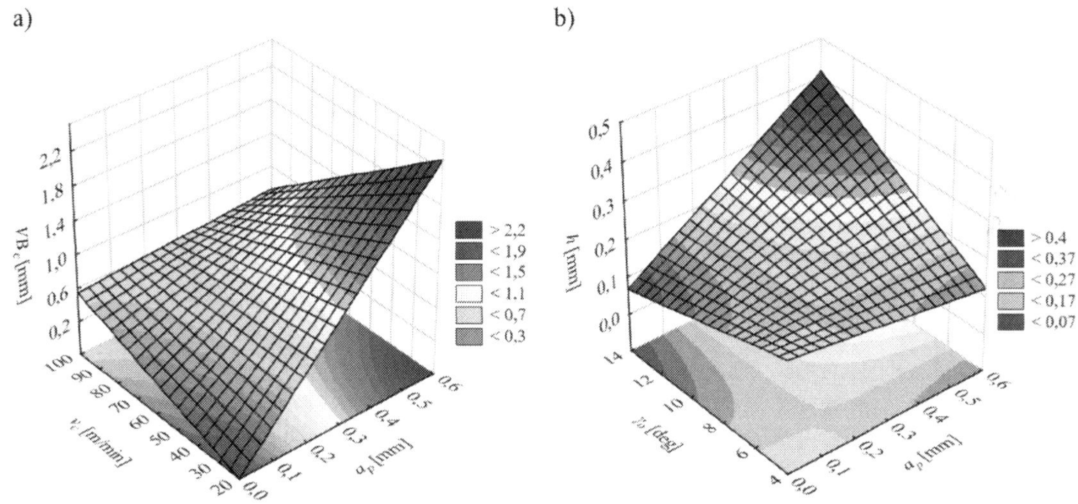

Figure 4. Plot functions of influence of a) cutting speed v_c and depth of cut a_p on cutting wear indicator VB_c, b) orthogonal rake angle γ_0 and depth of cut a_p on burr high h.

The next stage of conducted research focused on determination of cutting parameters values a_p and v_c and tool angle γ_0 impact on wear and burr high. The effects are presented in the Figure 4. The burr high vary from 0.1 mm to 0.4 mm. Smaller values of VB_c are obtained for small a_p values and higher values of γ_0 angle. It is possible to reduce the negative impact of depth of cut on burr high by using smaller γ_0 angle values.

The analysis of force variability in the interrupted turning is based on the analysis of the difference in mean values between the cutting phase and the entry into the spacing. The increasement in cutting force and thrust force was analysed in this way. The mean values of cutting force F_c and thrust force F_p were chosen for discussion. Cutting force good indicator of material decohesion velocity and plastic deformation changes in the cutting zone. The thrust force is directly connected with depth of cut values. It was found that there is no linear correlation between force increment, wear or burr high (Table 2). The dependence and influence between those parameters might be more complex. Figure 5 represent the quadratic plot functions of approximation between analysed factors. Low wear values and burr are generated at low increments of both cutting and thrust forces. To obtain small values of ΔF_c and ΔF_p the proper parameter selection should be conducted. Such basic optimization was shown in previously discussion. In further researches the temperature and dynamic aspect of interrupted cutting will be investigated to minimize the burr high and cutting tool wear.

Table 2. Matrix of correlations; N=12; $r \in \langle -1;1 \rangle$

Variable	h [mm]	VB_c [mm]	ΔF_c [N]	ΔF_f [N]	ΔF_p [N]
h [mm]	1,00	0,31	0,36	0,36	0,15
VB_c [mm]	0,31	1,00	0,19	0,26	0,58
ΔF_c [N]	0,36	0,19	1,00	0,96	0,81
ΔF_f [N]	0,36	0,26	0,96	1,00	0,86
ΔF_p [N]	0,15	0,58	0,81	0,86	1,00

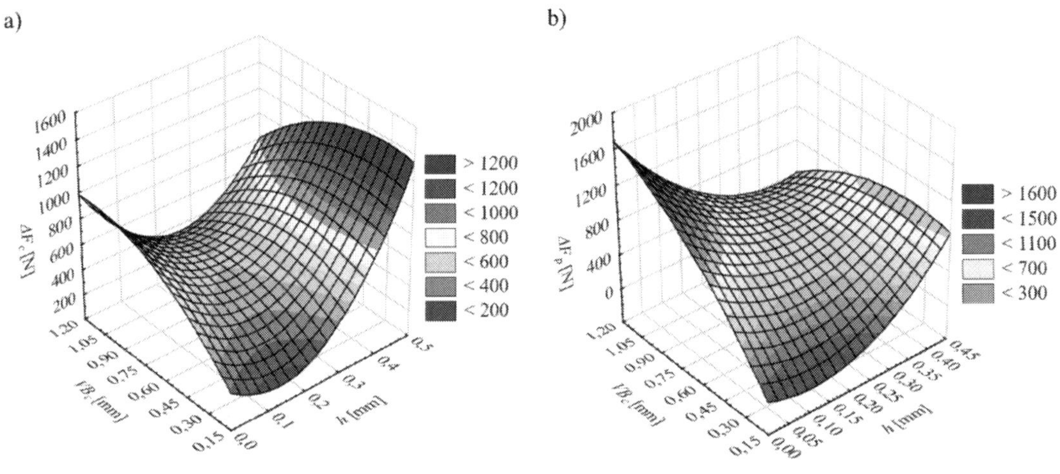

Figure 5. Quadratic plot function of force increasement of a) cutting force ΔF_c and b) thrust force ΔF_p and their dependency between cutting insert wear VB_c and burr high h.

4 Conclusions

On the basis of the carried out investigations, the fallowing conclusion are formulated:

- Tool geometry during interrupted cutting has lower impact on its performance then cutting parameters value.
- Plastic deformation on the edge of machined component is mainly influenced by the value of depth of cut. Feed per revolution and cutting speed have lower influence because decreased interaction area of rake face plane on analysed indicators.
- There is no significant correlation between cutting forces increase during interrupted turning of Inconel 718 on wear indicator VB_c and burr high h.

Acknowledgments

The presented research results, executed under the domestic project LIDER of No. 164/L-6/14/NCBR/2015 with grants for education allocated by the National Centre for Research and Development.

References

1. M. Rybicki, P. Szablewski, MATEC Web of Conferences, **121**, 03020 (2017)
2. D. Przestacki, T. Chwalczuk, MATEC Web of Conferences, **136**, 02006 (2017)
3. F. Jiang, Z. Liu, Y. Wan, Z. Shi, Journal of Materials Processing Technology, **213** (2013)
4. A.J. Oliveira de, A. E. Diniz, D.J. Ursolino, Journal of Materials Processing Technology, **209** (2009)
5. C.E.V. Venture, J., Kohler, B. Denkena, Journal of Manufacturing Processes, **19** (2015)
6. B. Denkena, J. Kohler, B. Breidenstein, A.M. Abrao, C.E.H. Ventura, Surface & Coatings Technology, **254** (2014)
7. W. Habrat, Solid State Phenomena, **261** (2017)
8. P. Twardowski , S. Legutko, G.M. Królczyk, S. Hloch, Advances in Mechanical Engineering, **6** (2015)
9. X. Cui, F. Jiao, P. Ming, J. Guo, Ceramics International, **43** (2017)
10. G. Urboain, L. N. Lopez de Lacalle, A. Fernandez, Journal of Soundand Vibration, **333** (2014)
11. I. Cascon, J.A. Saraua, Procedia, CIRP **31** (2015)
12. J. Ratava, M. Lohtander, J. Varis, Robotics and Computer–Integrated Manufacturing, **47** (2017)
13. A. Felusiak, P. Twardowski, Arch. Mech. Tech. Mater., **38** (2018)

The Effects of Laser Surface Modification on the Microstructure of 1.4550 Stainless Steel

Damian Przestacki[1,a], Aneta Bartkowska[1], Mateusz Kukliński[1] and Piotr Kieruj[1]

[1]*Poznan University of Technology, Marii Sklodowskiej-Curie 5 St., 60-965 Poznan, Poland*

Abstract. In this study a stainless austenitic steel 1.4550 was laser heat treated with diode laser. The influence a gouache coating on remelted steel substrate was carry out. The cooling system during laser melted was analysis as well. Melted layers were manufactured with different laser beam power between 0.6 kW and 1.4 kW, constant scanning laser beam speed v_l = 5.76 m/min and laser beam diameter equal d_l = 1.2 mm. The surface was treated at room temperature and under CO_2 cooling conditions and the results were compered. With the increase of the laser beam power, the dimensions of the laser tracks increase. The depth of laser tracks varies significantly than their width. The deepest melted layer was observed for a material that wasn't coated by any of absorbent paste and when there wasn't cooling system.

1. Introduction

One of new techniques improving machinability of material is laser assisted machining often reffered to as LAM [1-5]. The technique of laser assisted turning concentrates on improving machining of material due to increasing temperature in cutting zone which results from interaction with laser beam [1-5]. However, before undertaking attempts of hybrid machining (simultaneously heating and turning of machined material) it is significant to investigate what is happening what happens with material after only laser heating process [5-7]. Analysis of microstructure and microhardness significantly demonstrates these changes [1-15].

In many publications problems connected with laser heating of material and changes occurring in them are analyzed [1-19]. Mainly, they concern ferrous alloys [4,8,9], nickel alloys [6,7,13] or titanium alloys [2]. Group of hard-to-cut materials includes undoubtedly alloys reinforced with SiC and WC particles [10-12,15]. Recently, scientific research focuses on difficult-to-cut materials, which use are constantly increasing in industry [14-19]. Among ferrous alloys, steels containing extended content of nickel and chromium, ex. austenitic steel, may cause problems with machining.

Nowadays there is a tendency to manufacture products of higher quality in shorter time and with lower costs. To achieve some of mentioned objectives, it is necessary to use hard and resistant materials which usually have low machinability. Therefore, the objective of this research was to determine adequate parameters for laser assisted turning of commonly used austenitic steel.

2. Experiments

[a] Corresponding author: damian.przestacki@put.poznan.pl

Specimens of stainless austenitic steel 1.4550 were laser heat treated using diode laser TruDiode 3006 produced by Trumpf company. In the research, influence of impact of applied coating – gouache of thickness equal to approximately 20 μm and of CO_2 cooling system during interaction of laser beam with the material. All significant parameters of laser heat treatment are shown in table 1. During laser heat treatment constant laser beam scanning velocity and laser beam diameter were applied, equal to respectively v_l = 5.76 m/min and d_l = 1.2 mm. On the other hand, laser beam power was variable and equal to P = 600 W, P = 1000 W i P = 1400 W.

Table 1. Laser heat treatment parameters

P [W]	d_l [mm]	Coating	Cooling	v_l [m/min]
600	1.2	gouache coating	without	5.76
1000				
1400				
600			CO_2	
1000				
1400				
600		without coating	CO_2	
1000				
1400				
600			without	
1000				
1400				

Specimens of dimensions 26 mm x 10 mm x 6 mm after laser heat treatment were included in thermosetting resin. Then, specimens were grinded with abrasive papers of degressive gradation, polished using alumina oxide suspension and etched with glycerin reagent. To carry out microscopic observations light microscope OPTA-TECH of LAB 40 series was used both with OptaViewIS software. Moreover microhardness testing with Vickers method was carried out, using load of 100 g (HV0.1) on Zwick 3212B microhardness tester.

3. Results of experiment

Figure 1 presents influence of laser beam power without applying initial coating and without applying cooling system during laser heat treatment. It can be seen that with increasing laser beam power, dimensions of laser tracks also increase, in particular their depth. With applied parameters of laser heat treatment without initial coating and without applying cooling system during laser heat treatment, obtained laser tracks are free of pores and cracks.

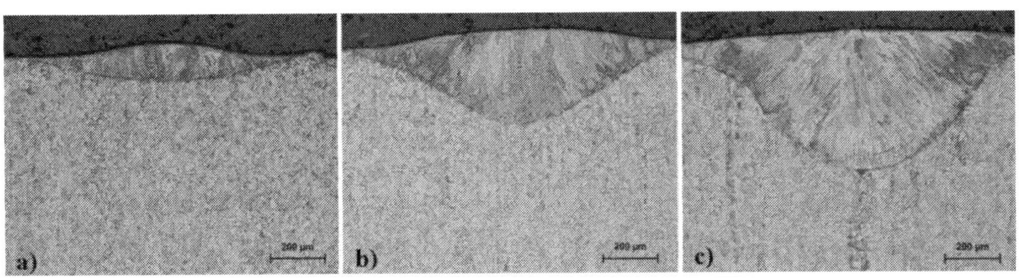

Figure 1. Microstructure of 1.4550 steel after impact of laser beam without initial coating and without cooling during laser heat treatment; a) 600 W, b) 1000W c) 1400W

Figure 2 presents influence of laser beam power if initial gouache coating is applied and without applying cooling system during laser heat treatment. It can be seen that increasing laser beam power dimensions of laser tracks also increase. Depth of laser tracks changes significantly. Also in this case applied laser heat treatment parameters resulted in obtaining layer free of cracks and pores.

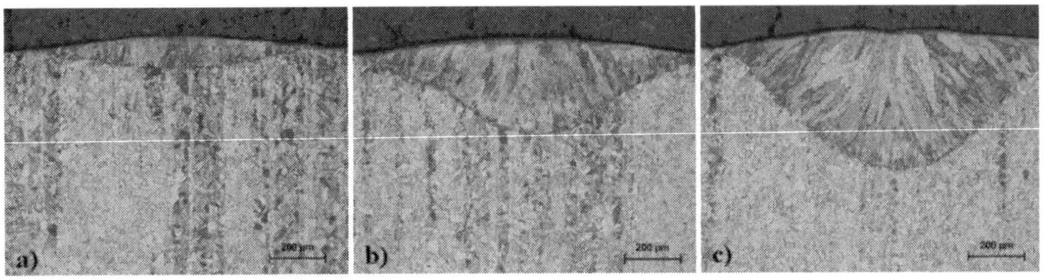

Figure 2. Microstructure of 1.4550 steel after impact of laser beam with initial gouache coating and without cooling during laser heat treatment; a) 600 W, b) 1000W c) 1400W

Figure 3 presents influence of laser beam power without applying initial coating but with applying CO_2 cooling system. Usage of increasing laser beam power results in obtaining larger dimensions of laser tracks. Especially changes can be seen if depths of laser tracks are analyzed. With applied laser heat treatment parameters without applying initial gouache coating and with applying CO_2 cooling system during the process obtained laser tracks are free of pores and cracks. Only near the surface fine crystals oriented perpendicularly to the surface can be seen and they were not visible on laser tracks obtained when cooling system was not applied.

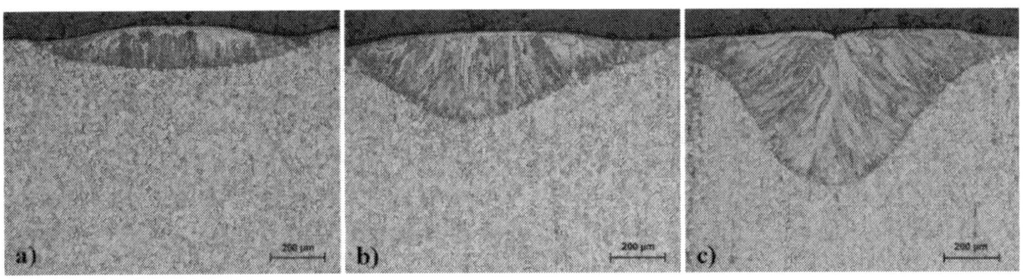

Figure 3. Microstructure of 1.4550 steel after impact of laser beam without initial coating and with CO_2 cooling system during laser heat treatment; a) 600 W, b) 1000W c) 1400 W

Figure 4 presents influence of laser beam power with applied both initial gouache coating and CO_2 cooling system during laser heat treatment. It can be seen that with increasing laser beam power the relation is maintained and the higher laser beam power, the larger dimensions of laser tracks.

If low laser beam power is applied (P = 600 W) with initial gouache coating and CO_2 cooling system, in obtained laser track cracks can be seen which are probably connected with high stresses caused by the cooling rate. Crack were not observed on laser tracks if laser beam power was increased to 1000 W or 1400 W.

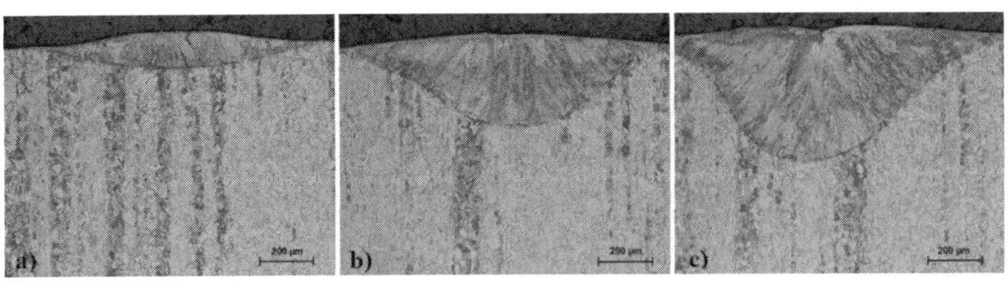

Figure 4. Microstructure of 1.4550 steel after impact of laser beam with initial gouache coating and with cooling using CO_2 during laser heat treatment; a) 600 W, b) 1000W c) 1400W

The influence of applied technological parameters on dimensions of laser tracks (depth, width) for 1.4450 steel after laser heat treatment are shown in table 2.

Table 2. Dimensions of laser tracks after laser heat treatment with applied parameters for 1.4550 steel

Track's depth [µm]	Track's width [µm]	P [W]	Coating	Cooling
116	901	600	gouache	none
327	1188	1000		
458	1193	1400		
127	959	600		CO_2
327	1207	1000		
472	1276	1400		
154	971	600	none	CO_2
305	1277	1000		
513	1362	1400		
134	929	600		none
332	1176	1000		
496	1355	1400		

Figure 5 presents microhardness profiles obtained in laser tracks and core of 1.4550 steel with applying variable parameters of treatment, i.e. laser beam power. Microhardness for each analyzed cases were similar. Here, for example, results for specimens with gouache coating and without applying cooling system during laser heat treatment are shown. Depth of re-melted zone is marked in diagram with dashed line. Microhardness of laser tracks, despite of differences in structure due to re-melting was comparable and equal to approximately 200 HV.

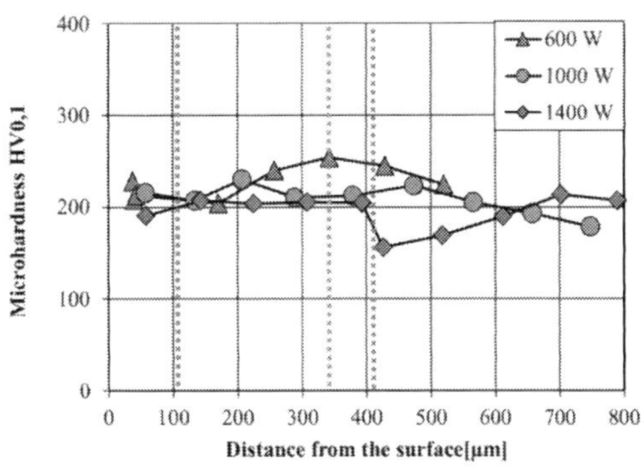

Figure 5. Microhardness profiles for 1.4550 steel covered with gouache coating and laser heat treated without cooling system during the process

On the basis of carried out research it can be seen that as a result of laser beam impact microstructure in re-melted zone is dendritic which is susceptible to machining. In re-melted zone direction of dendritic growth can be clearly seen and it has its beginning on boundary between re-melted zone and grains of the substrate. Obtained results of microhardness does not show significant differences between core and re-melted material, therefore, in further research, attempts connected with manufacturing multiple laser tracks will be taken and these will be machined in the next step.

4 Conclusions

On the basis of the carried out investigations, the following conclusion are formulated:

- with increasing the laser beam power dimensions of laser tracks also increase. Depth of laser tracks changes significantly in comparison with its width.

- The highest depth of re-melted zone was observed for material which was not covered with initial coating and if cooling system was not applied
- Microhardness in re-melted zone and in the substrate was similar

Acknowledgments

The presented research results, executed under the domestic project LIDER of No. 164/L-6/14/NCBR/2015 with grants for education allocated by the National Centre for Research and Development.

References

1. Z. Zhu, V.G. Dhokia, A. Nassehi, S.T. Newman, International Journal of Computer Integrated Manufacturing **26** (2013)
2. R.A. Rahman Rashid, S. Sun, G. Wanga, M.S. Dargusch, International Journal of Machine Tools & Manufacture **63** (2012)
3. Y. Jeon, C. M. Lee, International Journal of Precision Engineering and Manufacturing **13** (2012)
4. P. Dumitrescu, P. Koshy, J. Stenekes, M.A. Elbestawi, International Journal of Machine Tools & Manufacture **46** (2006)
5. C. R. Dandekar, Y.C. Shin, Int J Adv Manuf Technol **66** (2013)
6. D. Przestacki, M. Kuklinski, A. Bartkowska, INT J ADV MANUF TECHNOL, **93** (2017)
7. A. Bartkowska, M. Kukliński, P. Kieruj, MATEC Web of Conferences, MSE 121, **03006** (2017)
8. A. Bartkowska, A. Pertek, M. Popławski, D. Bartkowski, D. Przestacki, A. Miklaszewski, OPT LASER TECHNOL., **72** (2015)
9. A. Bartkowska; A. Pertek, Surface and Coatings Technology, **248** (2014)
10. D. Przestacki, M. Jankowiak, Journal of Physics: Conference Series **483** (2014)
11. M. Kawalec, D. Przestacki, D.,K. Bartkowiak, M. Jankowiak, ICALEO 2008 - 27th Inter.Congress on Applications of Lasers and Electro-Optics, Congress Proceedings, (2008)
12. D. Przestacki, T. Chwalczuk, S. Wojciechowski, INT J ADV MANUF TECHNOL, **91** (2017)
13. H. Ding, Y. C. Shin, Int J Adv Manuf Technol **64** (2013)
14. S. Sun, M. Brandt, M.S. Dargusch, International Journal of Machine Tools & Manufacture **50** (2010)
15. D. Bartkowski, A. Młynarczak, A. Piasecki, B. Dudziak, M. Gościański, A. Bartkowska, OPT LASER TECHNOL. 68 (2015)
16. Y. Xiong, W. Wang, R. Jiang, K. Lin, M. Shao, Materials, Vol.**11**(4)(2018)
17. K. Venkatesan, R. Ramanujam, P. Kuppan, Int J Adv ManufTechnol, Vol.**91**:3807-3821(2017)
18. D. Przestacki, T. Chwalczuk, MATEC Web of Conferences **136** (2017)
19. S. Wojciechowski, P. Twardowski, T. Chwalczuk, Journal of Physics: Conference Series, 20140000, Vol.**483**, pp.1-8(2014)

Computational Thermal Model of Unidirectional Composites with Random Fiber Array

Yuxuan Zhang[1], Hui Wang[1, a] and Wanqing Lin[1]

[1]*Department of Engineering Mechanics, Henan University of Technology, Zhengzhou 450001, China*

Abstract. The purpose of this work was to study the influence of microstructure on effective transverse thermal behavior of unidirectional fiber reinforced composites. Three types of microstructures are taken into account, including square periodic, hexagonal periodic and random arrangements of circular fibers. Unlike classical results at low fiber volume fractions and low thermal conductivity contrast between fibers and matrices, results provided by finite elements simulations for copper matrix composite reinforced with Carbon T-300 fibers have shown that random microstructures strongly affect the effective thermal properties of unidirectional composites for both high volume fractions and thermal conductivity contrast and can give closer predictions to the experimental results than the regular microstructures and the theoretical model.

1 Introduction

Unidirectional fiber reinforced composites [1] used for most engineering applications can be regarded as homogeneous materials with transversely isotropic/anisotropic properties from a macroscopic view. Moreover, it has been revealed that the macro-level behavior of unidirectional composite is usually controlled by its microstructure including constituent properties, fiber volume fraction, fiber shape as well as fiber arrangement. Usually, for the purpose of simplicity, the fiber arrangement in unidirectional composite follows an assumption that the fibers are dispersed in the matrix with regular pattern, i.e. the square pattern or the hexagonal pattern, so that a simple micromechanical unit cell model can be taken out from the composite for further homogenization analysis [2-4]. However, such assumption is different to the real distribution of fibers in the matrix material. It has been observed that the real microstructures are often random [5]. Therefore, how to assess the influence of randomly dispersed fibers to the effective thermal properties of composites have to be paid more attention [6].

Actually, at low fiber volume fraction and weak material property contrast between phases, the random composites can give rather similar predictions of overall properties of composite as the regular composites [7, 8]. But for high fiber volume fraction, i.e. 40% and more, and high thermal property contrast between matrix and fiber, there is still a doubt that microstructure influences effective thermal conductivity of composites.

Present work is a comparison between transverse effective thermal conductivity of unidirectional fiber reinforced composite respectively built with square, hexagonal and random distribution of non-overlapping circular fibers, for a large range of fiber volume fraction and thermal property contrast.

[a] Corresponding author : huiwang@haut.edu.cn

The homogenized effective thermal conductivities of these composites are evaluated by finite element simulations followed by an area averaging of local thermal properties.

2 Generation of random pattern of fibers

For an unit cell of unidirectional composite, the relation of the fiber volume fraction, the number of fibers, the fiber size and the unit cell size can be written as

$$v_f = \frac{n_f \times \pi r_f^2}{a \times a} \tag{1}$$

from which one can have

$$a = r_f \sqrt{\frac{n_f \times \pi}{v_f}} \tag{2}$$

where v_f, r_f and n_f are the fiber volume fraction, fiber radius and number of fibers in the unit cell. a is the side length of square unit cell.

Generally speaking, the fiber volume fraction and fiber radius are known parameters in practice for a known unidirectional composite, thus just n_f and a can be adjustable for establishing micromechanical unit cell model. From this point of view, the side length a of the square unit cell increases in terms of the number of fibers n_f in the unit cell, that is, how large the unit cell is.

To automatically generate random distribution of fibers in the unit cell to characterize their real case in composite as much as possible and simultaneously, avoid the fiber intersecting with the boundary of cell, the following algorithm is designed to produce the position of circular center by using random number

$$[x_c, y_c] = \text{rand}(1,2) \times (a - 2r_f) + r_f \tag{3}$$

where rand(1,2) is an inbuilt function in MATLAB to produce random point in the [0,1]×[0,1] region, and $[x_c, y_c]$ is the location of circular fiber center. Moreover, the fibers can't overlap actually, thus the distance between any two fibers should exceed two times of fiber radius.

Repeating such procedure until the prescribed number of fibers is generated in the unit cell. To demonstrate this algorithm, let's consider an example. If the fiber radius is taken as $3.5\mu\text{m}$, and the fiber volume fraction is set to be 50%, the random arrays with 100 and 400 fibers can be automatically produced by the present algorithm, as illustrate in Figure 1. It is found that the size of unit cell becomes bigger with the increase of the number of fibers, as expected from Eq. (2). Moreover, one can see that the 400 fibers are more densely packed than the 100 fibers.

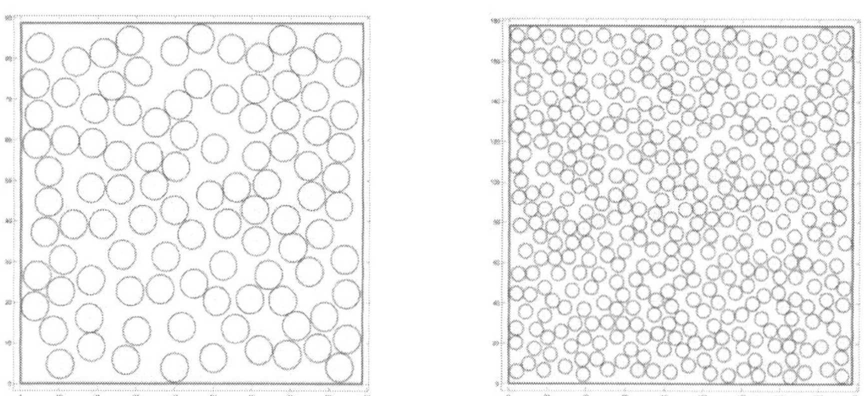

Figure 1. Two random arrays of fibers in the unit cell with 100 fibers (left) and 400 fibers (right)

3 Computational models

The composite under consideration is all reinforced with unidirectional circualr fibers of same diameters. Apart from the fiber volume fraction and the thermal property contrast between material phases, the only difference between samples studied is the fiber transverse distribution which could be distributed randomly or regularly in square or hexagonal pattern.

Determination of homogenized effective transverse elastic properties of these composites requires samples large enough to ensure statistical homogeneity. Due to its periodicity, the unit cell for the square patrtern and the hexagonal pattern of fibers is respectively reduced to a square cell surrounding a single fiber and a square cell embedding two fibers, as shown in Figure 2. While, for random case, the unit cell have to be chosen to be big enough, in other words, contain a sufficient number of fibers, so that the overall thermal conductivities are approximately independent of the applied boundary conditions [9].

In the following computational analysis, the number of fibers are 100 and 400, respectively for random composites, as illustrated in Figure 1. Corrspondingly, the regualr composite is assumed to have same fiber volume fraction for the purpose of comparison. The fiber is taken as T-300 Carbon fiber, which has diameter 7μm and transverse thermal conductivity 0.8W/mK. The matrix material is copper, which has thermal conductivity 380W/mK [10].

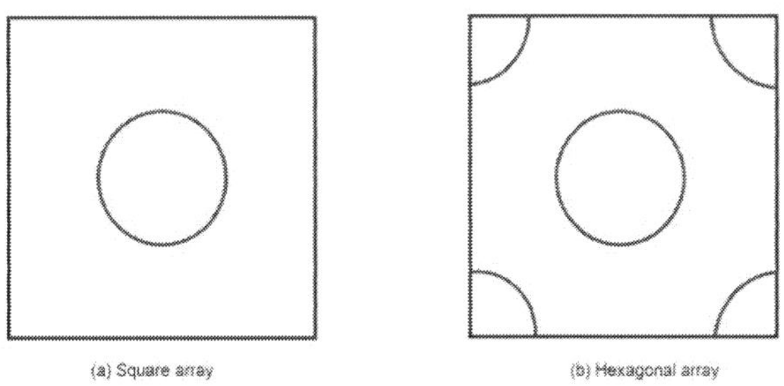

Figure 2. Regular arrays of fibers in the matrix

3.1 Effective thermal conductivity

For the unidirectional composite reinforced with long fibers, the transverse heat transfer can be regarded as the two-dimensional heat transfer in the xy plane. For such case, the temperature (T) at any local point in the unidirectional composite unit cell shown in Figure 3 must satisfy the governing equation in both the matrix and the fiber

$$\frac{\partial^2 T}{\partial x^2} + \frac{\partial^2 T}{\partial y^2} = 0 \tag{4}$$

The boundary conditions corresponding to Eq. (4) may be written as

$$\begin{aligned} & k\frac{\partial T}{\partial y} = 0 \ \text{ at } y = 0 \ \text{ and } \ y = a \\ & T = T_1 \ \text{ at } \ x = 0 \\ & T = T_2 \ \text{ at } \ x = a \end{aligned} \tag{5}$$

where k is the thermal conductivity of either the fiber or the matrix.

In addition to the boundary conditions given in Eq. (5), the conditions existing at the interface between the fiber and the matrix should be specified. For zero thermal resistance at the interface

$$\begin{aligned} & T_f = T_m \\ & k_f \left.\frac{\partial T}{\partial n}\right|_f = k_m \left.\frac{\partial T}{\partial n}\right|_m \end{aligned} \tag{6}$$

where n symbolizes the direction normal to the interface.

One can now solve the above problem consisting of Eqs. (4)-(6). Having obtained a solution to this problem, one would first calculate the local heat transfer rate at points along the surface $x = a$ and then an average heat transfer rate $\tilde{q}$ can be computed. Using this average value, a composite thermal conductivity will be now defined as [11]

$$k_x = \frac{\tilde{q}a}{\Delta T} \tag{7}$$

Here ΔT is the temperature difference between the isothermal surfaces $x = 0$ and $x = a$, as indicated in Figure 3.

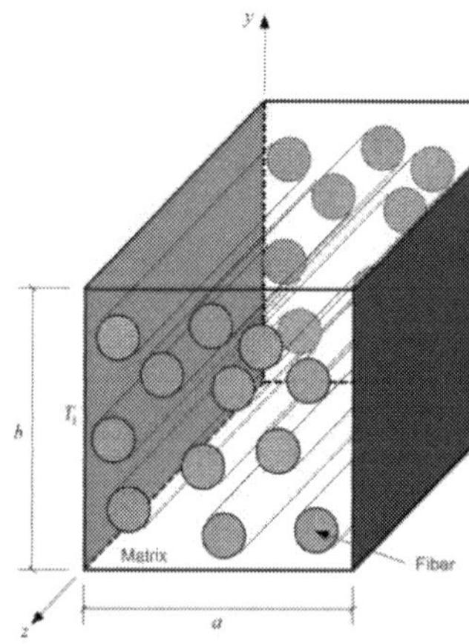

Figure 3. Transverse heat transfer in the unit cell

3.2 Meshing convergence

In the study, all composite samples are meshed by triangular-shaped linear finite elements. In order to ensure accurate results, even in high thermal gradient situations, the finite element mesh must be fine enough to avoid dependence of results with mesh scales. Different meshes are tested with regard to mesh densities. Figure 4 presents results of convergence for random unit cells with fiber volume fraction 50%. In both configurations the final mesh was fined enough to represent accurately the geometry of inclusions. It is observed that the required convergent number of elements dramatically increase for the random sample with 400 fibers, compared to the random sample with 100 fibers.

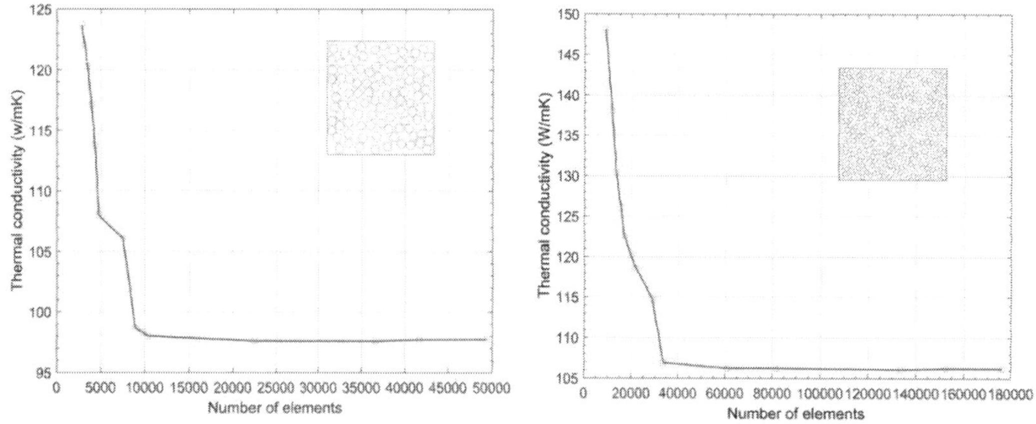

Figure 4. Effect of mesh density on the effective transverse thermal conductivity for the random microstructures with 100 fibers (left) and 400 fibers (right)

3.3 Results

In this subsection, the influences of fiber volume fraction and the thermal property contrast are investigated for both the regualr and random samples. Here, the regular arrays of fibers including the square array and the hexagonal array (see Figure 2) are compared. Besides, the theoretical prediction from the Gurtman model [10] which is based on a periodic and regular composite unit cell theory is also provided in Eq. (8) for performing comparison.

$$k_x = v_f k_f (1 + v_m A) + v_m k_m (1 - v_f A) \tag{8}$$

where

$$A = \frac{k_m - k_f}{(k_m + k_f)(1 + v_f)} \tag{9}$$

where k_m and k_f are thermal conductivity of fiber and matrix, respectively, and v_m=1-v_f.

3.3.1 Effect of fiber volume fraction

For various fiber volume fractions, i.e. changing from 5% to 50%, which cover the low value to the high value, the effective thermal conductivity of composite is evaluated. The number of fibers in the unit cell is chosen as 100 and 400 for random samples, 2 for regular hexagonal sample and 1 for regular square sample. Results in Figure 5 indicate that the random distribution of fibers gives different results to the theoretical prediction, while both the square and hexagonal arrays seem to predict well the transverse thermal conductivity of unidirectional metal matrix composites. From Figure 5, it is also resulting that the transverse thermal conductivity obtained from random samples is much lower than those from theoretical values and regular samples, and the discrepancy becomes more significant with the increase of fiber volume fractions. But, it is found that the results from random models are closer to the experimental results then the theoretical and regular predictions. Moreover, one can see that both the two random samples give slightly different predictions at high fiber volume fractions. Besides, it is observed from Figure 5 that the transverse thermal conductivity of composites nonlinearly decreases with the increase of fiber volume fraction. This can be attributed to the poor conductivity of T-300 fibers.

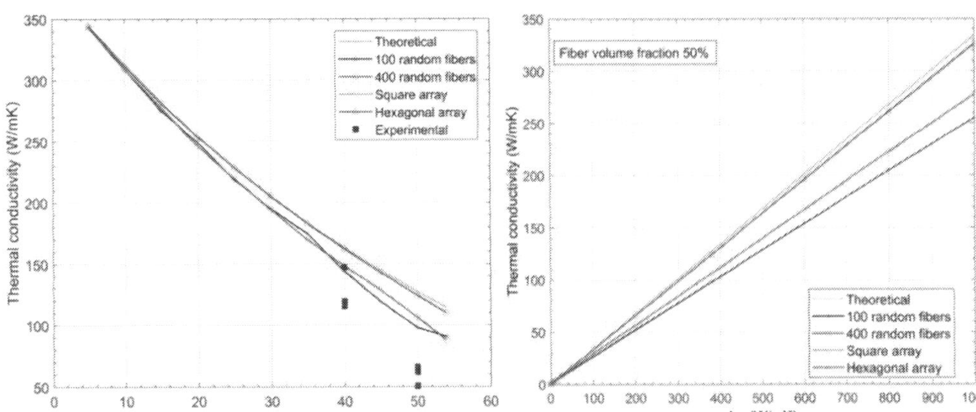

Figure 5. Effects of fiber volume fraction (left) and property contrast (right)

3.3.2 Effect of property contrast

To investigate the effect of property contrast of fiber and matrix materials, it is assumed that the thermal conductivity of fiber keeps 0.8W/mK unchanged while the thermal conductivity of matrix changes from 1000 to 0.1 W/mK. The fiber volume fraction is 50%. One can see from Figure 5 that all samples including the regular samples and the random samples give lower prediction than the theoretical equation(8), and the regular samples can give the closer results to the theoretical predictions, compared to the random samples, which have inhomogeneous fiber distributions. But the discrepancy of different predictions becomes larger as the k_m becomes bigger. In addition, in Figure 5, it is presented that the transverse thermal conductivity of unidirectional composites linearly varies in terms of the k_m.

4 Conclusions

The study investigates the influence of fiber distribution in the matrix by comparing random fiber arrays and regular fiber arrays. The random samples generated by a simple algorithm are used to approximate the real status of fibers in unidirectional composite. The results shows that the transverse thermal conductivity of unidirectional copper-matrix composites with Carbon fibers nonlinearly decreases from about 343 to about 87 W/mK when the fiber volume fraction increases from 5% to 54%, due to the poor thermal conductivity of T-300 fibers, whilst it linearly increases in terms of the matrix's thermal conductivity. More importantly, the obvious discrepancy can be observed between the random samples, the regular samples and the theoretical prediction. The random samples can give closer results of transverse thermal conductivity of unidirectional composites to the experimental results. This can be attributed to the inhomogeneous distributions of fibers. Denser packing of fibers unavoidably causes more complex behaviour of thermal transfer in composites, compared to the regular distributions.

Acknowledgements

The research work is partially supported by the Natural Science Foundation of China (grant no. 11472099) and the Program for Innovative Research Team of Science & Technology of Henan Province of China (Grant no. 19IRTSTHN020).

References

1. P.K. Mallick, *Fiber-Reinforced Composites: Materials, Manufacturing, and Design* (CRC Press, 2007)
2. H. Wang, Q.H. Qin, Appl. Math. Comput. **268**, 311-321 (2015)
3. H. Wang, Q.H. Qin, Y. Xiao, Int. J. Heat Mass Tran. **92**, 228-235 (2016)
4. H. Wang, X.J. Zhao, J.S. Wang, Compos. Sci. Technol. **118**, 117-126 (2015)
5. A.A. Gusev, P.J. Hine, I.M. Ward, Compos. Sci. Technol. **60** (4), 535-541 (2000)
6. B.Q. Kang, P. Yan, C.P. Jiang, Acta Metall. Sin-Engl. **29** (5), 140-145 (2012)
7. H. Wang, Q.H. Qin, Eng. Computation. **28** (8), 1079-1097 (2011)
8. D. Trias, J. Costa, J.A. Mayugo, J.E. Hurtado, Comp. Mater. Sci. **38** (2), 316-324 (2006)
9. T. Kanit, S. Forest, I. Galliet, V. Mounoury, D. Jeulin, Int. J. Solids Struct. **40** (13), 3647-3679 (2003)
10. J. Koráb, P. Štefánik, Š. Kavecký, P. Šebo, G. Korb, Compos. Part A. **33** (4), 577-581 (2002)
11. G.S. Springer, S.W. Tsai, J. Compos. Mater. **1**, 166-173 (1967)

The Research on a Confined Modeling Method of the Effect of Deformable Ground Soil and Crawler Board

LV Wei-wei[1,a], LI Zhong-xin and LOU Peng

School of Mechanical Engineering, Beijing Institute of Technology, Beijing 100081, China

Abstract: There is a great relationship between the passing capacity of vehicles on the ground deformation road and the properties of confined and shear of ground soil, so it is necessary to establish a model that can reflect the soil pressure of the deformation. Physical significance of classic soil pressure experience parameters model is indeterminate, it was generally obtained by a lot of specific soil test. In this paper a new analysis model was built though introducing the maximum stress value of the crawler board and soil surface on basis of a kind of existing analysis model. This model reveals the relations among the confined characteristics of soil, soil parameters and the geometry of confined crawler board. These soil parameters can be obtained through the conventional soil mechanics test without a large number of specific soil test. Through the contrast experiment and predictions results, this model can effectively predict the sinkage of ground soil under the load, and provide a theory basis for the prediction of vehicle mobility.

1 Introduction

The passing property of vehicles on the soft ground is closely related with soil pressure and shear resistance characteristics. In the study of vehicle ground mechanics, the empirical model is usually used to describe the confined characteristics of the soil. The most representative model is the model [1-2] made by M.G.Bekker. The model is based on the test of vehicle and soil parameters. The simplified equation is used to establish the relationship between the external load and the subsidence deformation of the soil, which provides a theoretical basis for vehicle performance testing. However, the physical meaning of the empirical parameters of the model is not clear. It is necessary to determine the parameters through a large number of expensive and time-consuming soil test tests, and the model is limited to a particular type of model. In recent years, many scholars have established a number of new empirical models [3-4], hoping that the model parameters are not dependent on specific soil tests, but the results show that these model parameters are still directly or indirectly related to the parameters of the Bekker model, and in essence the empirical parameters of the model need to be obtained through a large amount of specific pressure plate subsidence soil tests. Kacigin, Ageikind [5] establish an analytical model, including the external load of soil, the geometric size of the plate, the subsidence of soil and the mechanical properties of soil, which reflects the relationship between the amount of soil subsidence, the load of the soil and the stress in the soil, in which the model parameters can be obtained through the conventional soil mechanics test without a large number of specific soil test. It is only the model that considers the external load applied by the plate / wheel as the uniformly distributed load, which results in inaccurate prediction of the model.

[a] Corresponding author : lvweiwei@bit.edu.cn

On the basis of the analysis model, considering the influence of the soil stress on the maximum stress on the contact surface of the Caterpillar board and the soil, a new analysis model is established to establish the relation between the soil subsidence and the size of the load, the parameters of soil characteristics and the size of the plate, and the test method of the model parameters is given, and the test is tested. It is reasonable to predict the subsidence of soil by the model, which provides a theoretical basis for studying the subsidence deformation of deformable ground soil under the action of vehicle.

2 The influence of contact stress between track plate and soil on normal stress

Ageikind et al [6] believes that the normal stress at the z depth of soil is related to the geometric characteristics of the plate and the uniform load of the contact surface, as shown in formula (1) and (2).

$$\sigma_z = \frac{J \cdot p}{1+\left(\frac{z}{A_0+B}\right)^2} \tag{1}$$

$$\sigma_z = \frac{J \cdot p}{1+\left(\frac{z}{A_0+B}\right)^2} \tag{2}$$

σ_z—The internal positive stress of the soil in depth is Z; J—The dimensionless coefficient ,is related to the size of the plate, $J = \frac{0.03+\frac{L}{B}}{0.6+0.43\cdot\frac{L}{B}}$; p—the uniformly distributed load of the soil on the plate, kpa; z—The depth of the soil, m; A_0—The coefficient of quantity, is related to the width of the plate and the depth of the soil; B—The smallest straight diameter or width of the plate, m.

The recent research shows that the stress in the soil is related to the maximum stress of the contact surface, not the average stress of the contact surface. Therefore, the maximum stress of the contact surface is introduced in this paper to amend the analysis model proposed by Ageikind and to establish a new analytical model, such as formula (3).

$$\sigma_z = \frac{J \cdot p_{max}}{1+\left(\frac{z}{A_0+B}\right)^2} \tag{3}$$

p_{max}—The maximum stress on the contact surface of the plate to the soil, kpa.

According to the classical soil mechanics[7], the maximum stress on the contact surface between the plate and the soil is 3 times that of the average stress on the contact surface, and the relationship between them is shown by formula (4).

$$p_{max} = \xi \cdot p \tag{4}$$

ξ is a dimensionless constant, independent of the state of the soil, and only related to the width of the plate, $\xi = 1 + 1.86 \cdot \left(\frac{x}{2} + \frac{x^2}{3} - \frac{x^3}{5} \right)$, $x = 2.5 \cdot (1 - \exp(-3.74 \cdot B))$. By calculating the above formula, we can get the normal normal stress σ_z of the soil in the soil depth of Z under the load.

A circular plate with a diameter of $0.3m$ is placed horizontally in a soil trough with a soil depth of $1m$. The soil trough is filled with dry sand, the uniform load $p = 21.2, 42.5, 63.7, 85.0 and 106.2 kpa$ is applied on the circular plate, and the soil normal stress at the soil's depth of $0.2, 0.35 and 0.5m$ is tested. The results of the test and prediction results of the analysis model (1-2) are shown in Figure 1.

In Figure 1, the soil normal stress at depth of $0.2, 0.35 and 0.5m$ predicted by Beck's empirical model and analytical model are shown under $p = 21.2, 42.5, 63.7, 85.0 and 106.2 kpa$ in the dotted line and dashed line from left to right. The dotted line represents the results of the soil internal normal stress predicted by Beck's empirical model, and the solid line indicates the normal stress in the soil obtained by the new analytical model. The point in the map indicates the soil normal stress measured by the depth of $0.2, 0.35 and 0.5m$ the pressure sensor under the different uniform load test. It's known that the results of the analytical model are in good agreement with the test results from Figure 1, which indicates that the normal stress in the soil is related to the maximum stress value of the caterpillar plate and the soil contact surface, and can be used to modify the soil pressure characteristic model.

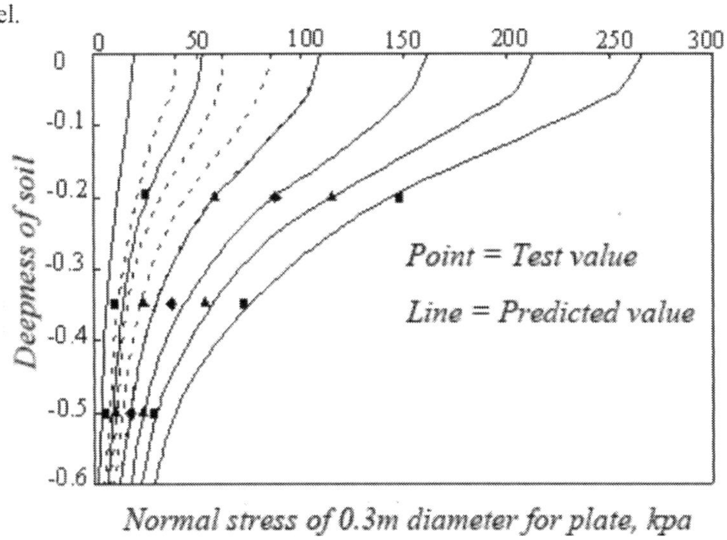

Figure 1. Press plate test

3 Soil pressure characteristic model

According to the dynamic soil theory, the force of the crawler and the soil is shown in Figure 2, and the normal pressure on the vertical crawler and the passive earth pressure σ_p the depth of Z below the a point can be expressed by the formula (5) [8].

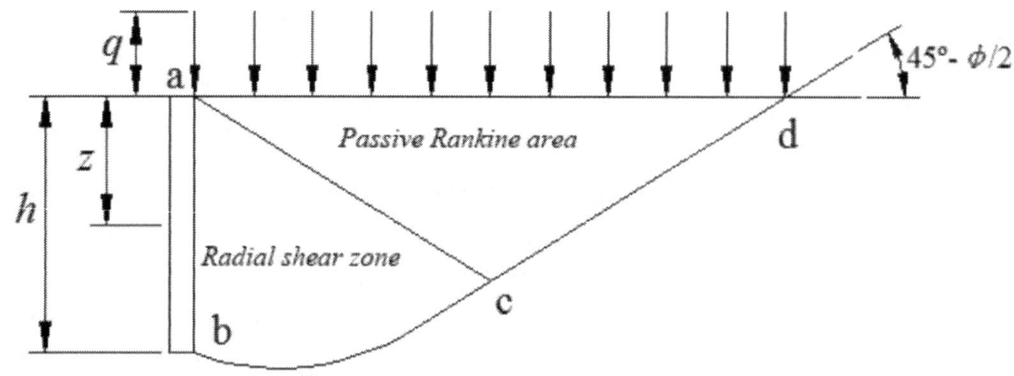

Figure 2. Caterpillar and soil action

$$\sigma_p = \left(N_\gamma \cdot n_\gamma \cdot B + N_p \cdot n_p \cdot z\right)\cdot \frac{\gamma \cdot g}{1000} + N_c \cdot n_c \cdot c \tag{5}$$

Where: $N_\gamma = \exp(6.65 \cdot \tan\phi\ -1.75)$; $N_p = \exp(4.76 \cdot \tan\phi + 0.15)$;

$N_c = \exp(3.32 \cdot \tan\phi + 1.5)$; $n_\gamma = 1 + 0.25 \cdot \frac{B}{L}$; $n_P = 1 + 1.5 \cdot \frac{B}{L}$; $n_c = 1 + 0.3 \cdot \frac{B}{L}$.

σ_p ——The bearing capacity of soil, kpa ; N_γ, N_p, N_c —— The coefficient of soil carrying capacity, dimensionless; γ ——The severe of soil, kg/m^3 ; n_γ, n_p, n_c —— Scale effect coefficient of plate; g —— Gravity acceleration $9.8m^2/s$; c ——Soil cohesion, kpa ; ϕ ——Soil friction angle.

According to the theory of soil mechanics limit equilibrium, the stress state of any point in the soil can be expressed by the Mohr stress circle in Figure 3. The shear strength law of the soil is expressed by the Kulun strength line. The conditions of the Mohr circle tangent to the Kulun strength line reflect the limit equilibrium condition of the soil very vividly.

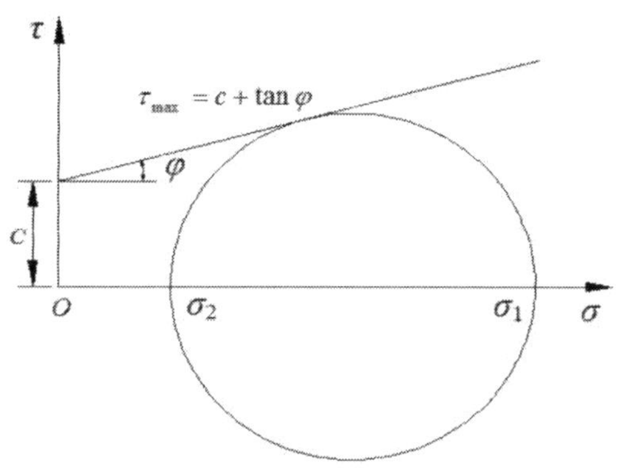

Figure 3. Mole Kulun damage criterion

According to the geometric relationship between the stress circle and the Coulomb strength line in the limit equilibrium state, the relationship between the principal stress of soil material and the maximum shear stress of soil is established.

$$\sigma_1 = \sigma_3 \tan^2\left(45^\circ + \frac{\phi}{2}\right) + 2c\tan\left(45^\circ + \frac{\phi}{2}\right) \tag{6}$$

By the limit equilibrium theory, the vertical stress σ_z on the unit area in the soil depth of z is the same, it is the minimum principal stress σ_3, but the horizontal stress σ_x is gradually increased and it is the maximum principal stress σ_1 when the limit is reached. It is the dynamic soil pressure strength σ_p.

The analysis model of soil pressure characteristics will be established by introducing σ_z and σ_p to the formula (6), as shown in formula (7).

$$p = \frac{\left(N_\gamma \cdot n_\gamma \cdot B + N_p \cdot n_p \cdot z\right) \cdot \frac{\gamma \cdot g}{1000} + N_c \cdot n_c \cdot c \ - 2c \tan\beta}{J \cdot \tan^2 \beta} \left[1 + \left(\frac{z}{A_0 + B}\right)^2\right] \tag{7}$$

Where $\beta = 45^\circ + \frac{\phi}{2}$.

The model includes uniform pressure, soil subsidence, plate size and soil properties parameters(cohesive force, internal friction angle and soil gravity). The parameters of the model can be obtained by conventional soil mechanics experiments without a large number of time-consuming and expensive specific soil experiments.

4 Test verification

The soil parameters are $w = 2.3$, $c = 0$, $\phi = 33^\circ$.It's filled with dry sand in the soil trough, and circular plates witha diameter of $0.3m$ $2.5cm,5cm,10cm,15cm$ and $30cm$ are separatelyput into the soil trough to carry out the pressure plate test. The solid line 1 to 5 in Figure 4 indicates circular platessinkage of the diameter of $0.3m$ $2.5cm,5cm,10cm,15cm$ and $30cm$ predicted by the analytical model respectively, and the point indicates the soil trough test data. It can be seen the consistency of the prediction results of the analytical model and the experimental results is good from Figure 4. The rationality of predicting the subsidence of soil by the model is verified by experiments, which provides a theoretical basis for the research on vehicle mobility.

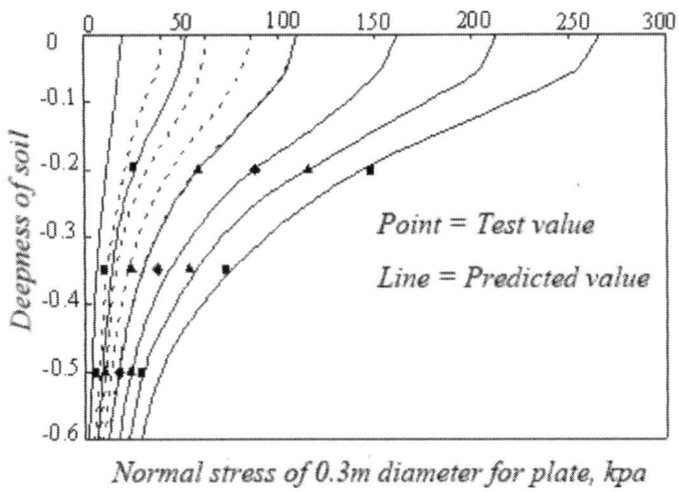

Figure 4. Comparison between experimental results and model prediction results

5 Conclusion

Based on the dynamic soil pressure and limit equilibrium theory, this paper analyzes the interaction relationship between the crawler plate and the ground soil, and establishes an analysis model of the soil pressure characteristics. The model establishes the relationship between the geometric parameters of the crawler plate, the soil subsidence, the external load and the vehicle driving parameters, and These soil parameters can be obtained through the conventional soil mechanics test without a large number of complex and specific soil test. It's discovered that the prediction results of the model are basically the same as those of the test results by comparison with test results, which provides a theoretical basis for predicting the mobility of vehicles.

References

1. Bekker MG. Theory of land locomotion. Ann Arbor (MI): Universityof Michigan Press; 1956.
2. Bekker MG. Off-the-road locomotion. Ann Arbor (MI): University ofMichigan Press; 1960.
3. Guskov VV. Optimum parameters of agricultural tractors. Moscow:Mashinostroenie; 1966 [in Russian].
4. Kacigin VV, Guskov VV. The basis of tractor performance theory. JTerramech 1968;5(3).
5. Ageikin JS. Off-the-road wheeled and combined traction devices:theory and calculations. New Delhi: Amerind Pub. Co.; 1987 [translation from 1972 Russian book].
6. Ageikin JS. Off-the-road mobility of automobiles. New Delhi:Amerind Pub. Co.; 1987 [translation from 1981 Russian book].
7. Tsytovich NA. Soil mechanics. Moscow: Gosstroizdat; 1963 [inRussian].
8. Modest Lyasko,LSA model for sinkage predictions. Journal of Terramechanics, 47 (2010) 1–19.

MATEC Web of Conferences **237**, 02012 (2018) https://doi.org/10.1051/matecconf/201823702012
D2ME 2018

A Numerical Method for Analysing Heat Conduction in Composites Containing Encapsulated Phase Change Materials

Hui Wang[1] and Qing-Hua Qin[2, a]

[1] *College of Civil Engineering & Architecture, Henan University of Technology, Zhengzhou 450001, China*
[2] *Research School of Engineering, Australian National University, Canberra, ACT 2601, Australia*

Abstract. In this study, a three-dimensional transient heat transfer model in a three-phase composite system is established to investigate effects of temperature reduction in a composite system due to the use of encapsulated phase change material (PCM). The entire composite system is composed of cement matrix material, PCM, and hollow metal microspheres (HMSs) which are introduced to accelerate the phase change efficiency of the PCM and to simultaneously hold the liquid phase of the PCM. The present transient heat transfer model is numerically solved via finite element technique for investigating the transient thermal performance of the three-phase composite system. The temperature distribution on the specific area is compared to that in the pure cement material and the two-phase composite system without metal shell for demonstrating the ability of temperature adjustment of the PCM. Finally, effects of the spatial distribution of HMS on the temperature variation in the three-phase composite system is further investigated to provide comprehensive understanding on energy adjustment of this composite system.

1 Introduction

Efficient thermal energy controlling has received considerable attention in recent years in building industries. In doing so, the cement material, a type of most popular building material, is employed to couple with PCMs with high storage capacities for improving its performance of thermal energy storage. PCMs can store massive amount of latent heat at a constant temperature (phase change temperature of the PCM) during melting and solidification processes, and thus is very suitable for reducing energy consumption of buildings [1, 2].

However, PCMs usually have very low thermal conductivities, e.g. around 0.2W/(mK) for octadecane and paraffin waxes and 0.5W/(mK) for inorganic salts. This prolongs the charging and discharging period of latent heat, and causes slow thermal response in practical applications. To overcome this disadvantage of PCMs, different heat transfer enhancement technologies have been developed through adding different high thermal conductivity materials in cement composites, such as metal foams [3, 4], metal powders [5], and nanostructures [6-8], to increase their thermal conductivity. However, most of the existing studies based on the experimental research and the reported outcomes

[a] Corresponding author: qinghua.qin@anu.edu.au

are not consistent due to the limitations of experimental stuffs and approaches. Moreover, the mechanisms behind the changes of thermal properties need to be further explored.

Motivated from the discussion above, this report provides a three-dimensional transient thermal analysis of cement material filled with PCM enclosed inside hollow metal microspheres to examine the feasibility of temperature adjusting using encapsulated PCM in cement materials. In this work, the encapsulation of PCM is used as it can perfectly hold the liquid phase of the PCM and more importantly it has beneficial outcome on mechanical strength of building materials. Additionally, the thermal conductivity of solid metal phase is generally several thousand times higher than that of PCM, hence the usage of hollow metal microspheres can enhance the heat transfer capability of PCM inside them and accelerate the phase change efficiency of PCM. To demonstrate the effect of hollow metal microspheres on heat transfer enhancement in PCM system, the predicted results are compared to those without PCM or without hollow metal microspheres.

2 Problem description

The physical problem to be investigated is shown in Figure 1. The PCMs are encapsulated inside the hollow metal microspheres with outer diameter d=50mm and the thickness of the metal wall t=5mm. They are, then, embedded into a cement material matrix. Five microspheres are arranged in a regular way as displayed in Figure 1. The distance between adjacent microspheres is denoted as D. Specially, when $D=2R=d$, the adjacent microspheres are in contact each other. It is assumed that the composite system consisting of the PCM, metal spheres, and cement is heated from the top surface through a constant heat flux G (Wm^{-2}). The other surfaces are assumed perfectly insulated boundaries, as depicted in Figure 1.

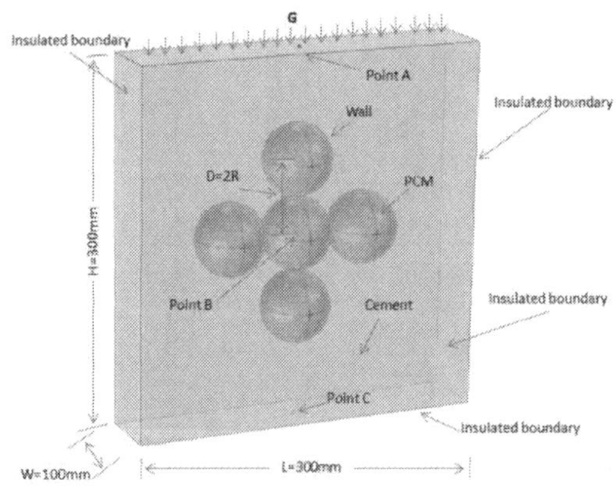

Figure 1: Illustration of cement composite with encapsulated PCM

In our computation, the octadecane is used as the PCM and the pure aluminium is chosen as the solid wall of the microspheres. The thermophysical properties of cement, aluminium and octadecane are tabulated in Table 1 for reference. Additionally, the initial conditions and the applied constant heat flux boundary condition are tabulated in Table 2.

Table 1: Thermophysical properties of octadecane, aluminium wall and cement

Property	Octadecane	Wall	Cement
Density ρ (kg/m^3)	800	2675	1450
Mass heat capacity $c_p(J/kg/k)$	1250	903	840
Latent heat of fusion $L_f(J/kg)$	125000	/	/
Melting temperature $T_m(K)$	303	/	/
Thermal conductivity k (W/m/K)	0.2	211	0.93

Table 2: Initial and boundary conditions for the three-phase composite model

Quantity	Value
Initial temperature $T_0(K)$	293
Incident energy intensity G (Wm^{-2})	750

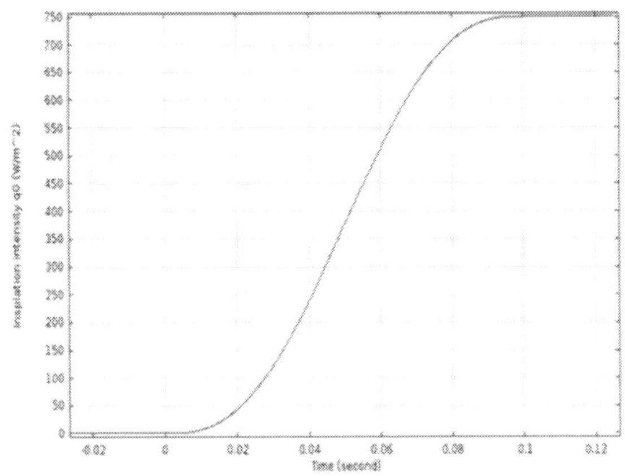

Figure 2: Smoothed incident energy intensity G

It is noted that the incident energy intensity, i.e. $G = 750Wm^{-2}$, is applied on the top surface of the system since $t > 0$. Obviously, it will cause a temperature discontinuity at the starting time. To fix this problem, a smoothed step function which increases the incident energy intensity from 0 to $750Wm^{-2}$ in 0.1s is introduced, as shown in Figure 2.

3 Numerical results and discuss

In this section, the enhanced heat transfer in the present composite system is numerically investigated. Because the heat transfer in such multi-phase system is highly nonlinear, analytical solutions are only valid for some very simple cases [9-15]. Generally, numerical methods such as the finite element method [16-20], the boundary element method [21, 22], and the finite difference method [3, 23] are resorted to provide significant flexibility for achieving high quality of solutions and avoiding expensive experiments. Among them, the temperature-based finite element technique is very suitable for such phase-change problem involving moving melting boundary during the phase-change process and multiple materials considered in the paper. Hence, the finite element simulation is carried out through the commercial software COMSOL in this study. To show the effect of heat transfer enhancement induced by the hollow metal microspheres, the two-phase composite system including the cement material and the pure octadecane microspheres with same dimensions and the pure cement sample (without the PCM and the hollow metal microspheres) are also analyzed respectively for comparison. Additionally, three key points displayed in Figure 1 are chosen to investigate the temperature variation with the time. It is noted that Points A, B and C are at the center of the top surface, the middle PCM microsphere, and the bottom surface, respectively.

In the following computation, the standard linear tetrahedron element is employed to model the pure cement sample, the two-phase composite system and the three-phase composite system, and the extra fine mesh scheme controlled by physical problem is used to guarantee the solution accuracy. The number of elements is 75351, 239647 and 312195 for the three different systems, respectively, as shown in Figure 3.

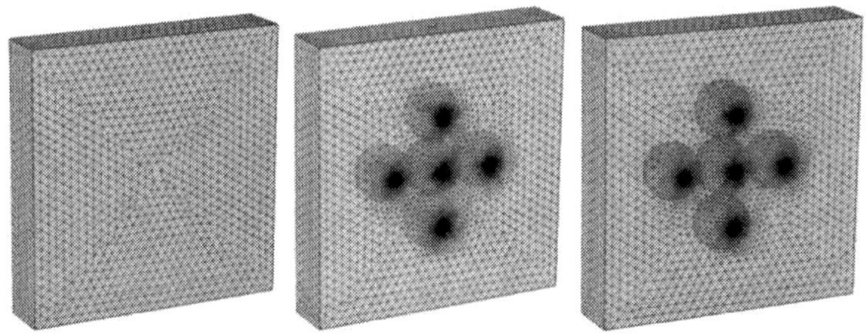

Figure 3. Mesh division for (a) the pure cement sample, (b) the two-phase composite system and (c) the three-phase composite system

3.1 Heat transfer in the pure cement

Firstly, the pure cement sample is considered for investigating the temperature distribution in terms of time. From the numerical results shown in Figure 4, it is found that the temperature at Point A, which locates at the top heating surface, increases more quickly than those at Point B and C, which are far away from the heating surface. Because of the low thermal conductivity of cement material, the provided thermal energy reaches Point B and C at about 5100s and 17000s, respectively, and then causes temperature to increase. Before this, the temperature at Point B and C keeps in the initial temperature.

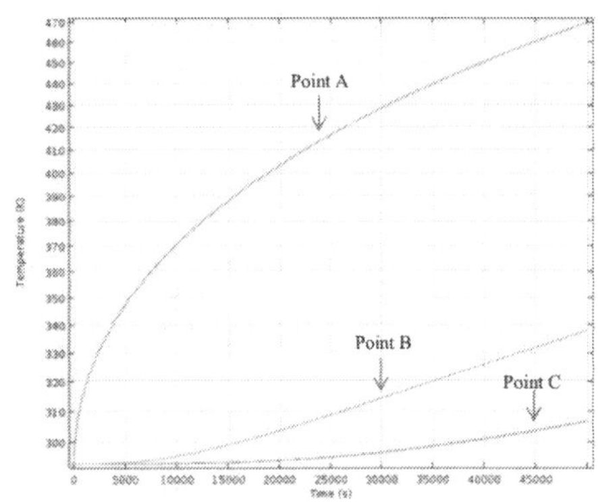

Figure 4. Temperature variation with time for the pure cement sample

3.2 Heat transfer in the two-phase system

Secondly, the heat transfer in the two-phase composite system consisting of the PCM microspheres and the cement material is considered. Results in Figure 5 indicate that the dramatic temperature increase at Point A is closely similar to that in the pure cement sample, whilst the temperature at Point B begins to slowly increase from the initial temperature at about 7200s, and it approximately reaches the melting temperature 303K of the PCM at around t=44040s. This can be attributed to the presence of the PCM microsphere above the middle one. After this, the temperature at Point B keeps almost horizontal until 46440s. This means that the provided heat is used for overcoming the latent heat of fusion of the PCM around Point B to make it change its phase from the solid state to the liquid state and doesn't bring the temperature rise. After the PCM ball around Point B fully melts, the temperature starts to rise again due to the heat transfer in the fluid PCM. Compared to that in Figure 4 for the pure cement sample, however, the temperature variation at Point B in the PCM/CEMENT system is apparently slower. This can be attributed to the lower thermal conductivity of the PCM than that of the cement. Also, it is seen from Figure 5 that the rise of temperature at Point C occurs later, compared to the pure cement sample, as was expected. This can be attributed to the phase change of PCM balls. To observe the temperature variation in the system clearly, Figures 6(a), (b), (c) and (d) in which the green interface represents 303K, which is also the melting temperature of PCM, display the variation of temperature at different time instances. It is observed that the thermal energy transfers into all PCM balls very slowly, due to the lower thermal conductivity of the PCM than that of the cement. Besides, due to the presence of PCM, the isotherm becomes curved at the region near the PCM balls. The melting front surface of the PCM moves downward as the time increases and all PCM balls fully becomes liquid state at about 77640s, as shown in Figure 6(d) that the minimum temperature in the whole region is 303.403K.

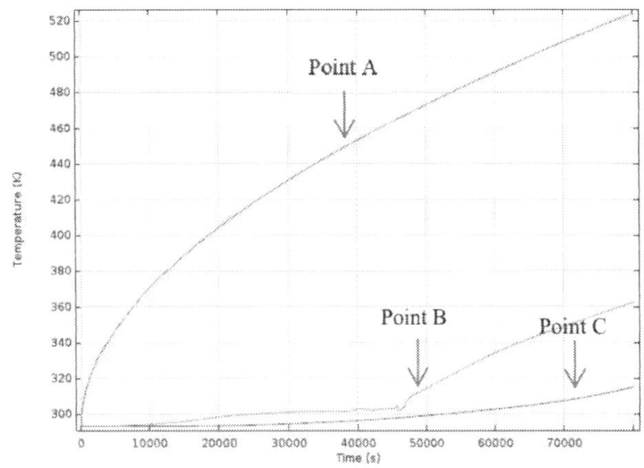

Figure 5: Temperature variation with time for the PCM/cement system

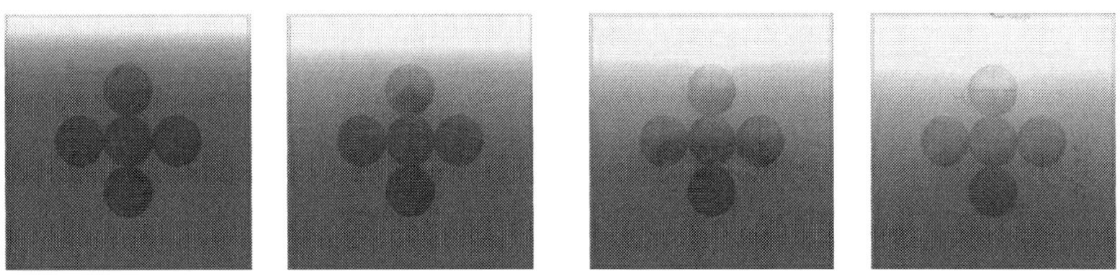

(a) 6600s (b)22200s (c)44040s (d)77640s

Figure 6: Temperature variation of the PCM/cement system at different time instances

3.3 Heat transfer in the three-phase system

Thirdly, the three-phase composite system is taken into consideration. Interestingly, as shown in figure 7, because the thermal conductivity of the aluminium wall is more than one thousand times higher than that of the PCM, the heat energy can rapidly transfer into the PCM and causes the temperature rise at point B quickly. At about 16980s, the temperature reaches the melting temperature and after this instance the temperature keeps unchanged due to the occurrence of phase change. Figures 8(a), (b), (c) and (d) clearly show the temperature variation with the time in the entire domain. It is clearly seen that all PCM balls encapsulated with aluminium spherical shell melt more rapidly than the pure PCM balls in the two-phase composite system, and all PCM balls fully becomes liquid state at about 43500s.

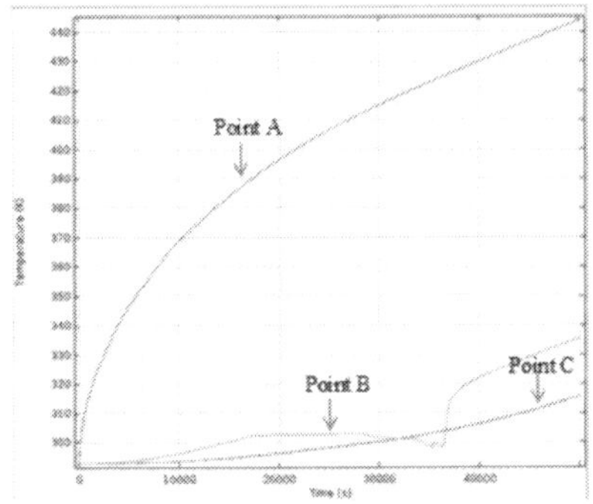

Figure 7: Temperature variation with time for the three-phase system

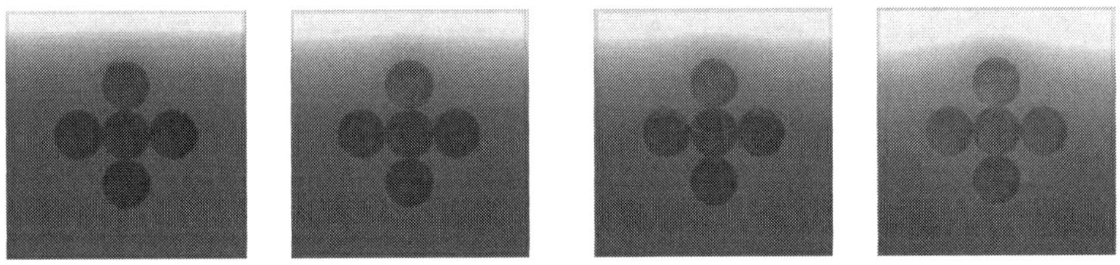

(a) 8100s (b)14100s (c)16980s (d)43500s

Figure 8: temperature variation of the three-phase system at different time instances

3.4 Effect of the distribution of encapsulated PCMs

From the computation above, it is found that the presence of hollow metal microsphere can improve the heat transfer efficiency in the PCM and leads to rapid phase change from the solid state to the liquid state. However, the adjacent PCM microspheres always keep in contact with each other in the above simulation. Actually, they can disperse in the cement material in any form. In order to investigate the effect of distribution of the five microspheres, a different three-phase composite model is considered, in which the four balls around the middle ball move outward by setting D=150mm.

In Figure 9, the temperature platform at Point B appears at about 19680s which is evidently later than that in Figure 7. This is reasonable that the heating energy can't rapidly reach the middle metal microsphere embedded with the PCM because of the block of cement material between adjacent PCM microspheres. While for the model in Figure 7, the contact of adjacent metal microspheres makes the heating energy reach the middle metal microsphere more rapidly. This phenomenon can also be observed in Figure 10, which displays the temperature variation at different time instances. It is found that the large distance between neighboring encapsulated PCM microspheres alleviates the speed of thermal energy flowing into the PCM.

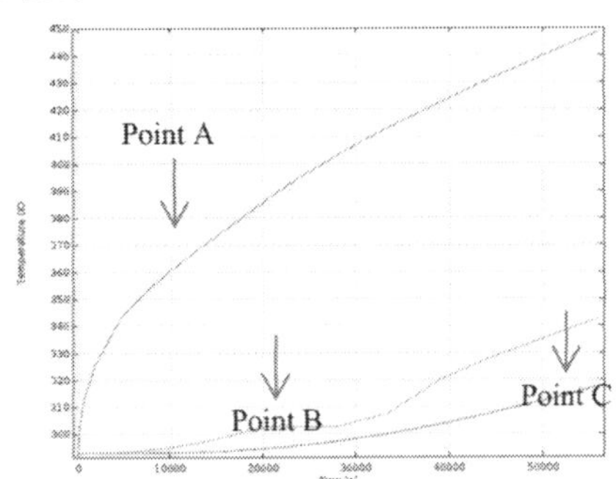

Figure 9: Temperature variation with time for the discrete three-phase system

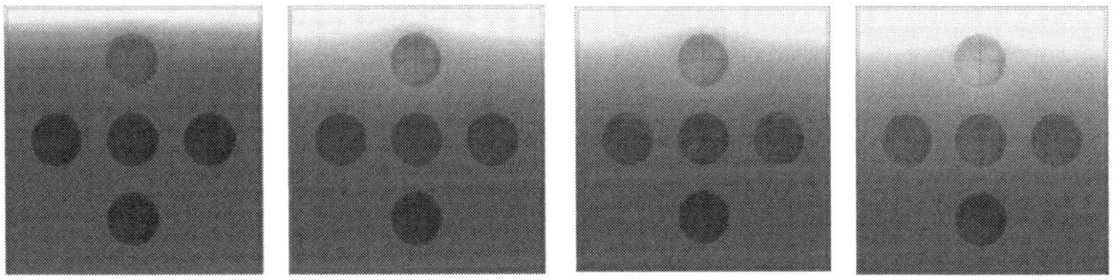

(a)8100s (b)15000s (c)19680s (d)42000s

Figure 10: Temperature variation of the discrete three-phase at different time instances

4 Conclusions

In this study, the influence of heat transfer enhancement is numerically studied in three-phase composite system, in which the PCM are enclosed into the hollow metal microsphere and then are embedded into the cement matrix. Compared to the results from the pure cement sample, a two-phase system without the hollow metal microspheres, the effects of metal wall and distance between neighbouring PCM microspheres on solid/liquid phase change heat transfer are very significant. The introduction of hollow metal microsphere shortens the melting time of the whole system, as well as enhances the melting rate of PCM tremendously. Besides, the large distance between neighboring encapsulated PCM microspheres alleviates the speed of thermal energy flowing into the PCM system.

References

1. Zalba, B., J. M. Marin, L. F. Cabeza and H. Mehling, Applied Thermal Engineering, **23**(3): 251-283.(2003)
2. Zhou, D., C. Y. Zhao and Y. Tian, Applied Energy, **92**: 593-605.(2012)
3. Zhao, C. Y., W. Lu and Y. Tian, Solar Energy, **84**(8): 1402-1412.(2010)
4. Tian, Y. and C. Y. Zhao, Energy, **36**(9): 5539-5546.(2011)
5. Mettawee, E. S. and G. M. R. Assassa, Solar Energy, **81**(7): 839-845.(2007)
6. Xia, L., P. Zhang and R. Z. Wang, Carbon, **48**(9): 2538-2548.(2010)
7. Ji, P., H. Sun, Y. Zhong and W. Feng, Chemical Engineering Science, **81**: 140-145.(2012)
8. Şahan, N., M. Fois and H. O. Paksoy, Solar Energy Materials and Solar Cells, **137**(137): 61-67.(2015)
9. Yu, S. and Q. H. Qin, Theoretical and Applied Fracture Mechanics, **25**(3): 279-288.(1996)
10. Qin, Q. H. and S. W. Yu, International Journal of Solids and Structures, **34**(5): 581-590.(1997)
11. Qin, Q. H., Y. W. Mai and S.-W. Yu, International Journal of Fracture, **91**(4): 359-371.(1998)
12. Qin, Q. H. and Y. W. Mai, International Journal of Solids and Structures, **36**(16): 2463-2479.(1999)
13. Qin, Q. H., Y. W. Mai and S. W. Yu, International journal of solids and structures, **36**(3): 427-439.(1999)
14. Qin, Q. H., International Journal of Solids and Structures, **37**(39): 5561-5578.(2000)
15. Qin, Q. H. and J. Q. Ye, International Journal of Solids and Structures, **41**(9-10): 2447-2460.(2004)
16. Dhanasekar, M., J. Han and Q. H. Qin, Finite Elements in Analysis and Design, **42**(14-15): 1314-1323.(2006)
17. Qin, Q. H. and H. Wang (2008). Matlab and C programming for Trefftz finite element methods, CRC Press.
18. Wang, H. and Q. H. Qin, European Journal of Mechanics-A/Solids, **29**(5): 801-809.(2010)
19. Wang, H. and Q. H. Qin, Archives of Mechanics, **62**(6): 493-510.(2010)
20. Wang, H. and Q. H. Qin, Computational Mechanics, **48**(5): 515-528.(2011)
21. Qin, Q. H., International journal of solids and structures, **30**(22): 3101-3111.(1993)
22. Qin, Q. H. and Y. W. Mai, Engineering Fracture Mechanics, **69**(5): 577-588.(2002)
23. Takeuchi, Y., Y. Yoshimoto and R. Suda, Journal of Computational and Applied Mathematics, **320**: 101-119.(2017)

Effect on the wettability, hardness and shear strength properties of 3%-nano Titanium Oxide (TiO_2) added Sn-3.8Ag-0.7Cu (SAC)/Copper (Cu) solder joint

Amares Singh[1, a] and Rajkumar Durairaj[2]

[1]*Center of Mechanical and Materials Engineering, Faculty of Engineering and Built Environment, SEGi University No. 9, Jalan Teknologi, Taman Sains Selangor, Kota Damansara PJU 5, 47810 Petaling Jaya, Selangor, Malaysia*
[2]*Lee Kong Chian Faculty of Engineering and Science, Universiti Tunku Abdul Rahman, Jalan Sungai Long, Bandar Sungai Long, 43000 Kajang, Selangor, Malaysia.*

Abstract. Demand towards smaller electronic devices prompts the electronic interconnection to have less density and minimal weight percentage and such of this concern tips to applications of nanoparticles incorporation to the solder alloy. In this work, the effect of titanium dioxide (TiO_2) nanoparticles (<100nm) additions on the performance of wettability micro hardness, and shear strength of the Sn-3.8Ag-0.7Cu (SAC) solder alloy soldered on the Copper (Cu) substrate was investigated. The SAC added 3% TiO_2 nanoparticles had a slight decrease in the micro hardness compared to the plain SAC in this study but the value still surpasses many other common solder alloy's hardness value. The wettability test was conducted by taking account the contact angle achieved by soldering the SAC added 3% TiO_2 solder alloy to the Cu substrate and contact angle produced was less than 40°. As for the shear strength, there was a clear increment of 13% in the shear strength compared to the plain SAC solder alloy. Quite observant, the additions of TiO_2 nanoparticles displayed significant influence on these properties that contributes to excellent performance. Much detailed discussion are elaborated in the content below aiding with results and theory behind the result achieved.

1. Introduction and Literature

Lead free solders are the current affection in the electronic industry as many governed body have banned the usage of Plumbum (Pb). Solder alloys are known as the medium that holds the integrity of an interconnection in ana electronic deivce, thus, its properties mainly mechanical strength are much debated. Various lead free solder alloys such as SnCu [1], SnAg [2], SnZn [3], SnBi [4], SnAgZn [5], SnAgCu [6] are researched and the outcome based on the results are profound. Nevertheless, some major drawbacks are also being listed upon testing these solder alloys, namely the poor wetting (SnZn) [7], thick intermetallic (IMC) layers (SnCu, SnAg) [8], low shear strength (SnAgZn, SnAg) [9], prone to corrosion (SnZnAg, SnZn) [10] and many more.

Adding to that, the shrinking in sizes of the electronic chips and devices demands less mass and density consumption of a solder alloy. This then creates a concern as only by additions of elements, the mechanical strength of any solder alloys can be enhanced, but at the same time, it will be consuming quite a sum of mass. Such trend prompts the elements to be added to be in smaller

[a] Corresponding author: amaressinghgill@segi.edu.my

particles dimensions to enhance the mechanical strength and also avoid increase in mass consumption [11]. Nanoparticles additions are the solution for this issue and its effect to the prior solder alloys should be beneficial and not deteriorating the initial properties, which did occur in some studies. Nanoparticles additions can be divide in to two types, i) metal nanoparticles (Ag, Cu, Ni, In and etc) and ii) oxide/ceramic nanoparticles (TiO_2, SiC, $SrTiO_3$ and etc.), with all these depending on the outcome of the applications [12]. These nanoparticles can influence by effecting the solubility of the solder alloy and taking part in the diffusion process or contrast to that, being a discrete particles and influence in increasing the mechanical strength. It is crucial to provide and understand the influences of these nanoparticles to a solder alloy to allow application in the electronic industry.

The SnAgCu solder is widely debated by its enhancement in the advantage of having a low melting temperature (217°C) [13], near to the soldering temperature (230-250°C) [14], and possessing a comparable mechanical strength, yet some concerns arises on the thick intermetallic layer produced by this solder over sometime that degrades its initial high strength [15]. Sn and Cu elements in this solder alloy have high affinity among them and with the availability of Cu in the substrate creates an opportunity for the Sn to react and produce thick IMC layer (Cu_6Sn_5) between the solder and substrate [16]. It then worsen as this layer reacts with Cu and creates another layer between the Cu_6Sn_5 and the substrate, the Cu_3Sn layer [17]. A schematic diagram in Figure 1 shows the prior and after reflow reaction of the above explanation. This paper involves the additions of nanoparticles in the SnAgCu solder system and investigate the influence of the nanoparticles to the solder alloy.

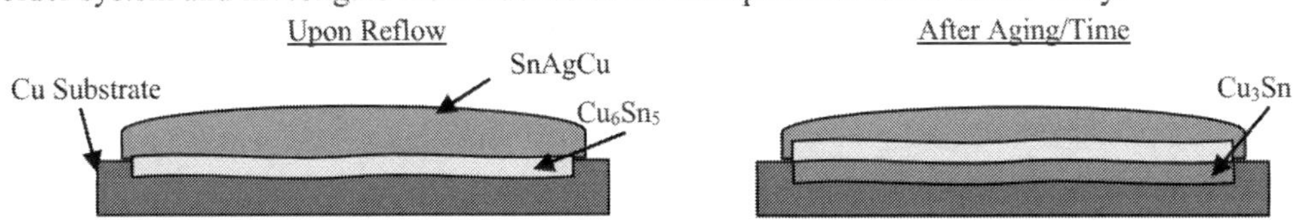

Figure 1. Intermetallic layer formation between solder and substrate (upon reflow and after aging).

2. Materials and Experimental Procedure

The preparation of bare Sn-3.8Ag-0.7Cu (SAC) solder alloys were based on the weight percentage of 95.5% Tin (Sn) Alfa Aesar , 3.8% Argentum (Ag) and 0.7% Copper (Cu) and all the materials are readily available and purchased from Alfa Aesar Inc. with 99.9% purity. An overall weight of 20g of SAC was used and this small amount of weight composition is selected to replicate the usage of small consumption of solder paste used in the electronic industries. These elements were then melted using an induction furnace for $1^{1/2}$ hour at a constant temperature of 1100°C to ensure a homogeneous mixture of all elements. The TiO_2 nanoparticles (size < 50nm, Alfa Aesar Inc) of 3% addition from the overall weight of the solder alloy (20g) gives a total mass of the nanoparticles of 0.6 gram and the nanoparticles were was mechanical mix in the SAC solder alloy and melted in the furnace at 1100°C for hour to allow the mixing of the nanoparticles in the solder alloy. To ensure all nanoparticles are properly mixed, the SAC added with the TiO_2 (SAC + 3% TiO_2) was re-melted using a hot plate with stirring for 30 minutes. The SAC + 3% TiO_2 solder alloy were then compacted into billets form, (5mm × 1mm) for experiment purposes. The billets were polished and etched in Nital for 60 seconds to avoid debris before tests and produce clean surface finish. Wettability test was done by measuring the contact angle (calculated using the VIS Pro software) between the SAC + 3% TiO_2 and Cu substrate (10×10mm), upon soldering at 250°C for 30s. The hardness of the SAC + 3% TiO_2 solder alloy was based on the Vickers microhardness test with 1kgf load dented to the solder billet. Shear test meanwhile, was tested using the Universal Tensile Machine (Instron) with application of single shear lap method with the SAC solder billets soldered on the Cu substrate (30mm × 10mm × 0.5mm) at 250°C. Cross section sample was analysed using the Scanning Electron Calorimetry (SEM) to investigate the interfacial property between the solder and substrate.

3. Results and Discussion

3.1 Wettability of SAC + 3% TiO_2/Cu.

The wetting characteristics of the SAC + 3% TiO_2 solder alloy was studied using the contact angle parameter. Wetting behavior could also be studied by applying the surface tension equation (1) that shall give the wetting force [18] to evaluate the wetting characteristic but this aspect is widely discussed in literatures and was not included in this study.

$$\gamma_{sv} = \gamma_{ls} + \gamma_{lv} \cos\theta \tag{1}$$

γ_{sv} = *surface tension of solid,* γ_{ls} = *interfacial energy,*
γ_{lv} = *surface tension of liquid,* θ = *contact angle*

It is defined that a smaller contact angle produces better joint and strength [9, 19] and researches recommends contact angles less than 50° is advantageous [14]. The results of the contact angle for the SAC + 3% TiO_2 solder alloy was an average of 32.7° and is considered beneficial taking account the proposed sufficient contact angle of below 50° by [14]. Being compared with other literatures, Sn3.8Ag0.7Cu (24.81°), Sn3.5Ag0.5Cu (33°), Sn6Zn4Bi (35°), Sn40Pb (8), Sn3.5Ag (38), Sn58Bi(35°), Sn2.5Ag0.7Cu0.1(Ce,La) + 0.1%Ni (35°), Sn3.5Ag4Ti (Ce,Ga) + Al_2O_3 (51°), Sn0.7Cu0.2Ni + 0.3%In (59°) and Sn0.7Cu + 0.2%S (35°) [7, 14, 18, 20-24], this solder tops in providing lower contact angle compared to other plain and nanoparticles added solder alloys. Yet, comparing with the plain SAC and the traditional SnPb solder alloy, the contact angle of latter was lower compared to SAC + 3% TiO_2 in this study. Point has to be taken in to account that these bare solder alloys (SAC and SnPb) are not added with nanoparticles. Additions of nanoparticles will slightly increase the viscosity of the solder alloy that increases the contact angle. It is a typical characteristics of the nanoparticles to move to the leading edge of the molten solder during soldering, as they do not take part in the diffusion process [25], and at an instant of 250°C of soldering temperature, the heat would have no influence to melt the TiO_2 (T_M = 1843°C). This was similarly reported by [26]. This means that the nanoparticles behaves as discrete particles and moves to the site which has higher energy [20, 21], at the edge of the molten solder alloy during soldering. Though, a proper mixing would allow proper separation of nanoparticles evenly throughout the molten solder rather than occurrence of agglomeration that will cause catastrophic effect on wettability because the agglomerated nanoparticles will pin the leading edge and contributes to an ineffective wetting (large contact angle). This was the comment made by [22] upon addition of high amount of Al_2O_3 (> 3%) nanoparticles to the Sn3.5Ag that later decreased the joint strength as a cause of large contact angle (crack initiator). This aspect was considered in this research and a minimal of 3% additions was chosen to be added in SAC. This was also aided with proper mixing of the nanoparticles and contributes to a lesser increase (about 8%) in the contact angle compared with the plain SAC. As aforementioned, producing contact angle below 50° reflects better joint between solder and substrate and was the outcome for this SAC + 3% TiO_2 solder joint.

3.2 Microhardness and Shear Strength of the SAC + 3% TiO_2 solder.

The Vickers hardness of the SAC + 3% TiO_2 solder alloy produced an average of 25.9Hv which is considered high compared to other solder alloys in recent available literatures, Sn58Bi (27.5Hv) [27], Sn0.7Cu0.05Ni (9.82Hv) [28], Sn0.3Ag0.7Cu (8.7Hv) [29], Sn-9Zn-In (15Hv) [30], Sn0.3Ag0.7Cu + 1% CNT (9.05Hv) [29], Sn3.5Ag (14.8Hv), Sn3.5Ag0.7Cu (15.8Hv), and Sn3.5Ag0.7Cu + 0.5TiO_2 (17.6Hv) [31]. There was a clear increase in the hardness values of the SAC + 3% TiO_2 solder alloy. Firstly, the increase is due to the nanoparticles in the solder alloy acting as site for heterogeneous nucleation because they do not take part in the diffusion process [25, 29]. This will increases the eutectic area of the solder alloy, and eutectic areas are known to have higher surface are per unit volume, contributing to the increase in load to penetrate through this phases [18]. According to energy absorption theory [32], the molten solder alloy of SAC being exposed with more nanoparticles (TiO_2) will tend to extract the nanoparticles to the grain boundary (high energy site) and this will decrease the growth kinetic of atoms in the molten solder alloy, and as consequent, reduces the grains sizes [10]. Smaller grains are known to have higher surface area to withstand stress exposed. Similar statements were provided by [24]. The TiO_2 nanoparticles at this instant likewise behaves as a solid solution that adds up as another barrier for the stress to overcome, thus increasing the stress to deform the structure and increases the hardness [33]. The concern here is that to limit the additions of the nanoparticles (3% of TiO_2), where an over additions might have no positive grain refinement effect in the molten solder alloy because a content of high amount of these nanoparticles will attract to each other as a cause of Van Der Waals force between them [9].

An interfacial between solder and substrate is an important factor to provide integrity between the solder and substrate [1, 15]. This provides the basis of obtaining the shear strength of a solder joint. The shear strength of the SAC + 3% TiO_2 solder joint was recorded to be 51.5MPa compared with other recent results of shear strength of other common solder alloys, SnPb (49MPa), Sn3.5Ag (25), Sn3.0Ag0.5Cu (43MPa), Sn3.8Ag0.7Cu (44.9MPa), Sn0.7Cu (27.9MPa), Sn0.7Cu + 1% Si_3N_4 (28.1MPa) and Sn3.0Ag0.5Cu + 3%ZrO_2 (44MPa) [14, 34-36]. The increase of the shear strength is directly attributed by the intermetallic (IMC) layer produced between the solder and substrate. This is where the elements from the solder and substrate comes into contact and diffusion takes place to produce the IMC layers [35]. A typical IMC layer of Cu_6Sn_5 was noticed in between the solder and substrate as observed from Figure 2. Normally, with presence of adequate Sn and Cu, this layer would grow thicker and react with Cu and produce the Cu_3Sn layer as explained in Section 1. However, there was absence of this Cu_3Sn layer in this research pointing out that further diffusion process was blocked by the TiO_2 nanoparticles. The characteristics of the TiO_2 nanoparticles to segregate along the grain boundary will block the diffusion process between solder and substrate and vice versa [37], prohibiting thick Cu_6Sn_5 IMC layer and production of Cu_3Sn layer. Aiding with this, thin IMC layer, 2.32μm in this research enhances the shear strength of the joint. This is another factor contributing to a lower contact angle produced in this research as discussed earlier. Furthermore, the presence of TiO_2 nanoparticles along the grain boundary contributes to Orowan strengthening mechanism, whereby the stress induced during loading has to break through or bypass the nanoparticles at the end of dislocation, and this phenomenon increases the stress [10]. Table 1 shows points G (solder side), H (solder and interfacial) and I (interfacial) that were taken to analyse the element existence at these areas, and it was quite obvious the TiO_2 nanoparticles preferentially accommodated at points H and I. This disables the Cu to pass through beyond point H to react with Sn and produce thicker IMC layer by rapid diffusion and verifying the earlier justification.

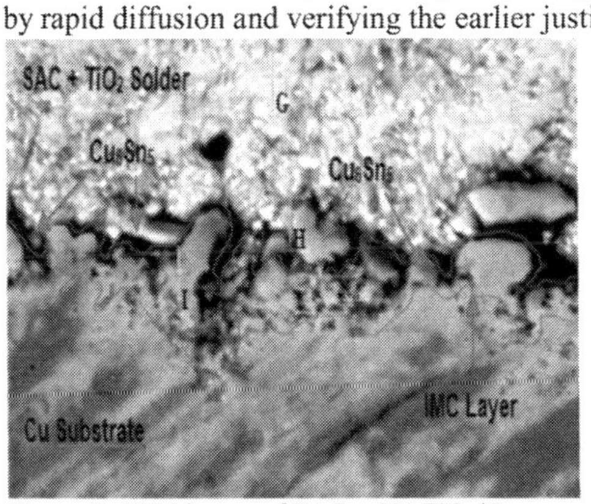

Figure 2. SEM image of SAC + 3% TiO_2/Cu joint.

Table 1. Element of SAC + 3% TiO_2/Cu joint.

Points	Elements (wt. %)					
	Sn	Ag	Cu	Ti	O	C
G	39.9	44.6	-	0.1	11.9	3.32
H	30.1	0.07	17.2	0.1	9.96	42.3
I	37.5	-	34.2	-	-	28.2

4. Conclusion

In this research, the outcomes were obvious to focuses on the interconnection (mechanical) of the nanoparticles added solder alloy. Key aspects of wettability, hardness and shear strength were discussed to provide the relevancy among these important parameters. As to improve the properties of the SAC solder alloy, this research introduced the 3% TiO_2 nanoparticles to the solder alloy. The additions of the nanoparticles provides low contact angle that is desirable to be applied in the electronic industry. The surface tension was not discussed in this paper as the information are widely available yet the information (1) was provided to serve as literature aspect. Microhardness result (25.9Hv) shows an increment as the TiO_2 possesses a high resistance for deformation compared to many other researched solder alloys. The presence of TiO_2 as discrete particles at the interface between solder and the substrate was vital as they block a vast diffusion process to maintain a thin IMC layer and added with these TiO_2 nanoparticles' presence, the shearing strength was increased quite noticeably (51.5MPa). Literally, the presence of thee nanoparticles are they key of such beneficial properties enhancement, yet preparation of the nanoparticles additions should be also given key attention, as improper mixing could result in deterioration rather than improvement.

References

1. M.A.A. Mohd Salleh et al. Materials Science & Engineering A **556** (2012)
2. L. Zhang et al. Microelectronics Reliability **52** (2012)
3. Chen Wenxue et al. Rare Metal Materials and Engineering **39** (2010)
4. L. Shen et al. Materials Science & Engineering A **558** (2012)
5. L. Zhang and K.N. Tu. Materials Science and Engineering R **82** (2014)
6. S.Y. Chang et al. Materials and Design **32** (2011)
7. Ervina Efzan M.N. and Tan S.Y. Applied Mechanics and Materials. **315** (2013)
8. Y. Yao et al. Journal of Alloys and Compounds **682** (2016)
9. Ervina Efzan Mhd Noor and Amares Singh. Soldering & Surface Mount Technology **26** (2014)
10. G. Ren et al. Journal of Alloys and Compounds **665** (2016)
11. Y. Tang et al. Materials and Design **55** (2014)
12. Anattiya Wattanakornphaiboon et al. *Materials Today: Proceedings* **5** (2018)
13. G. Chen et al. Journal of Alloys and Compounds **685** (2016)
14. Amares Singh et al. Journal of Mechanical Engineering **5** (2018)
15. Y. Yang et al. Microelectronics Reliability **53** (2013)
16. Santhosh Kumar et al. JPEDAV **32** (2011)
17. R. Tian et al. Materials Science & Engineering A **709** (2018)
18. Y. Wang et al. Materials and Design **119** (2017)
19. Satyanarayan and K.N. Prabhu. Advances in Colloid and Interface Science **166** (2011)
20. L.C. Tsao et al. Materials and Design **31** (2010)
21. S Amares et al. IOP Conf. Ser.: Mater. Sci. Eng. **205** (2017)
22. R. Kolenak et al. Materials and Design **32** (2011)
23. L.F. Li et al. Materials and Design **64** (2014)
24. H. Huang et al. Microelectronics Reliability **74** (2017)
25. S. Chellvarajoo et al. Materials & Design **82** (2015)
26. S. Chellvarajoo et al. Materials and Design **67** (2015)
27. L. Yang et al. Materials Science and Engineering A **667** (2016)
28. N Saleh et al. IOP Conf. Ser.: Mater. Sci. Eng **238** (2017)
29. Zhu et al. Materials Science and Engineering A **727** (2018)
30. Wang et al. *IEEE International Conference on Nano/Micro Engineered and Molecular Systems* (2013)
31. S.Y. Chang et al. Materials and Design. **32** (2011)
32. K. Mehrabi et al. Journal of Alloys and Compounds **688** (2016)
33. W.-q. Xing et al. Materials Science & Engineering A **678** (2016)
34. S.-j. Jeon et al. Microelectronic Engineering **91** (2012)
35. Choudhury, L. Ladani. Journal of Alloys and Compounds **680** (2016)
36. X. Hu et al. Materials Science & Engineering A **673** (2016)
37. Lu et al. Journal of Alloys and Compounds **765** (2018)

Influence of Process Parameters of Printing on Mechanical Properties of Plastic Parts Produced by FDM 3D Printing Technology

Petr Vosynek[1,a], Tomas Navrat[1], Adela Krejbychova[1] and David Palousek
[1]*Brno University of Technology, Technicka 2, 61669 Brno, Czech Republic*

Abstract. Fused Deposition Modelling (FDM) is a fast-growing 3D printing technology. This technology expands rapidly even in households. Most users set print parameters only according to their own experience, regardless of the final mechanical properties. In order to predict the mechanical behaviour of the FDM-printed components, it is important to understand not only the properties of the printing material but also the effect of the printing process parameters on the mechanical properties. Components manufactured by FDM technology have an anisotropic structure, therefore the filling angle, fill shape, air gap, print orientation, and print temperature affect the resulting mechanical properties. This work deals with the change of mechanical properties depending on the setting of the filling angle, the shape of the filling, the orientation of the parts during printing, the influence of the material and pigment manufacturer.

1 Introduction

In this work, 3D Fusion Deposition Modelling (FDM) technology is used. The essence of this technology is to melt plastic strings through a hot-dip nozzle on a building mat. The height of the applied layer is not the same as the width of the pushed fibre behind the nozzle but is due to the thickness of the nozzle, which is usually 0.4 mm. In the case of more complex shapes of the resulting object, support structures are used. The building and support materials are transported to the print head by system of tubes. The print head comprises a melting chamber where the material is heated and pushed through the nozzle onto the building pad. A drive is used to push the material into the print head so that the material pushes the rotating gear wheels towards the nozzle. The material immediately solidifies on the substrate and, once one layer is applied, the substrate is reduced by the thickness of the layer. Upon completion of printing, the resulting product is broken off from the pad, the supports may be removed mechanically or by a chemical or ultrasonic bath. Among the basic materials used for 3D FDM printing are ABS (Acrylonitrile Butadienstyrene), PLA (Polyactic Acid), PA (Polyamide), PC (Polycarbonate), PET-G (Polyethylene Terephthalate - Glycol) and others, [1].

For our purposes, FLASHFORGE Dreamer was used for printing. This printer has two printheads, so it can print with two printing materials. Printing takes place in a closed chamber that isolates the internal temperature and reduces external temperature influences. The manufacturer promises an accuracy of 0.0025 mm on the Z axis and 0.01 mm on the XY plane. The maximum print size of this printer is 231 x 150 x 140 mm, [2].

[a] Corresponding author: vosynek@fme.vutbr.cz

The GCODE file format is used to print the model. The code is generated in FlashPrint, which determines also the print parameters. Several parameters that affect mechanical properties needs to be considered when printing, specifically temperature, print speed, part orientation, infill density, direction, width and raster, overlay size. The parameters are mostly set only based on user experience regardless of the mechanical properties of the printed part. Unlike the injected parts, the FDM-printed parts are anisotropic. If the strength of a component is required, it must be printed in such a way that the mechanical forces are distributed longitudinally along the length of the fibres, [3].

The influence of the direction of the grid and the influence of size of the overlap on the ultimate tensile strength was experimentally investigated in [4] by S. Ahn. The result of this research is that the optimal parameter for joining fibres regarding to mechanical properties is the negative value of the air gap. This gap primarily affects the resulting print strength. Study [4] shows that FDM-printed samples achieved lower tensile strength than samples injected into the mould. This is due to non-homogeneity in the connection between the fibres. This inhomogeneity greatly reduces the effective cross section of the samples. The smallest tensile strength had samples printed with a 90° infill angle. This is due to the fact, that the force is transmitted only by the connections between the individual fibres, not the fibres themselves. Samples for this experiment were made from ABS P400.

The influence of the orientation of the part during printing and the filling angle on the ultimate tensile strength and Young's modulus is described in [5] (I. Durgun). During the experiment he worked with ABSplus-P430. Samples were printed with five different fill angles (0°, 30°, 45°, 60° and 90°) with three different orientations (horizontally, perpendicularly and vertically). The worst strength results were obtained when the orientation of printed fibres was perpendicular to the pad. The highest tensile strength was achieved by a sample printed at a vertical position with a fill angle of 0°, where the force acts longitudinally along the length of the fibres.

A comprehensive experiment on the mechanical properties of the samples was performed by F. Rayegani [6]. When printing samples, it combined the different orientation of the component (0° and 90°), the angle (0° and 45°) and the fill width (0.2032 mm and 0.5588 mm) and the air gap (-0.00254 mm and 0.5588 mm). Maximum strength was achieved with sample orientation of 0°, fill angle of 45°, fill width of 0.2032 mm and air gap of -0.00254 mm. From the results, the air gap can also be greatly influenced by the resulting strength at a positive level much less than that of the negative air gap.

Mechanical properties of parts manufactured by FDM technology may vary according to supplier, colour pigment and additives. Finding a suitable material with high mechanical properties is essential for use in production. At the same time, the process parameters of the 3D printing process are important. For this reason, this article deals with the comparison of the mechanical characteristics of samples made from materials from 2 suppliers, and other influences such as the orientation of the parts during the construction and its filling.

2 METHODS

A comparison of mechanical properties was made for PET-G material from two manufacturers Devil Design (DD) and Plastics Mladeč (PM). Mechanical properties were determined by tensile tests.

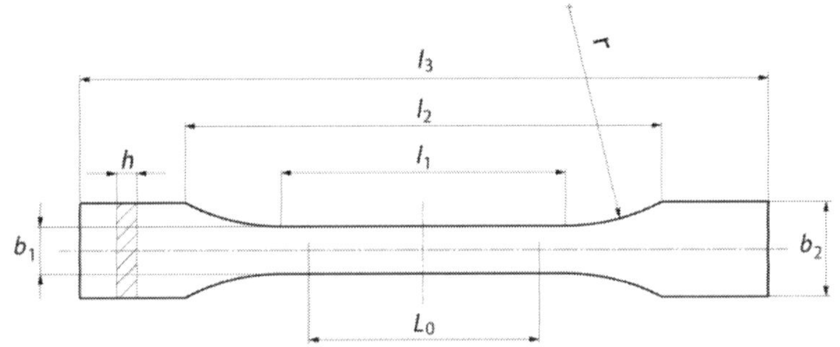

Figure 1. Geometry of specimens

The shape and dimensions of the specimens were chosen according to ČSN ISO 527-2. With respect to the printer's workspace, a small test specimen (fig. 1) was chosen, with dimensions $l_3 \geq 75$ mm, $l_1 = 30$ mm, $r \geq 30$ mm, $l_2 = 58$ mm, $b_2 = 10$ mm, $b_1 = 5$ mm, $h \geq 2$ mm, $L_0 = 25$ mm.

Two different types of printing strings were available from each manufacturer. From the manufacturer DD, the material is a transparent colourless (natural - NAT) and a transparent ruby red (RR) material. For the second manufacturer PM: transparent colourless (NAT) and transparent blue (BL). All these materials have a print diameter of 1.75 mm. The samples were designed for comparison the mechanical properties of the material with respect to each manufacturer, the orientation of the printed part, the angle of the raster (fig. 2) and to check the influence of the pigment. All samples were made up of 100 % infill density to be as similar as the injected (casted) parts. In practice, the hexagonal raster (fig. 2) is the most commonly used, with a smaller percent filling to save the material. Therefore, samples with hexagonal raster and 50 % infill density were printed from each manufacturer. For each comparison set, 5 equal pieces were created – in total 12 x 5 = 60 samples. The list of the measured samples and sets are shown in tab. 1.

Table 1. List of specimens.

Set	Manufacturer	Colour	Infill shape	Raster angle	Specimen orientation	Infill density
1.	DD	NAT	line	45° / -45°	XY horizontal	100 %
2.	DD	RR	line	45° / -45°	XY horizontal	100 %
3.	DD	RR	line	45° / -45°	XZ horizontal	100 %
4.	DD	RR	line	45° / -45°	XZ vertical	100 %
5.	DD	RR	line	0° / 90°	XY horizontal	100 %
6.	DD	RR	hexagon	-	XY horizontal	50 %
7.	PM	NAT	line	45° / -45°	XY horizontal	100 %
8.	PM	BL	line	45° / -45°	XY horizontal	100 %
9.	PM	BL	line	45° / -45°	XZ horizontal	100 %
10.	PM	BL	line	45° / -45°	XZ vertical	100 %
11.	PM	BL	line	0° / 90°	XY horizontal	100 %
12.	PM	BL	hexagon	-	XY horizontal	50 %

Sets of individual samples were printed at one time so that all samples had the same ambient conditions as temperature or humidity. The specified print parameters are summarized in tab. 2.

Table 2. Print parameters.

Extruder temperature	220 °C
Platform temperature	65 °C
Thickness of the printed layer	0.18 mm
Thickness of the first layer	0.27 mm
Print velocity	60 mm/s

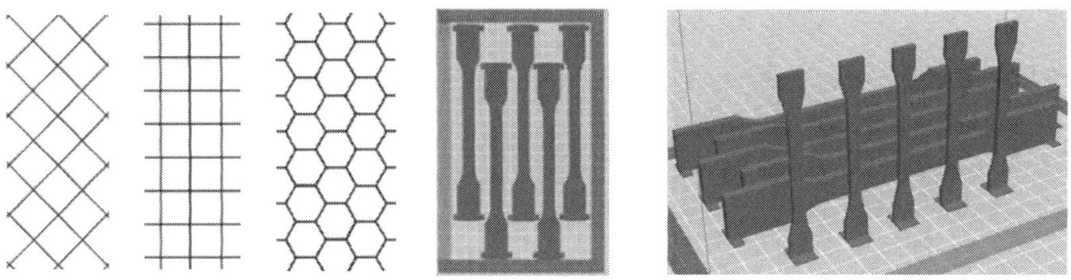

Figure 2. From left to right: types of inner infill (45°/-45°, 0°/90°, hexagonal), specimen orientation XY horizontal and XZ horizontal + vertical

3 Results and discussion

The Zwick universal test machine with a load cell head up to 22 kN was used for the measurement. Sample loading was carried out at a speed of 3 mm/min. The force was measured up to break or up to form a neck (for both cases it means significant decrease of the force). Ultimate tensile strength was calculated from the maximum force and the Young's modulus was determined by regression from the stress-strain curve.

In the fig. 3 and fig. 4, there are a comparison of the ultimate tensile strength and Young's modulus for both manufacturers. Each column represents the average strength for all sample sets denoting standard error of the mean. The green columns are represented by Devil Design and Blue Columns by Plastics Mladeč. Alongside each other, there are always samples with the same print parameters:

1) Transparent material, raster 45°/-45°, print plane XY; 2) Colour material, raster 45°/-45°, print plane XY; 3) Colour material, raster 45°/-45°, print plane XZ horizontal; 4) Colour material, raster 45°/-45°, print plane XZ vertical; 5) Colour material, raster 0°/90°, print plane XY; 6) Colour material, raster hexagonal.

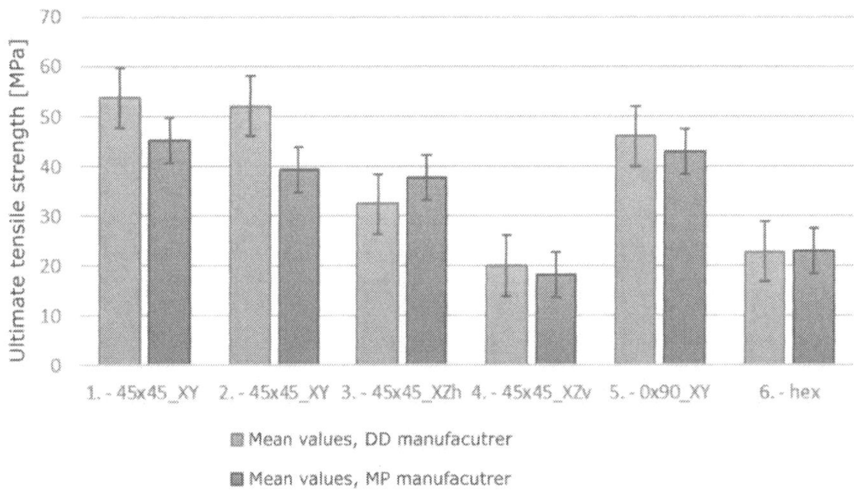

Figure 3. Comparison of ultimate tensile strength between DD a PM.

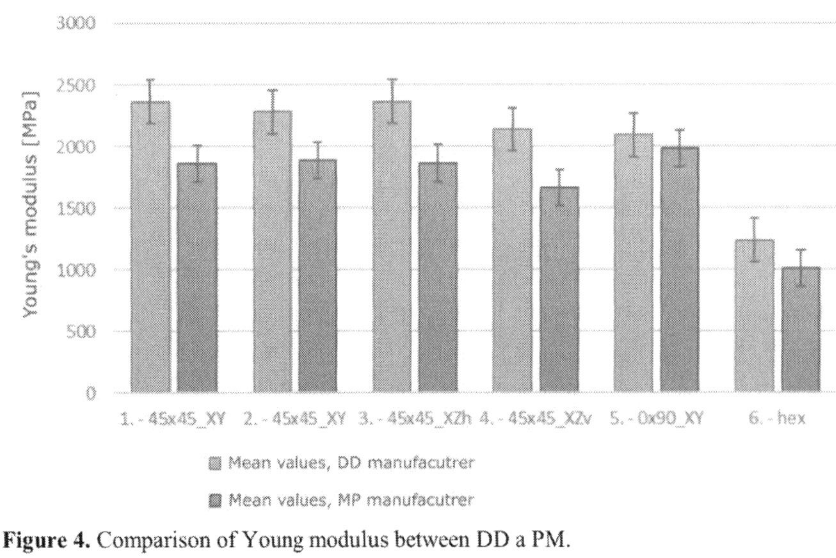

Figure 4. Comparison of Young modulus between DD a PM.

For samples printed at a 45°/45° infill angle, the neck was created. The neck was formed at an angle of 45° in the direction of the fibres and also in the shear plane of the sample. Exceptions were the blue samples from the manufacturer Plastics Mladeč (PM_BL_45x45_XY). Here, the necking were not significand and the end of the test were followed by rupture. This behaviour was probably due to pigment, because natural specimens from the same manufacturer with the same print parameters (PM_NAT_45x45_XY) shows also necking at the end of tensile test.
For all samples other than those mentioned above, there was a rupture. For specimens with a 0°/90° infill angle and hexagonal infill samples, the angle of fracture was 90°. Similarly, for samples printed vertically in the XZ plane.

The results of the measurements were statistically processed using the Minitab program. At 5% significance levels, we can assume that the data follows normal distribution (Anderson-Darling normality test). The individual sets are independent of one another (unpaired). To compare the different sets, we used the Two Sample t-test parameter test. Table 3 summarizes the percentage differences in mean values of ultimate tensile strength and Young's modulus.

Table 3. Comparison of mean values of ultimate strength and Young modulus for given sets of samples (with p-values):

		Δ (%)	
Sets of samples – comparison between manufacturers		**Young modulus**	**Ultimate strength**
DD_NAT_45x45_XY	PM_NAT_45x45_XY	-21,1 p < 0,0005	-16,1 p < 0,0005
DD_RR_45x45_XY	PM_BL_45x45_XY	-17,2 p < 0,0005	-24,6 p = 0,002
DD_RR_45x45_XZh	PM_BL_45x45_XZh	-21,1 p < 0,0005	-
DD_RR_hex	PM_BL_hex	-18,4 p = 0,011	-
Sets of samples – pigment influence			
DD_NAT_45x45_XY	DD_RR_45x45_XY	-	-3,0 p = 0,029
PM_NAT_45x45_XY	PM_BL_45x45_XY	-	-12,9 p = 0,028
Sets of samples – infill angle and infill shape influence			
DD_RR_45x45_XY	DD_RR_0x90_XY	-8,3 p = 0,001	-11,7 p < 0,0005
DD_RR_45x45_XY	DD_RR_hex	-45,8 p < 0,0005	-56,3 p < 0,0005
PM_BL_45x45_XY	PM_BL_hex	-46,6 p < 0,0005	-41,6 p = 0,001
Sets of samples – sample orientation influence			
DD_RR_45x45_XY	DD_RR_45x45_XZh	-	-37,8 p = 0,001
DD_RR_45x45_XY	DD_RR_45x45_XZv	-	-61,7 p < 0,0005
DD_RR_45x45_XZh	DD_RR_45x45_XZv	-	-38,4 p = 0,003
PM_BL_45x45_XY	PM_BL_45x45_XZv	-11,8 p = 0,004	-53,6 p < 0,0005
PM_BL_45x45_XZh	PM_BL_45x45_XZv	-10,8 p = 0,022	-51,7 p < 0,0005

4 Conclusions

From the analysis of the results it is obvious that the greatest influence on the resulting strength is the orientation of the part during printing. The highest tensile strength with respect to the orientation of the component is achieved by XY-printed samples and the lowest samples printed vertically in the XZ plane, which do not reach even 50 % of the tensile strength value of samples printed in the XY plane. Furthermore, the tensile strength significantly affects the shape of the infill, where the hexagonal shape is less, but the percent filling of the samples plays a large role here. Hexagonal infill samples are printed with 50% infill density, and 23% less material is used than other samples with 100% infill density.

For Devil Design manufacturer, the effect of raster orientation was noted, with samples built with the 45°/45° raster ultimate tensile strength being 12% higher than those with a 0°/90° raster. For Plastice Mladeč, this influence was statistically insignificant. Better mechanical properties were measured on samples from Devil Design, where the tensile strength and the Young's modulus were about 20 % higher than those of Plaster Mladeč's manufacturer.

The influence of the pigment was different according to the manufacturer. For the natural (colourless) material from Devil Design, the tensile strength was 3 % higher than that of the red material, at Plastice Mladeč, the difference between the tensile strength between natural and blue material was nearly 13 %.

Acknowledgement

Financial support through the project NETME CENTRE PLUS (LO1202) by the Czech Ministry of Education, Youth and Sports under the "National Sustainability Programme I" is gratefully acknowledged.

References

1. FDM technologie (Fused Deposition Modeling). 3D inventor blog, online, (2016)
2. Flashforge Dreamer Dual Extrusion 3D Printer. FLASHFORGE USA, online, (2015)
3. What is Design for Additive Manufacturing. ENGINEERS rule: Technology for Design and Engineering online, (2016)
4. S. H. Ahn, M. Montero, D. Odell, S. Roundy, P. K. Wright, Anisotropic material properties of fused deposition modeling ABS. Rapid Prototyping Journal, 8(4): 248-257, (2002)
5. I. Durgun, R. Ertan, Experimental investigation of FDM process for improvement of mechanical properties and production cost. Rapid Prototyping Journal, 20(3), 228-235, (2014)
6. F. Rayegani, G. C. Onwubolu. Fused deposition modelling (FDM) process parameter prediction and optimization using group method for data handling (GMDH) and differential evolution (DE). The International Journal of Advanced Manufacturing Technology, 73(1-4), 509-519, (2014)

Liquid-Phase Sintering and Properties of PIMed 10-20 vol.% SiC_p-Reinforced Aluminium Composites

Tapany Patcharawit[1,a], Arada Ngeekoh[1], Usanee Kitkhamthorn[1] and Nutthita Chuankrekkul[2]

[1]*School of Metallurgical Engineering, Suranaree University of Technology, 111 University Avenue, Muang, Nakhon Ratchasima, 30000, Thailand*
[2]*The Metallurgy and Materials Science Research Institute, Chulalongkorn University, Phayatai Road, Patumwan, Bangkok, 10330, Thailand*

Abstract. Liquid-phase sintering and properties of powder injection moulded aluminium composite reinforced with SiC_p has been investigated via thermal-phase analyses and hardness property. Mixing of 10 and 20 vol.% SiC_p and aluminium powder was done via ball milling prior to feedstock preparation at 55% solid loading. Powder injection moulding was carried out at 170°C, followed by solvent and thermal debinding. The brown samples were sintered in a nitrogen atmosphere at 650-660°C, prior to age-hardening. Experimental result shows relatively high % linear shrinkage of 9.9-21.9 % and high % volumetric shrinkage of 34.5-46.5%. Thermal analysis indicates the formation of the liquid around 632 °C, facilitating liquid phase sintering. The composite microstructures consists of uniformly distributed SiC_p throughout the densified aluminium matrix. SiC_p addition scales up SiC_p clustering, accompanying with gap porosity. The composite reinforced with 10 vol.% SiC_p possesses higher sintered density at 2.63 g/cm^3, yielding 93.1% theoretical. However the maximum macro Vickers hardness was measured at 164.2 H_v for 20 vol.% SiC_p– reinforced aluminium composite in the age-hardened condition. Liquid phase sintering was found effective for the microstructure evolution of the PIMed composite, rendering densification. Hardening is influenced by SiC_p addition and the roles of AlN, Al_2Cu and Mg_2Si precipitates.

1 Introduction

Aluminium matrix composites have been increasingly recognized for engineering applications such as in the automotive, electronics, aerospace and biomedical fields for examples [1-3]. Their properties are dependent on selected materials utilized such as the matrix alloys and reinforcements. Particulate reinforcement such as Al_2O_3, SiC, Si_3N_4, and AlN are of great interest for improvement of hardness, tensile, modulus and wear properties. Powder injection moulding (PIM) is one of the most effective solid-state fabrication techniques available for composite production in which tailored properties can be achieved by selection of the matrices and reinforcements in the powder mixing stage. The PIM process involves powder pre-treatment, feedstock preparation, injection moulding, followed by debinding, and finally sintering along with post-treatment if required. Particularly in the sintering stage, strength development is due to bonding of adjoining powders via solid/liquid phase sintering. Solid phase sintering at a temperature range just below the melting point of the powder renders strong

[a] Tapany Patcharawit: tapany@sut.ac.th

solid bonds via material transport mechanisms such as lattice diffusion, grain boundary diffusion, viscous flow, surface diffusion, gas phase transport for example [4]. Grain boundary is the most important source of materials for diffusion to take place. However, as the sintering temperature is raised to just above the solidus of the components; hence, some extent of the liquid forms, which contributes to densification and shrinkage. The viscous flow is enhanced favourable for fast sintering. The assisting requirements are 1) increasing solid solubility in the newly formed liquid melt, 2) solubility of solid in the liquid and 3) low liquid solubility in the solid [5]. Additives which lower liquidus and solidus temperatures are also favourable.

As for the PIMed aluminium composites investigated in this current research, the formulated composition involved deformable, lower melting point Al-Cu alloy powder as the matrix and hard-abrasive, high melting point SiC_p of a smaller size as the reinforcement, aiming for hardness and wear improvement. Hardening is expected via second phase precipitation and particulate reinforcement. Liquid phase sintering was applied to improve densification on the expense of shrinkage and dimension stability. The influences of SiC_p addition along with microstructure evolution during liquid phase sintering is also discussed.

2 Experimental

The commercial Al-Cu powder has its composition as detailed in Table 1 and its average size is 73 μm. The silicon carbide particulate (SiC_p) of averagely 20 μm size was added at 10 and 20 vol.%. Feedstocks was prepared at 55 % solid loading for powder injection moulding at 170°C. The obtained green samples of 20x20x5 mm^3 dimensions were subjected to solution debinding. In order to determine the sintering temperature of the composite, Simultaneous Thermal Analysis (STA) was selected using a NETZSCH STA 449 F3 Jupiter operating at a 40 K/min heating rate over 25-900°C in a nitrogen atmosphere. Microstructure and phase analysis were carried out after sintering and age-hardening using a JEOL-JSM 6061LV SEM coupled with EDX spectroscopy, and a D8 XRD over a 2θ range of 20°-80°. Bulk density and macro Vickers hardness was examined to relate microstructure and property of the composites.

Table 1. Chemical composition of Al-Cu powder by XRF (wt.%).

Al	Cu	Mg	Si	Fe
Balanced	4.77	1.11	1.01	0.12

3 Experimental Results

3.1 Thermal Analysis and Microstructure Examination

3.1.1 Thermal Analysis of the Composite

Differential Scanning Calorimetry (DSC) and Thermogravimetric (TG) analyses are graphically demonstrated in Figs. 1 a) and b) for green compacts of 10 vol.% and 20 vol.% SiC_p additions respectively. It can be seen that the endothermic reactions observed on the DSC curves occurring over 42-156°C are in accordance with the melting ranges of paraffin wax, stearic acid and polypropylene respectively. Accordingly, the moulding temperature was set at 170°C in order to enable a decent flow of feedstock into the mould cavity. Another endothermic reaction at approximately 458°C is possibly incident for the decomposition or vaporization of polypropylene while the TG curves signify mass losses at the onset of 226 °C towards 471-483°C. This therefore implies a complete thermal debinding of the multi-component binder around 500°C prior to entering liquid phase sintering. Moreover, at approximately 632°C, endothermic reaction is visible due probably to the liquid phase formation [6].

https://doi.org/10.1051/matecconf/201823702015

It is therefore recommended that the sintering temperature should be in a range of 650-660°C to achieve liquid phase sintering for such PIMed composites.

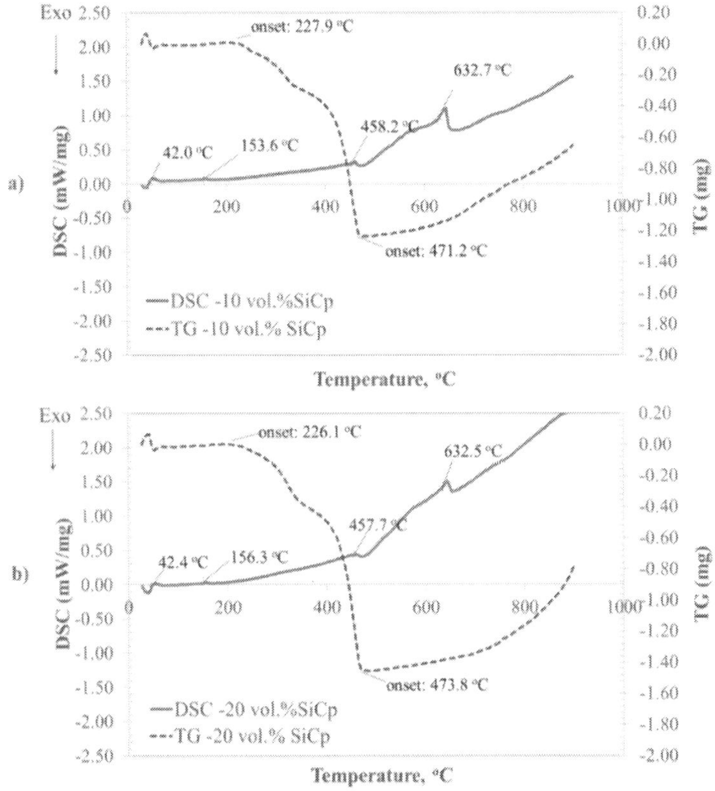

Figure 1. DSC and TG analyses of 10 and 20 vol.% SiC_p green compacts.

3.1.2 Microstructure Analysis

Optical micrographs of 10 and 20 vol.% SiC_p – reinforced aluminium composites in the as-sintered condition are comparatively shown in Figs. 2 a) and b) respectively. Fine SiC_p (dark) is distributed throughout the aluminium matrix (light). There is no excessive grain growth of the matrix. Al-SiC_p interfaces exhibit good bonding. A greater degree of SiC_p agglomeration was observed at higher SiC_p content, giving greater amount of SiC_p-SiC_p interfaces and a higher degree of porosity.

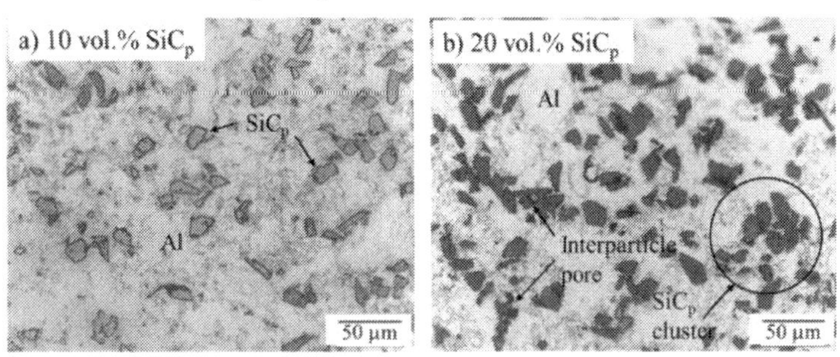

Figure 2. Optical micrographs of 10 and 20 vol.% SiC_p reinforced aluminium composites.

3.1.3 XRD Analysis

XRD results are illustrated in Figs. 3 a) - d) for 10 and 20 vol.% SiC_p – reinforced aluminium composites in the as-sintered and age-hardened conditions. Existence of phases are quite similar for both composites reinforced with 10 and 20 vol.% SiC_p. The main peaks of Al were detected as the matrix. The SiC peaks were also identified as the reinforcement and it is more evidently shown especially for composite possessing higher SiC_p addition. AlN peaks are also apparent, observed in a greater extent than those of SiC_p. Some research [7] classified that AlN is desirable for densification, while some [8] identified the role of AlN as detrimental. Precipitates such as Al_2Cu and congruent Mg_2Si were found both in the as-sintered and age-hardened conditions. Sufficient Mg content of 1.11 wt.% allows congruent Mg_2Si to be formed, which is believed to be beneficial for additional hardening of the composites aside from that of Al_2Cu as the main precipitate phase. Al_2O_3 and Si peaks were also detected in a comparatively lesser extent than other phases' peaks.

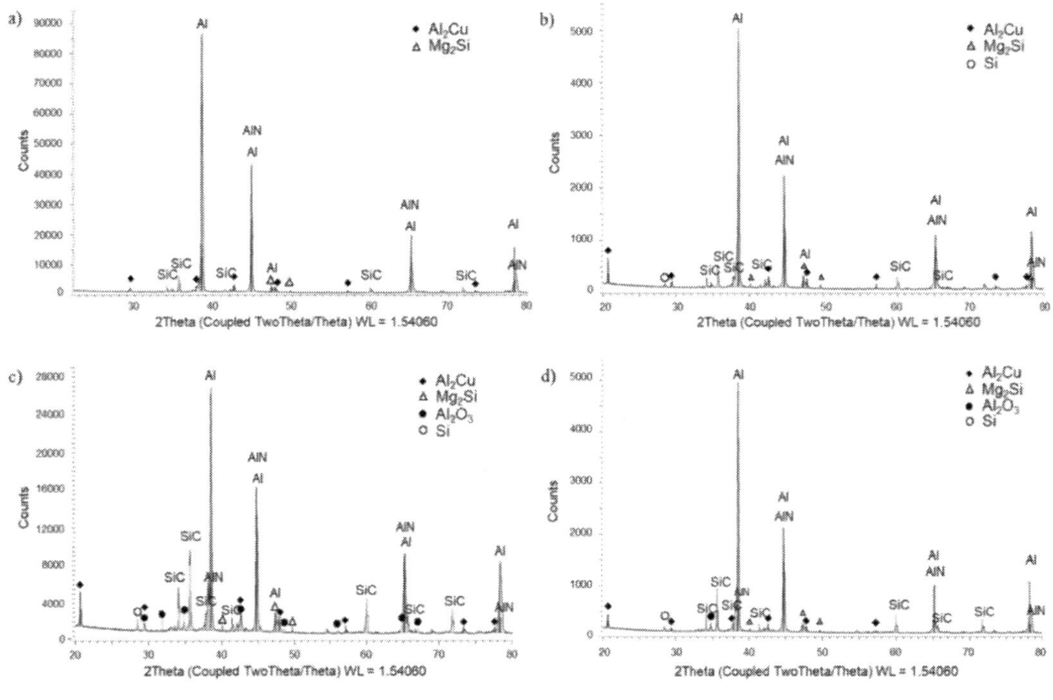

Figure 3. XRD analyses of 10 vol.% SiC_p reinforced aluminium composites in a) as-sintered, b) age-hardened conditions, and 20 vol.% SiC_p reinforced aluminium composites in c) as-sintered, d) age-hardened conditions.

3.1.4 SEM and EDS Analysis

Microstructure investigation at higher magnifications by a secondary mode electron microscopy shows a reasonably good interface between SiC_p and aluminium matrix in the case of 10 vol.% SiC_p addition. However, as mentioned previously, SiC_p clusters are routinely observed when the SiC_p increases to 20 vol.%, as illustrated in Fig. 4 a). This resulted in interparticle or gap porosity accompanying with SiC_p clusters. Figure 4 b) locates EDS analyses at 5 different locations, which confirm the existence of phases found in the as-sintered condition as detailed by EDS peaks of particular elements in Figs. 4 c)-g) for spectrums 1-5 respectively. Spectrum 1 indicates a composition related to that of the aluminium alloy matrix, giving 91.86 wt.% for Al and 5.16 wt.% for Cu, with small amount of 2.98 wt.% O, as listed in the composition table. Spectrum 2 shows a eutectic composition in relevant to its morphology seen on the SEM micrograph. SiC_p was clearly detected in spectrum 3. The Al_2Cu is identified as precipitate in spectrum 5, giving the composition of 45.04 and 54.96 wt.% for Al and Cu respectively. Secondary mode micrograph of the age-hardened composite shows a similar result to that of as-sintered condition as displayed in Fig. 5 a). Five different locations for EDS analyses are exhibited in Fig. 5 b) where relevant EDS peaks are illustrated in Fig. 5 c)-g) for spectrums 1-5 respectively. The analysed composition of 5 different locations is inclusively summarized within. Spectrum 1 clearly indicates composition of SiC_p while eutectic composition reflects in composition analysed for spectrum 2. The matrix composition is detected as 96.74 wt.% Al and 3.26 wt.% Cu in Spectrum 3 while EDS analysed in spectrum 4 as a mixed composition of mainly Cu, Al and C. Spectrum 5 evidently shows a composition of Al_2Cu precipitate. Such EDS analyses on existing phases are in good agreement with XRD and XRF analyses previously mentioned, with an exception for AlN, which cannot be resolved by SEM used in this current research. Future work is then required to identify such a significant phase.

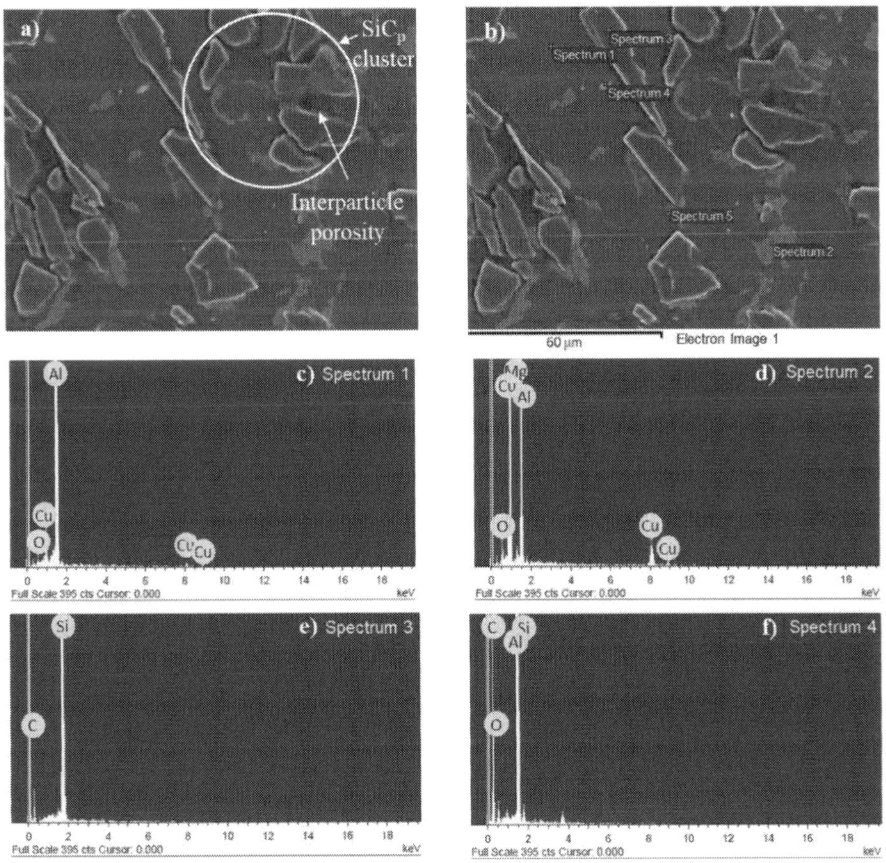

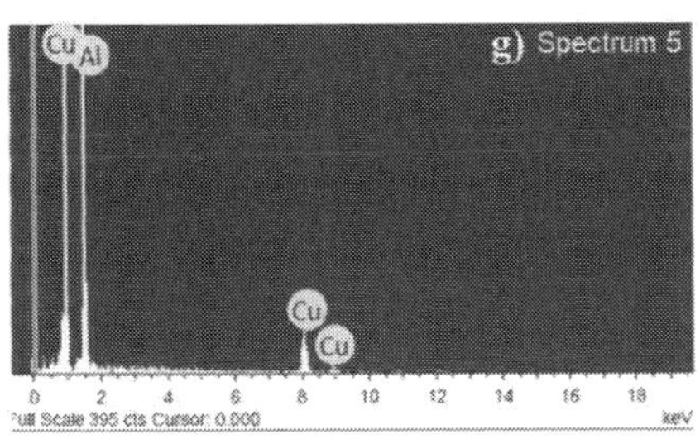

Composition of spectrums 1-5 (wt.%)					
Ele.	SP 1	SP 2	SP 3	SP 4	SP 5
Al	91.86	43.35	-	31.16	45.04
Cu	5.16	49.62	-	-	54.96
Si	-	-	59.79	1.11	-
Mg	-	1.73	-	-	-
O	2.98	5.30	-	5.21	-
C	-	-	40.21	62.52	-

Figure 4. SEM and EDS analysis of as-sintered 20 vol.% SiC_p reinforced aluminium composites.

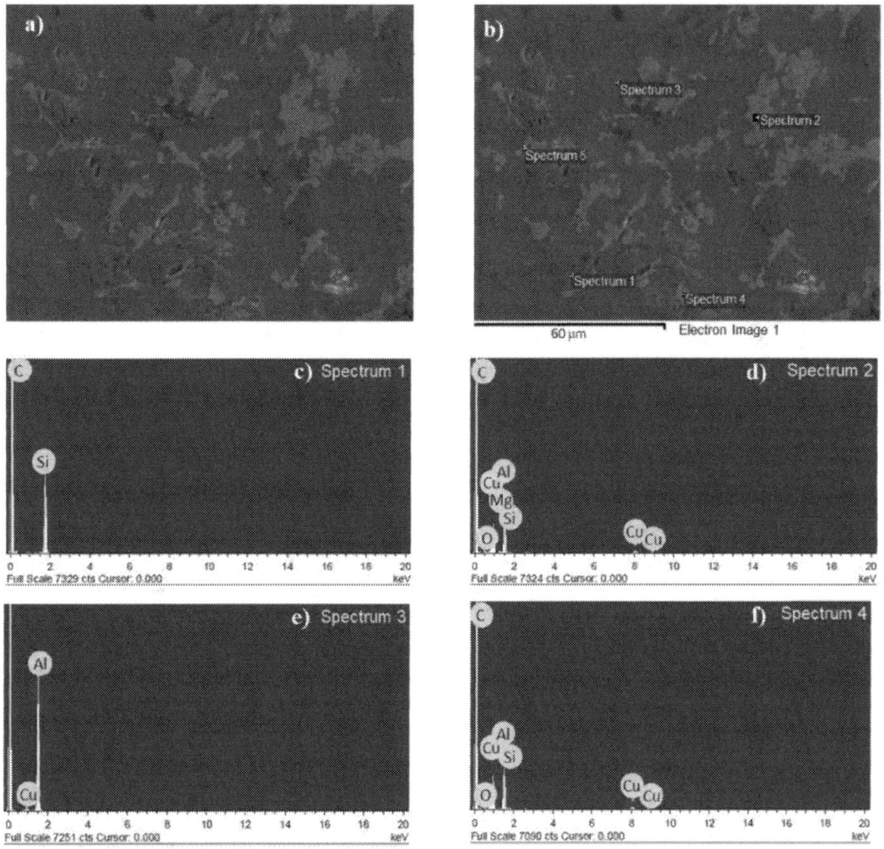

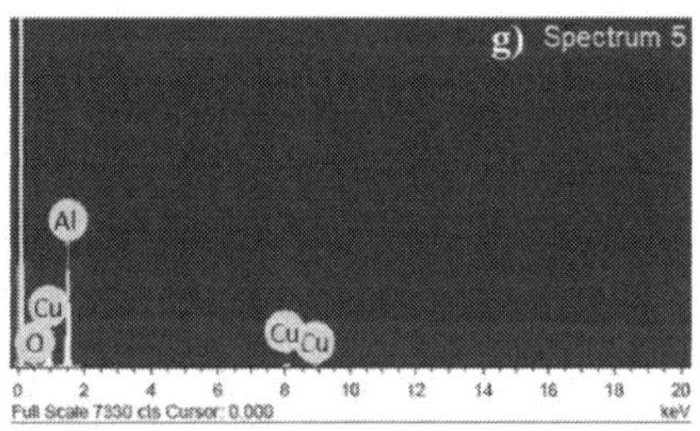

Composition of spectrums 1-5 (wt.%)					
Element	SP 1	SP 2	SP 3	SP 4	SP 5
Al	-	32.93	96.74	37.37	64.12
Cu	-	46.98	3.26	48.2	33.62
Si	56.88	0.79	-	1.02	-
Mg	-	2.55	-	-	-
O	-	4.23	-	1.77	2.26
C	43.12	12.53	-	11.65	-

Figure 5. SEM and EDS analysis of age-hardened 20 vol.% SiC_p reinforced aluminium composites.

3.2 % Shrinkage and Bulk Density

Percentage of shrinkage in thickness, length and width are listed in Table 2, giving the % volumetric shrinkage likewise. It is seen that % linear shrinkages are rather large, which are in a range of 9.88 - 21.95%. Increasing SiC_p content scales down the % shrinkage, owing to its high thermal stability. Further, by assuming a simple calculation of linear shrinkage (Y_s) as expressed in equation (1) [9] as a function of solid loading of the feedstock (ϕ), theoretical density of the mixed Al powder and SiC_p (ρ_{th}) according to equation (2), and the final sintered density (ρ), the Y_s can be obtained as summarized in Table 2. It is noted that the calculated linear shrinkage is significantly smaller than the measured ones. The volumetric shrinkage is essentially high, due to high amount of binder used at 45 vol.% plus densification by liquid-phase sintering involved. Increasing SiC_p content from 10 to 20 vol.% therefore results in a significant reduction in the volumetric shrinkage from 46.62% to 34.54%.

$$Y_s = 1 - \{\phi/(\rho/\rho_{th})\}^{1/3} \quad (1)$$

Table 3 summarises green and sintered densities of the composites, giving the optimum sintered density of 2.63 g/cm^3 for 10 vol.% SiC_p-reinforced aluminium composite. This accounts for 93.1% theoretical sintered density according to the rule of mixture. Equations (2) and (3) express theoretical density for the green compact ($\rho_{green,\, th}$) and sintered sample ($\rho_{sinter,\, th}$) respectively.

$$\rho_{green,\, th} = \rho_{Al}V_{Al} + \rho_{SiC}V_{SiC} + \rho_{binder}V_{binder} \quad (2)$$

$$\rho_{sinter,\, th} = \rho_{Al}V_{Al} + \rho_{SiC}V_{SiC} \quad (3)$$

where ρ_{Al}, ρ_{SiC}, and ρ_{binder} are densities of aluminium alloy (2.78 g/cm^3) [10], SiC (3.20 g/cm^3) and multi-component binder (0.94 g/cm^3). Accordingly, V_{Al}, V_{SiC} and V_{binder} are volume fractions of aluminium alloy, SiC_p and multi-component binder respectively, where $V_{Al} + V_{SiC} + V_{binder} = 1$ for equation (2) and $V_{Al} + V_{SiC} = 1$ for equation (3). Increasing SiC_p content therefore deteriorates the sintered density, due possibly to agglomeration behaviour of high surface energy SiC_p [11, 12] during feedstock preparation. These SiC_p clusters and interparticle porosity still remained even after the samples experiencing the liquid phase sintering.

Table 2. % Average shrinkage in thickness, length, width, and volume.

SiC_p (vol.%)	% Shrinkage				% Calculated linear shrinkage (Y_s)
	Thickness	Length	Width	Volume	
10	21.95	17.63	16.99	46.62	16.12
20	9.88	15.18	14.47	34.54	13.45

Table 3. Average green and sintered density, and macro hardness properties.

SiC_p (vol.%)	Bulk green density (g/cm^3)		Bulk sintered density (g/cm^3)		Macro Vickers hardness, (H_v)	
	Measured	%Theoretical	Measured	%Theoretical	As-sintered	Age-hardened
10	1.83	93.37	2.63	93.1	132.6	160.8
20	1.84	92.93	2.43	85.0	125.0	164.2

3.3 Macro Vickers Hardness

Macro Vickers hardness result shows increased values after age-hardening for both composites with 10 and 20 vol.% SiC_p additions. The maximum macro hardness value was measured at 164.2 H_v for 20 vol.% SiC_p - aluminium composite after age-hardening, as listed in Table 3. The precipitation of fine Al_2Cu and Mg_2Si during age-hardening are possibly the cause of hardening effect.

4 Discussion

4.1 Densification of the PIMed SiC_p-Reinforced Aluminium Composites

As previously shown that the calculated linear shrinkage of the PIMed composite is 13.45-16.12%. However, the measured linear shrinkage of the composites is rather high, essentially for the volumetric shrinkage (34.54-46.62%), which is related to high binder content and liquid phase sintering involved. If the shrinkage related to eliminating of the binder is excluded, effect of the remaining solid phases, i.e., aluminium and SiC_p on shrinkage is now considered. After the binder was rid of, the sample then entered the liquid phase sintering. Additional shrinkage is due mainly to aluminium powder, which partly melted to facilitate particle rearrangement, and then shrank upon solidification. In this current research work, densification occurred without any significant warpage or dimension instability fortunately. Comparative work of previous research [11] investigated sintering of 20 vol.% SiC_p-reinforced aluminium composite, sintered at a higher temperature of 740°C over 1 hour reported the density of 2.52 g/cm^3 and hardness of 95.9 and 136.9 H_v in the as-sintered and age-hardened conditions respectively without any warpage. By replacing the aluminium powder with one of a smaller size (39.6 μm), a greater degree of densification prevailed, due to shorter diffusion distance and larger curvature stresses [12]. Though, gaps between SiC_p-SiC_p and SiC_p-Al interfaces were sealed in this case, dimension stability is nevertheless sacrificed. Therefore, in this current work, the sintering scheme has been changed by lowering sintering temperature 650-660 °C. Extending the sintering period is then obligatory to attain effective sinterability of the composite, offering relatively greater hardness values of 125.0 and 164.2 H_v in the as-sintered and age-hardened conditions as mentioned previously. Lower sintering temperature is to avoid dimension instability and extended sintering time is to aid diffusion process to take place when the sintering temperature is approaching the melting temperature.

As the liquid phase sintering proceeded, higher diffusion rate allowed [12] 1) rearrangement of aluminium powder and SiC_p favourable for particle packing, 2) solution-precipitation suitable for pore filling and through the 3) final stage of densification. In the early stage of liquid phase sintering of this type of composite, as the aluminium oxide skin was consumed [13], Mg played a key role on wetting of the liquid aluminium on the surface treated SiC_p. [14]. Spreading of the liquid phase then dissolved the sintered bonds and rearranged the aluminium powder and SiC_p by providing capillary force to pulls all solid grains together [12]. Except for the SiC_p, the aluminium powder is of course easily soluble in the liquid melt, which enhanced the transport rate, resulting in grain coarsening and densification. Further, pore isolation and annihilation is associated with reduction of surface energy in this final stage. Reasonably good sintered density of 2.63 g/cm^3 at 93.1% theoretical is then achieved for 10 vol.% SiC_p reinforced aluminium composite. Nevertheless, increasing SiC_p content from 10 to 20 vol.% slowed down the mass transport due to its lower solubility in the liquid melt. More effort was then required to pull more solid grains of SiC_p and the remaining aluminium powder and, resulting in less % shrinkage and inferior densification.

4.2 Hardening of the PIMed SiC_p-Reinforced Aluminium Composites

To consider hardening mechanisms in the PIMed SiC_p-Reinforced Aluminium Composites, roles of SiC_p addition and age-hardening are now discussed. First, hard and abrasive SiC_p of a finer size was used to reinforce aluminium-copper alloy. Its intrinsic hard-abrasive property rendered the increment in macro hardness as the hardness indention stretching over an area accounting for hundreds of SiC_p. The influence of interparticle porosity existing in the SiC_p clusters found in the 20 vol.% SiC_p addition is then retarded, though this interparticle porosity reflects the inferior sintered density as mentioned earlier in section 3.2. However, previous work [15] has reported exceeding amount of SiC_p addition up to 30-40 vol.% has proof to give deleterious density and hardness.

Secondly, roles of precipitates such as AlN, Al_2Cu and Mg_2Si are considered. Though, AlN peaks are quite evident from the XRD results, confirmation by SEM and EDS is still an issue in this current work. It was reported elsewhere [16, 17] on sub-micron sizes of AlN found in aluminium composite produced by a liquid-route process at 900-1000 °C. However, in the solid-route process of powder metallurgy [18], it was also documented that AlN shell was formed enveloping the pure aluminium powder by nitridation at 560 °C for 6-18 h in N_2 flow prior to sintering. TEM characterization of this AlN shell revealed an aluminium matrix with a large number of AlN nanoparticles in a 10-50 nm. Therefore, it might be presumed that the AlN formation is possible over the 650-660 °C sintering temperature in this current work. However, due to the shorter exposure time to N_2, AlN might be formed as a thin-shell structure surrounding such aluminium powder and make it difficult to recognize. Further work is then required to identify morphology and preferential sites of AlN in order to explain its function on the strengthening effect of the composite. In addition, the formations of Al_2Cu and Mg_2Si are confirmed by both XRD and EDS point analysis results. Though, peaks of Al_2Cu and Mg_2Si were also detected both in the as-sintered and after age-hardening, the latter is theoretically fine and coherent. These fine coherent precipitates formed after age-hardening offer hardness increment in a range of 23.5 - 31.4%.

5 Conclusions

According to the experimental results, it can be concluded that the PIMed aluminium composites reinforced with 10 -20 vol.% SiC_p provided uniform microstructure, giving reasonably good sintered density of 93.1% theoretical at 2.63 g/cm^3 and maximum macro Vickers hardness of 164.2 H_v after age-hardening. Densification and high values of linear and volumetric shrinkage are due to liquid phase sintering. Hardening is influenced by SiC_p addition and the roles of AlN, Al_2Cu and Mg_2Si precipitates.

References

1. S. Rawal, JOM **53** 14-17 (2001)
2. C. Yan, W. Lifeng, R. Jianyue, Chin. J. Astron. **21** 578-584 (2008)
3. M. Dyzia, Materials, **11** 1-13 (2018)
4. S.-J. L., Kang, *Sintering: densification, grain growth & microstructure* (2005)
5. R. M. German, *Sintering: From Empirical Observations to Scientific Principles* (2014)
6. A. Arockiasamy, R. M. German, P. Wang, M. F. Horstemeyer, W. Morgan, S. J. Park Sintering, Powder. Metall. **54** 354-359 (2011)
7. T. W. Kim, Mater. Sci. Eng., A **483-484** 648-651 (2008)
8. I. A. MacAskill, R. L. Hexemer Jr, I. W. Donaldson, D. P. Bishop, J. Mater. Process. Technol. **210** 2252-2260 (2010)
9. D. F. Heaney, *Handbook of Metal Injection Molding* (2012)
10. J. M. Martín, F. Castro, J. Mater. Process. Technol. **143-144** 814-821 (2003)
11. A. Ngeekoh, V. Thongnoppakoon, A. Changlor, T. Jindangam, N. Phongreed, T. Patcharawit, N. Chuankrerkkul, the 41st Cong. Sci. and Tech. of Thai. (2014)
12. R. M. German, P. Suri, S. J. Park, J. Mater. Sci. **44** 1-39 (2009)
13. A. Kimura, M. Shibata, M. Katayama, T. Kanie, H. Takada, Appl. Phys. Lett. **70** 3615-3617 (1997)
14. J. Hashim, L, Looney, M. S. J. Hashmi, J. Mater. Process. Technol. **119** 324-328 (2001)
15. T. Patcharawit, P. Tongkhon, A. Wiangsamut, P. Auisungnoen, N. Chuankrerkkul, Appl. Mech. Mater. **376** 17-22 (2013)
16. C. Borgonovo, D. Apelian, M. M. Makhlouf, JOM **63** 57-64 (2014)
17. S. Kumari, U. Pillai, and B.C. Pai, J. Alloys Compd. **509** 2503-2509 (2011)
18. M. Balog, P. Yu, M. Qian, M. Behulova, P. Svec Sr, R. Cicka, Mater. Sci. Eng., A **562** 190-195 (2013)

Effects of Dopant Types and Contents on Morphology and Strength Properties of PANI/PAN Electrospun Nanofibers

Taohai Yan[1,2,3], Jinhua Jiang[1] and Nanliang Chen[1,a]

[1]*College of Textiles, Donghua University, Shanghai 201620, P.R. China.*
[2]*Clothing and Design Faculty, Minjiang University, Fuzhou,Fujian 350121, P.R. China*
[3]*Fujian Key Laboratory of Novel Functional Fibers and Materials, Minjiang University, Fuzhou Fujian 350121, P.R. China*

Abstract: In this paper, the electrospinning of polyaniline under different dopant conditions were studied, the experimental phenomena was introduced, analyzed the relationship between the nanofiber surface morphology of DBSA-PANI/PAN, HCl-PANI/PAN and PTSA-PANI/PAN membranes and dopant type and content. The average diameter of DBSA-PANI/PAN electrospinning fiber membrane showed a decreasing trend when the content of doped polyaniline increased 1%-4%, and reached a minimum value of 780nm when the concentration was 4%. The strength of DBSA-PANI/PAN fiber membrane was the highest when the concentration of doped polyaniline was 1%, and the maximum intensity was 1.8×10 5 Pa. The average diameter of HCl/PANI/PAN fiber membranes decreases with the increase of doped polyaniline concentration, and it fluctuates up and down within a small range. When the content of doped polyaniline was increased to 4%, the average fiber diameter reached a minimum of 420 nm. The strength of the HCl-PANI/PAN fiber membrane was the highest when the concentration of doped polyaniline was 4%, and the maximum intensity was 1.08×10 5 Pa. The average diameter of the PTSA-PANI/PAN electrospun fiber membrane was the smallest at 2% doped polyaniline content and was 430 nm, the strength of PTSA-PANI/PAN fiber membrane is the highest when the content of doped polyaniline is 2%, and the maximum intensity is 3.9×10 4 Pa. In this paper, the basic research of doped polyaniline electrospinning nanofibers has made a useful supplement to the existing research of polyaniline nanofibers, and has a reference value for the application of polyaniline.

1 Introduction

In 1862, polyaniline was first used as a dyestuff in H.Letheby's research. There have been three phases of polyaniline research: the discuss on "Aniline Black" as a dyestuff, the research of organic semiconductor, and as well as extensive studies of conducting polymer[1]. Due to the lack of profound theoretical knowledge of macromolecular at that time, the further study of polyaniline had been dragged on for nearly a century before 1960s. In 1976 MacDiarmid found that polyaniline boasted unique properties such as electric conduction. This led to polyaniline being considered a conductive polymer with widespread applications. Polyaniline is easily obtainable as a raw material, is easily manipulated by synthetic processes, has a high conductivity, a unique doping performance, an excellent electromagnetic microwave absorbing performance, is environmental stabile, chemically stabile, and has optical properties[2]. Efficient doping of the macromolecular functional protic acids

[a]Corresponding author. Email: nlch@dhu.edu.cn

with PANI molecular chains, such as dodecylbenzenesulfonic acid(DBSA), camphorsulphonic acid(CSA), paratoluenesulfonic acid (TSA), poly (acryl ic acid)(PAA), Sodium pyrophosphate(TSPP), naphthalenesulfonic acid (NSA) have been performed. The acid acts as a template or steric stabilizer to form a doped PANI limb dispersion. Previous studies have found that functional proton sulfonic acids that contain aromatic rings and 10 to 15 carbon chain alkyl groups are excellent dopants for PANI. The dodecyl benzene sulfonic acid (DBSA) is the most commonly doped macromolecule, which could effectively improve the dispersibility of the product and does not weaken the performance of polyaniline [3]. Electrospinning is a process that uses electrostatic forces to form fibers[4]. This process uses a high voltage supply, which induces the charge of a certain polarity into a polymer solution or a polymer melt that is accelerated to a collector that has the opposite polarity. The obtained electrospun fibers have a small diameter, in the micrometer to nanometer range[5]. The polymer content of the electrospun polyaniline composite fiber influences the fiber morphology by changing the content of the PANI or the polymer to make a stable jet during electrospinning, which forms a stable fiber. The diameter and the electrical conductivity of the polyaniline composite fibers have increased and the mechanical properties have been enhanced under specific conditions [6].

Five groups of doped polyaniline solutions with different mass fractions were designed, 1 wt.%, 2 wt.%, 3 wt.%, 4 wt.%, 5 wt.%, PAN content in solution remains at 10 wt.%, the molar masses of the three dopants are the same as those of polyaniline, the three dopants are p-Toluenesulfonic acid (PTSA), Dodecyl benzene sulfonic acid (DBSA), hydrochloric acid(HCL), the relationship between nano-composite fiber membrane morphology and dopant species was analyzed.

2 MATERIALS AND METHODS

2.1 Materials and equipment

Polyaniline (PANI), p-Toluenesulfonic acid (PTSA) and polyacrylonitrile (PAN) (MW 150,000) was purchased from Beijing Kuol Chemical Technology Co., Ltd.. (Beijing, China). Dodecyl benzene sulfonic acid (DBSA) was purchased from Zhanfan Chemical co., LTD. (Guangzhou, China). Hydrochloric acid was purchased from Damao Chemical Reagent Factory (Tianjin, China). N,N-Dimethylformamide (DMF) was purchased from Tianjin Zhiyuan Chemical Reagent Co., Ltd (Tianjin, China). All chemicals were analytical reagent and were used without alteration. The microcomputer type conductivity meter DDBJ-350 was purchased from Qiwei instrument co., LTD. (Hangzhou, China). The high pure water system EXPLORER was purchased from Chenhuang Scientific Instrument co., LTD (Shanghai, China). The high voltage DC power supply DW-P503-1ACDF was purchased from Dongwen High Voltage Power Supply co., LTD (Tianjin, China). The two-channel trace injection pump WZS-50F6 was purchased from Smith Medical Instrument co., LTD (Zhejiang, China).

2.2 Methods

Electrospinning PAN, 6%, 10%, and 12% pure PAN nanofilms, it was found that 10% PAN nanomembranes had the best spinning effect, then add 1%, 2%, 3%, 4%, 5% equimolar mass of polyaniline and dopant in 10% PAN solution in DMF. The solution was first dispersed in an ultrasonic cleaning apparatus for 10 minutes, and then the solution was returned to a magnetic stirrer and stirred for about 12 hours to obtain a uniform and stable solution. The doped polyaniline and PAN solution was put into a needle. The distance between the metal needle and the receiving plate was 10cm, the voltage was 15.00kv, and the pump flow rate was 1ml/h.

2.3 Morphological characterization of doped-PANI/PAN electrospun fibers membrane

A scanning electron microscope (Hitachi S-4800) was used to quantify the morphology and the surface of the electrospun fibers. The samples were sputtered with gold and pictures were taken. The diameters of more than 50 electrospun fibers were measured via scanning electron microscopy (SEM) images using Image J software. The average diameters and standard deviations were calculated.

2.4 Fiber membrane strength test

The equipment used for this fiber membrane strength test was a 3365-INSTRON strength tester. For each pattern, take 6 sets of rectangular specimens of 1cm×10cm×0.1mm to obtain fiber membrane strength data, calculate the average fiber strength based on the formula.

Calculate the average fiber strength according to the formula:

$$\sigma = P \div A \tag{1}$$

In the formula: σ − Breaking strength or volumetric strength (Pa), P − Breaking force (N), A − Cross section area (m2).

3 Results and Discussion

3.1 Electrospun PAN nanofiber membrane

3.1.1 SEM analysis of PAN electrospun nanofiber membrane

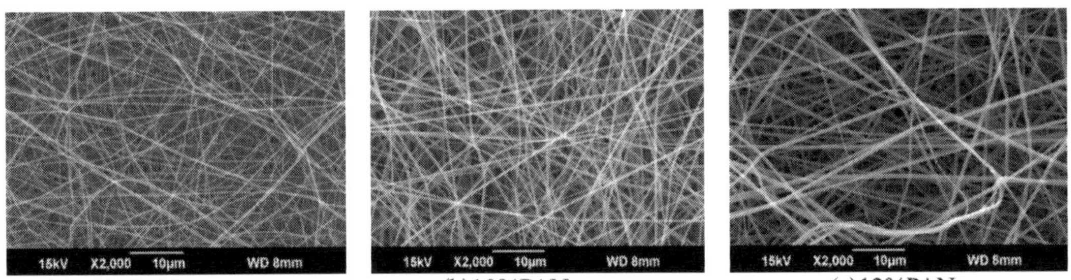

(a) 8%PAN (b)10%PAN (c)12%PAN

Figure 1. SEM of polyacrylonitrile nanofiber membranes with different concentrations

Figure 1 shows the morphological structure of the nanofiber membranes under different concentrations of PAN spinning solution obtained after 2,000-fold magnification. The fibers of the three nanofiber membranes are uniform and smooth. The diameters of nanofibers with concentrations of 8%, 10%, and 12% were 121 nm, 173 nm, and 299 nm respectively. It is easy to see that the fiber diameter increases with the increase of PAN concentration.

3.2 DBSA-PANI/PAN NANOFIBER MEMBRANE

3.2.1 SEM analysis of DBSA-PANI/PAN nanofiber membrane

(a)PAN10%+PANI-DBSA1% (b)PAN10%+PANI-DBSA2% (c)PAN10%+PANI-DBSA3%

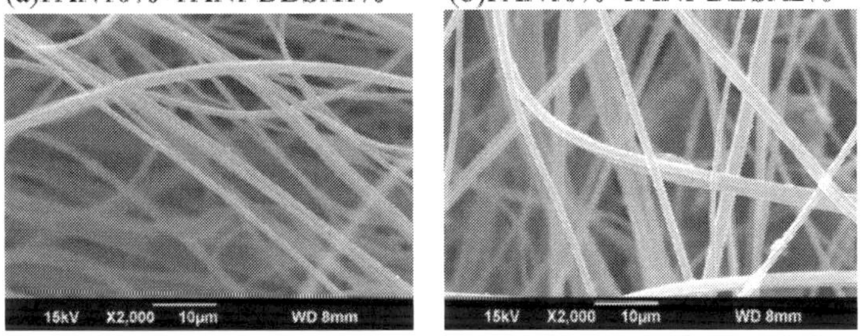

(d)PAN10%+PANI-DBSA4% (e) PAN10%+PANI-DBSA5%

Figure 2. SEM of fiber membranes at different concentrations of PANI-DBSA

In Figure 2(a), (b), (c), (d) and (e) represent SEM of 1%, 2%, 3%, 4%, and 5% DBSA-PANI electrospun membranes, the average fiber diameters are 910 nm and 890 nm, 830 nm, 780 nm, and 880 nm respectively. Electrospun nanofibers have a relatively smooth dry phase and a uniform fiber diameter. With the increase of polyaniline and dodecyl benzene sulfonic acid concentration, the average fiber shows a tendency to decrease, the average diameter of the fiber reaches a minimum value when the concentration of polyaniline and dodecyl benzene sulfonic acid increases to 4%. And the concentration reaches 5%, the mean diameter of the fiber increases and the average diameter and concentration of the fiber are almost the same as the average diameter at 2%.

3.2.2 Strength test of DBSA-PANI/PAN nanofiber membrane

Strength of 1%, 2%, 3%, 4%, and 5% DBSA-PANI electrospun membranes are2.12×10^5Pa, 1.42×10^5Pa, 7.4×10^4 Pa, 1.8×10^5Pa and 6.1×10^4Pa respectively. As the content of doped polyaniline increases, the fiber strength gradually decreases, but at a concentration of 4%, the strength reaches a peak and then decreases. Its strong value rule is not very obvious and has a certain degree of randomness.

3.3 HCL-PANI/PAN NANOFIBER MEMBRANE

3.3.1 SEM analysis of HCl-PANI/PAN nanofiber membrane

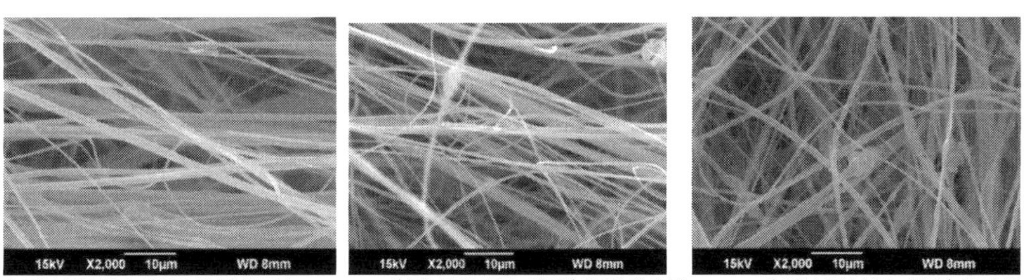

(a)PAN10%+HCl-PANI1% (b)PAN10%+HCl-PANI 2% (c)PAN10%+HCl-PANI 3%

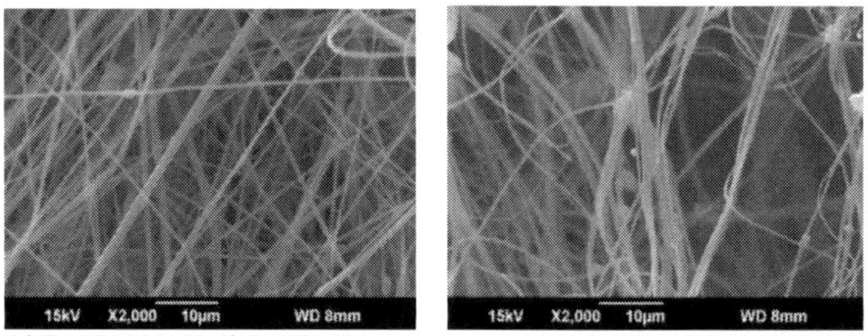

(d)PAN10%+HCl-PANI 4% (e)PAN10%+HCl-PANI 5%

Figure 3 SEM of fiber membranes at different HCl-PANI concentrations

In Figure 3(a), (b), (c), (d) and (e) represent SEM of 1%, 2%, 3%, 4%, and 5% DBSA-PANI electrospun membranes, the average fiber diameters are550nm, 450nm, 480 nm, 420nm and 610nm respectively. The fiber diameter is uniform, the strips are clear, there are no obvious impurities, and the film formation is relatively good. With the increase of HCl-PANI concentration, the diameter of nanofiber membranes showed a decreasing trend overall and fluctuate up and down within a small range. When the concentration of polyaniline and hydrochloric acid was increased to 4%, the average diameter of the fibers reached a minimum. And when the concentration reached 5%, the average diameter of the fibers rapidly increased.

3.3.2 Strength test of HCl-PANI/PAN nanofiber membrane

Strength of 1%, 2%, 3%, 4%, and 5% HCl-PANI electrospun membranes are 6.6×10^4Pa, 5.7×10^4Pa, 6.3×10^4Pa, 1.08×10^5Pa and 5.3×10^4P respectively. As the content of doped polyaniline increases, the fiber strength gradually decreases, but at a concentration of 4%, the strength reaches a peak and then decreases. Its strong value rule is not very obvious and has a certain degree of randomness. With the increase of the concentration, the strength of the fibrous membrane showed a small fluctuation at first and then decreased. When the concentration increased to 4%, the strength of the fibrous membrane reached a maximum, and then the strength of the fibrous membrane showed a downward trend. Because that the solution concentration increases polyaniline content in the composite fiber system increases, the fiber membrane strength increased significantly, but increased to a certain concentration, the solubility of solute in the solution decreased, the fiber membrane strength instead showed a downward trend.

3.4 PTSA-PANI/PAN NANOFIBER MEMBRANE

3.4.1 SEM analysis of PTSA -PANI/PAN nanofiber membrane

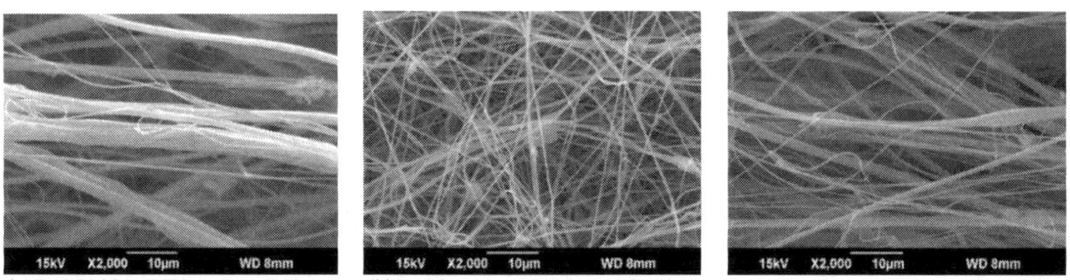

(a)PAN10%+PTSA-PANI1% (b)PAN10%+ PTSA-PANI 2% (c)PAN10%+ PTSA-PANI 3%

Figure 4. SEM of fiber membrane at different PTSA-PANI concentrations

In Figure 4(a), (b) and (c) represent SEM of 1%, 2% and 3% PTSA-PANI electrospun membranes, the average fiber diameters are580 nm, 430nm and 530nm respectively. With the increase of the concentration of doped polyaniline, the average fiber diameter showed a trend of decreasing first and then increasing. There was obvious adhesion between the fibers, and a small amount of spindle-shaped spherical particles appeared. With the increase of polyaniline concentration, the viscosity of the solution increased, and the adhesion was more obvious.

3.4.2 Strength test of PTSA-PANI/PAN nanofiber membrane

Strength of 1%, 2%, 3% and 4% PTSA-PANI electrospun membranes are 2.4×10^4Pa, 3.9×10^4Pa, 1.1×10^4Pa and 2.1×10^4Pa respectively. With the increase of the concentration of doped polyaniline, the strength of the fiber membrane showed a trend of increasing first and then decreasing. At a concentration of 2%, the strength of the fiber membrane reached a maximum value, and the strength of the fiber membrane decreased afterwards.

4 Conclusion

Under the same conditions, the mean diameters of the fibrous membranes of HCl/PANI/PAN and PTSA/PANI/PAN were similar and the smallest, and the average diameter of the DBSA/PANI/PAN membrane was the largest. Under the same conditions, the fiber membrane strength of DBSA/PANI/PAN was the highest, the fiber membrane strength of HCl/PANI/PAN was the second, and the strength of PTSA/PANI/PAN fiber membrane was small. Hydrochloric acid as an inorganic small molecule protonic acid, its polyaniline-doped solubility is low, the small molecule acid is easily dedoped, resulting in poor mechanical properties of the doped polyaniline fiber membrane. Dodecyl benzene sulfonic acid has a large molecular mass and a large size, the binding capacity of the molecular chain is improved after doping, so that the final fiber membrane exhibits a higher strength, resulting in a higher strength of the final fibrous membrane. The nanostructured polyaniline fiber has unique physical and chemical properties. When the size of the polyaniline material is at the nanometer level, due to its structural advantages, it has a larger size, while it has a smaller size effect and a tunneling effect than the conventional fiber. The application prospect of doped polyaniline nanofiber film is broad.

References

1. Jung, Y., Baeg, K. J., Kim, D. Y., Someya, T., Park, S. Y.. Synth. Met., 159 (2009)
2. Engin, F. Z., Usta, İ. J. Text. I., 106(2015)
3. Lakshmidevi, V., Yelamaggad, C. V., Venkataraman, A.. ChemistrySelect, 3 (2018)
4. Dural-Erem, A., Erem, H. H., Ozcan, G., Skrifvars, M.. J. Text. I., 106(2015)
5. Nurwaha, D., Han, W., Wang, X.. J. Text. I., 104(2013)
6. Sujith, K., Asha, A. M., Anjali, P., Sivakumar, N., Subramanian, K. R. V., Nair, S. V., Balakrishnan, A.. Mater. Lett. , 67(2012)

Wheel Loader Driving Intention Recognition with Gaussian Mixture - Hidden Markov Model

Guoxiang Cao[1], Anlin Wang[1,a] and Donghuan Xu[1]

[1] *Research Institute of Mechanical and Electronic Engineering, School of mechanical and energy engineering, Tongji University, Shanghai 200029, China*

Abstract. Accurate recognition of driving intentions can delay upshifts under the intention of quick acceleration to maximize vehicle power performance; avoid frequent gear changes in automatic transmissions for rapid deceleration intention and make all power to flow to the bucket in the desire for fast motion of cylinders. However, due to the ambiguity of the human intentions and multiple meanings of depressing on the accelerator pedal in wheel loader, it is difficult to recognize driving intention. Nevertheless, the driver's intentions are directly reflected in the accelerator pedal, brake pedal and hydraulic valve control handle. By detecting these observable signals such as the signals of acceleration pedal's displacement and velocity, brake pedal's displacement and velocity and valve status Gaussian Mixture - Hidden Markov Model(MGHMM) can recognize the unobservable driving intentions. The experiment is done in Simulink and the results show that MGHMM can recognize driving intentions as expected.

1 Introduction

Wheel loader often works in construction site and the bumpy road condition brings shakes to the driver, which always fatigue the driver. In addition, frequent shifts also intensify fatigue of the driver. The application of automatic transmission in wheel loader, however, relieve driver's fatigue, but the automatic transmission makes the decisions of gearshift only when the throttle opening and vehicle speed satisfy the shift requirements. The method is effective in car, but the condition in wheel loader is different. The wheel loader has mainly two missions digging and transporting while the car only has the latter, which leads that pushing the acceleration pedal down may be not to accelerate the loader body but accelerate to tip bucket. Therefore, in order to prevent the automatic transmission from issuing wrong instructions, it is necessary to accurately recognize the driving intention. On the other hand, classic two-parameter gearshift schedule cannot avoid the problem of frequent gearshift in rapid deceleration. However, if gearshift intention can be recognized accurately, the problem of frequent gearshift in rapid deceleration would be avoid by forcing downshift.

Human behaviour intentions have uncertainties and the increasing the throttle opening has multiple meanings. The above two points make the recognition of driving intention more difficult. For the recognition of driving intention man has raised many methods. Since the problem of driving intention recognition is basically pattern recognition and condition monitoring problem, Neural Network [1-4] and other machine learning methods are suitable to solve the problems. These methods are based on a great deal of experimental data. Fuzzy Logic Control [5] is also applied in driving intention

[a] Corresponding author: wanganlin@tongji.edu.cn

recognition. Similarly, Fuzzy Logic Control need rich experience and a large amount of time to adjust the controlled system. The above methods cannot efficiently solve the problem of wheel loader driving intention identification. However, some recent endeavours in this direction in car using a popular methodology Gaussian Mixture - Hidden Markov Models (MGHMM) [6] such as work from H Hou [7], H Berndt [8] et al is fruitful. It can recognize the implicit parameter of temporal data patterns and build the recognition model, and the driving intention recognition itself is a problem of Hidden Markov because human behaviour is made up of a sequence of internal "mental" states which is not directly observable, but it can be inferred from observable signals, such as acceleration pedal's displacement and velocity as well as brake pedal's displacement and velocity.

In this paper five input signals are considered including signals of acceleration pedal's displacement and velocity, brake pedal's displacement and velocity and valve status. And seven output results are Fast Slowdown, Slowdown, Natural, Keep, Acceleration, Rapid Acceleration and Valve Action. Through the trained MGHMM models driving intention can be recognized online in MATLAB Simulink.

2 Gaussian Mixture - Hidden Markov Model

The Hidden Markov Model is an extension and development of the Markov Model. It is a probabilistic model used to describe the statistical properties of stochastic processes. And it is a double stochastic process. The Hidden Markov Model showed as figure 1 consists of a markov chain and a general random process.

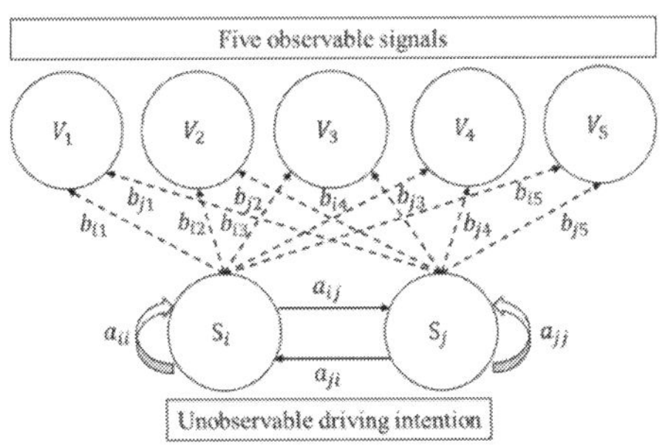

Figure 1. The Hidden Markov Model of unobservable driving intentions and observable signals

The transition probability matrix of markov chain describes the transition between two states. In general, the relationship between the state and the observed sequence is described by observation probability matrix.

HMM is defined as follows:

1. **X** represents a set of states.

$$\mathbf{X} = \{S_i | 1 \le i \le N, i \in Z\} \tag{1}$$

where N is the number of states.

2. **O** represents a set of observable sequences.

$$\mathbf{O} = \{V_i | 1 \le i \le M, i \in Z\} \tag{2}$$

where M is the number of observations that may be output from each state.

3. The state transition probabilities are

$$\mathbf{A} = \{a_{ij}\}, \tag{3}$$

where $a_{ij} = P[q_{t+1} = j | q_t = i], 1 \le i,j \le N$.

4. The observation probability matrix of State j is

$$\mathbf{B} = \{b_j(k)\} \tag{4}$$

where $b_j(k) = P[O_t = V_k | q_t = S_j], 1 \le j \le N, 1 \le k \le M$. B is the probability of output value.

5. The initial state distribution is

$$\boldsymbol{\pi} = \{\pi_i\} \tag{5}$$

where $\pi_i = P[q_i = S_i], 1 \le i \le N$.

Hence HMM can be characterized by equation:

$$\boldsymbol{\lambda} = (X, O, \pi, A, B) \tag{6}$$

HMM mainly solves the problem of evaluation, decoding and learning. The corresponding algorithms are Forward method, Viterbi method and Baum–Welch method.

Driving behaviour is a continuous behavior of time. For continuous observation sequences, the Gaussian Mixed Model is adopted. Then the matrix **B** in equation 4 is not a matrix but a probability distribution function of observation values.

$$b_j(o) = \sum_{k=1}^{M} c_{jk} \mathcal{N}\left(o, \mu_{jk}, U_{jk}\right), 1 \le j \le N \tag{7}$$

where c stands for the mixture coefficient for the kth mixture in the jth state. $\mathcal{N}$ stands for the probability distribution function of a normal distribution with mean μ and covariance U measured at observation o. Mixture coefficient c satisfies the following constraints:

$$\sum_{k=1}^{M} c_{jk} = 1 \quad 0 \le c_{jk} \le 1, 1 \le j \le N, 1 \le k \le M \tag{8}$$

Therefore, MGHMM can be expressed as

$$\boldsymbol{\lambda} = (X, O, \pi, A, c, \mu, U_{jk}) \tag{9}$$

The data to be learned is input into the model. When the model converges, it is considered that the model training has been completed.

3 The driving intention recognition model

3.1 The structure of driving intention recognition model

The driving intention of the driver in the time series is a set of decisions for each minute period under each specific work environment. That is, the driving intention recognition of each tiny period can show the change of the driver's overall driving intention and MGHMM can realize the long-term driving intention recognition by analysing short-term driving behaviour data. The figure 2 shows MGHMM recognition model structure. Considering the driver's driving intention is directly reflected in the accelerator pedal, brake pedal and hydraulic valve handle. The signals of accelerator pedal's displacement and velocity, brake pedal's displacement and velocity and hydraulic valve's switch state are chosen as input data of MGHMM. In each working cycle of the program, the seven trained models, Fast Slowdown(FS) MGHMM, Slowdown(S) MGHMM, Natural(N) MGHMM, Keep(K) MGHMM, Acceleration(A) MGHMM, Rapid Acceleration(RA) MGHMM and Valve Action(VA) MGHMM, will rapidly process the input-data with Maximum Likelihood Estimation(MLE) and the name of MGHMM corresponding the maximum value is output as the current short-term driving behaviour.

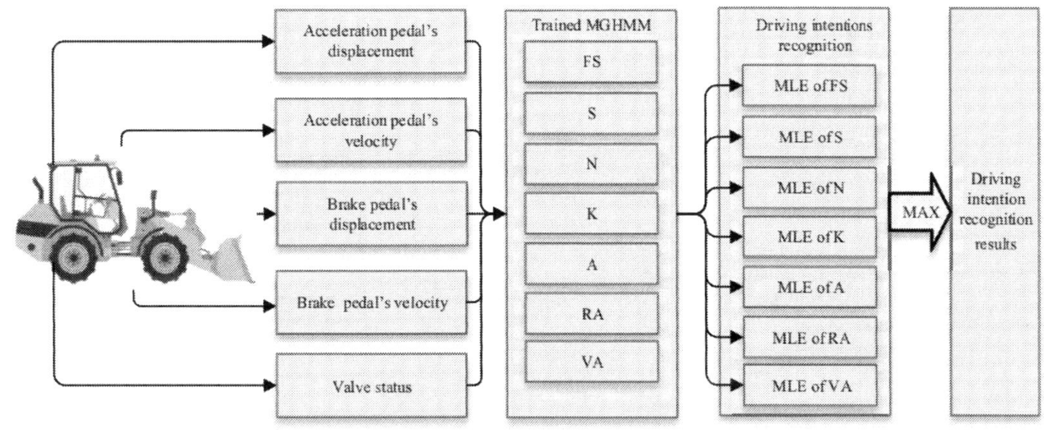

Figure 2. The structure of driving intention recognition model

3.2 The training of MGHMM model

All the driving behaviour MGHMMs are trained independently in MATLAB Simulink using the Baum-Welch algorithm which is an 'iterative update' algorithm to construct an HMM fitting given observation sequences. Thus, the parameters of a certain short-term driving behaviour model are optimized gradually for given learning data until convergence within a certain range. The following figure 3 shows the methods of learning data processing and MGHMMs model training processing.

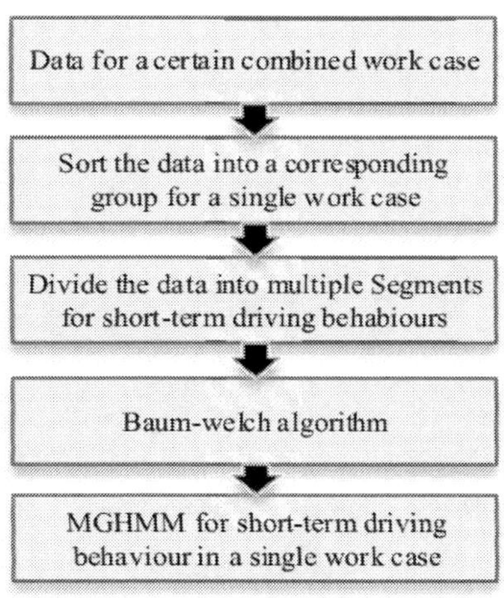

Figure 3. The method of learning data processing and MGHMMs training processing

First of all, data for a certain combined work case is collected. Then, the certain combined work case data is divided into a plurality of single work case data, and then a single work case data is divided into a plurality of time series segments. Finally, Baum-Welch algorithm is used to train each single work case MGHMM. In the experiment seven MGHMMs are finally trained, for example, When the speed of the accelerator pedal is greater than 0.2/s, the current driving intention, in this paper, is considered to be Fast Acceleration. And all the observed signals of Fast Acceleration work cases are divided into a group for Fast Acceleration MGHMM training.

4 Results of driving intention recognition

 https://doi.org/10.1051/matecconf/201823703001

After all the MGHMMs are trained, all the optimal parameters are imported into the workspace of MATLAB. The typical work conditions shown in the following figure 4 for model verification are selected. After the data to be recognized is imported into the trained models, the maximum likelihood estimators are output by the all trained models. The model name corresponding to the maximum of all MLEs is the current recognition result.

In order to improve the recognition accuracy, the input data is filtered and the recognition result is shown in the following figure 5.

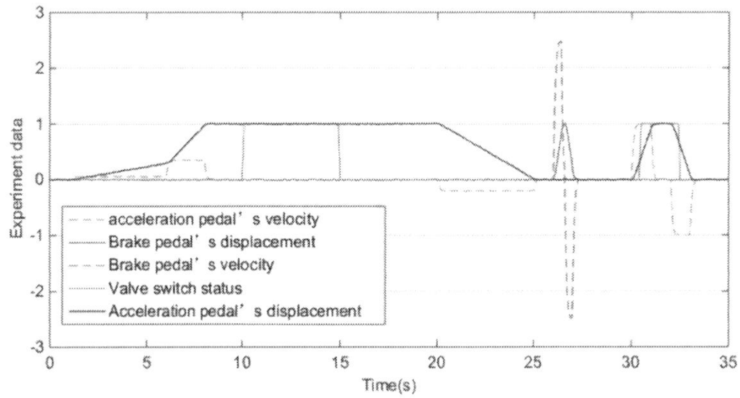

Figure 4. The typical work conditions of wheel loader

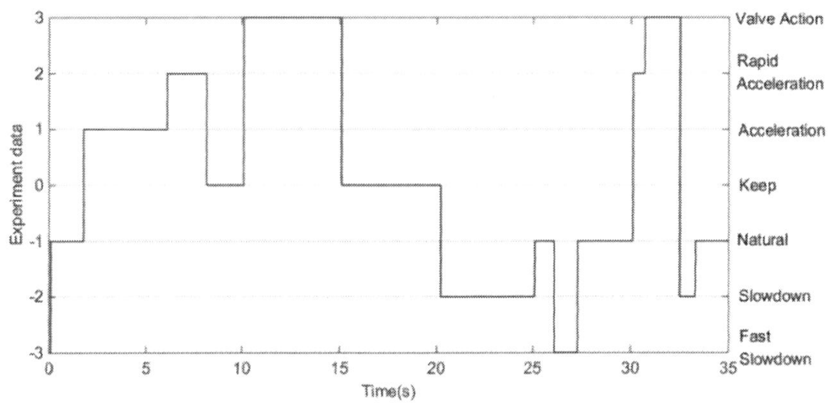

Figure 5. The results of driving intention recognition with MGHMM

The solid line in the figure 5 represents the recognition results of the driving intention. From the figure, it can be concluded that the MGHMM method can accurately identify the driver's driving intention, for example, in a period of 1st second to 6th second, the accelerator pedal speed is 0.06/s and other input signals almost zero. At the same time, the driving intention recognition result is Acceleration. In another period of 10st second to 15th second, the driving intention recognition result is Valve Action because the valve state is 1. Many experiments prove that the time step of 0.1s can achieve the highest recognition accuracy and meet the real-time requirement of online recognition.

5 Discussion and Conclusion

Because of the multiple meanings of throttle, depressing the gas pedal of a car represents the desire of acceleration, while depressing the gas pedal of wheel loader may be not to accelerate vehicle but to accelerate the tipper or lifting work. Hence, it is difficult to recognize driving intention of wheel loader. However, due to the driver's intention directly reflected in the accelerator pedal, brake pedal and hydraulic valve control handle, MGHMM method can be used to recognize unobservable driving intentions with these observable signals. In order to improve the recognition accuracy and satisfy the real-time requirement of online recognition, the input data is divided into small fragments of 0.1s. Thus, this allows the TCU (Transmission Control Unit) to quickly control the transmission hydraulic system to shift gears based on the identified driving intention. Ultimately the controllability of the gearshift can be ameliorated and work efficiency improved.

In this paper, the driver's intention recognition model uses only five signals that are directly related to the driver. If further research is carried out and vehicle information and traffic information, such as current vehicle speed and current road gradient information, are considered. The MGHMM can recognize more accurately driving intentions to provide better driving assistance.

Acknowledgement

This research is supported by National Key Research and Development Program(2017YFC0704000).

References

1. Q Wang, X Tang, P Wang, et al. *Driving intention identification method for hybrid vehicles based on neural network*. Transactions of the Chinese Society for Agricultural Machinery, **43**(8):32-36(2012).
2. C Wu, Y Li, X Ma, et al. *A Recognition Model for Lane Change Intention Based on Neural Network*, International Conference on Transportation Information and Safety. 125-130(2013).
3. C Zong, C Wang, D Yang, et al. *Driving intention identification and maneuvering behavior prediction of drivers on cornering*, International Conference on Mechatronics and Automation. 4055-4060(2009).
4. R Zhang, X Yan, C Wu. *A Recognition Model for Acceleration Intention of Automobile Drivers Based on Fuzzy Clustering*, International Conference on Transportation Information and Safety. 1938-1947(2011).
5. J Heine, M Sylla, I Langer, et al. *Algorithm for Driver Intention Detection with Fuzzy Logic and Edit Distance*, IEEE, International Conference on Intelligent Transportation Systems. IEEE, 1022-1027(2015).
6. Rabiner, L.R. *A tutorial on hidden Markov models and selected applications in speech recognition*. Proc. IEEE,**77**(2):257-286(1989).
7. H Hou, L Jin, Q Niu, et al. *Driver Intention Recognition Method Using Continuous Hidden Markov Model*. International Journal of Computational Intelligence Systems, **4**(May 2011):386-393(2011).
8. H Berndt, J Emmert, K Dietmayer. *Continuous Driver Intention Recognition with Hidden Markov Models*, International IEEE Conference on Intelligent Transportation Systems. IEEE, 1189-1194(2008).

MATEC Web of Conferences **237**, 03002 (2018) https://doi.org/10.1051/matecconf/201823703002
D2ME 2018

Assessment on Structural Integrity of In-service Machine Using De-noised Vibrational Modal Data and Artificial Neural Network

Zhi Chao Ong[1,a], Ee Teng Yap[1], Zubaidah Ismail[2] and Shin Yee Khoo[1]

[1] *Department of Mechanical Engineering, Faculty of Engineering, University of Malaya, 50603 Kuala Lumpur, Malaysia*
[2] *Department of Civil Engineering, Faculty of Engineering, University of Malaya, 50603 Kuala Lumpur, Malaysia*

Abstract. The recent oil price drop creates a demand for swift action within oil and gas industry to shift focus from increasing daily production rates, to optimizing existing assets in achieving growth. Industrial machinery, one of the industry's key asset many times failed due to high amplitude vibration that contributes to accelerated wear and tear and subsequently results in high cycle fatigue failure. As such there is a need to develop a structural integrity assessment for in-service machinery for continuous and safe operation. Vibration-based method such as Experimental Modal Analysis (EMA) is widely used for damage detection on civil and piping system under stationary environment. However, in industrial applications, system shutdown is very costly. EMA is also undesirable in this case due to the dominant ambient and system disturbances on the in-service system. An alternative method called Impact-Synchronous Modal Analysis (ISMA) is developed to perform modal analysis under noisy environment. Applying the ISMA technique in de-noising the non-synchronous disturbances at upstream could generate a cleaner and static-like modal data downstream for analysis. Artificial Neuron Networks (ANN) is then applied extensively in structural damage identification purposes based on changes in modal data due to its excellent pattern recognition ability. By leveraging on the latest technologies, i.e. ISMA and ANN as proposed, it allows real-time monitoring of assets, in this case, the machines, as well as the ability to transform continuous streams of data into useful information to predict damages.

1 Introduction

The recent drop in oil price creates a demand for swift action to shift focus from increasing daily production rates to optimizing existing assets in achieving growth. Development of a structural integrity assessment could allow continuous monitoring the health of the system and notify the end-user prior to any failure and to promote high health, safety, environment and quality (HSEQ) and cost efficiency to achieve the organizations operational and financial objectives.

Detection of damage is one of the most important factors in maintaining the integrity and safety of structures. Vibration-based methods are a global and non-destructive damage identification approach

[a] Corresponding author : alexongzc@um.edu.my

that has been considered for use on the dynamic properties of structure to identify the location and severity of damage without prior information of the type or level of damage. Many researchers focused on damage detection of civil structures and piping system where clean vibrational modal data could be obtained using the conventional Experimental Modal Analysis (EMA) under stationary environment [1-5]. However, the suitability of the modal parameters in damage detection is heavily affected by the ambient conditions and the consistency and reliability of the testing procedures. Downstream modal parameters could be contaminated by measurement error as well as modal extraction error. In addition, in industrial applications, shutting down the system could induce high downtime cost to the organizations. EMA is also undesirable in cases where it is to be performed on the in-service systems where the ambient and system disturbances dominate. It will be an uphill challenge when the same algorithms are applied in a noisy environment of in-service systems.

Most of vibration-based damage identification techniques are considered as some form of pattern recognition since they look for differences between two categories, e.g. before and after a structure is damaged or differences in damage levels. Artificial Neural Networks (ANNs) have been applied extensively in recent years due to their excellent pattern recognition ability, which is useful for structural damage identification purposes based on the change in modal analysis data [6-8]. Previous researches show that ANN and pattern recognition techniques have been mostly applied on stationary civil structures, beam-like and pipe-like structures damage detection. However, the nature of noisy ambient and system disturbances from in-service systems remain the main challenge for the researchers because it produces noisy modal data for damage detection or structural integrity assessment.

An alternative modal analysis method known as Impact-Synchronous Modal Analysis (ISMA) is developed which allows modal analysis to be performed in a noisy environment, when structure is in service and yet could provide a static-like vibrational modal data through de-noising processes [9, 10]. ISMA is considered a more practical technique to adopt in obtaining the modal data for the use of damage detection. Therefore, this study investigates a different form of parameters at the upstream of modal data, i.e. real time de-noised frequency response vectors combined with ANN with the aim to develop a damage detection scheme under noisy environment in a more practical way. Applying the ISMA technique which is a practical approach in de-noising the non-synchronous disturbances at the upstream could generate a cleaner and static-like modal data at downstream for a more accurate damage detection scheme under ANN environment.

2 Methods

2.1 Experimental Set-up

A rectangular Perspex plate with a dimension of 48cm x 20cm x 0.9cm (width x height x thickness), weighting 1.1kg, was used as the test specimen as shown in Figure 1 to simulate a similar vibration behaviour of an automobile. The plate was ground supported using screws at each of the four corners. It was connected to the steel plate and aluminium supports at every corners. The ground supports acted as the suspension/spring components of a typical automobile wheels.

The plate was divided into nine (9) measurement points. Since the plate is treated as an automobile, no sensor can be mounted on the wheel (located at the edges of the plate; point 1, 3, 7 and 9). By mounting it on the body it is still possible to find the damages or loss of stiffness at the wheel or suspension. Therefore, five (5) accelerometers were mounted in symmetric order on the plate as shown in Figure 2. The accelerometers used were Wilcoxon Research model S100C accelerometer which has a built-in charge amplifier. The mounting method used on the accelerometer is by cyanoacrylate adhesive. A manual PCB model 086C03 impact hammer was used to create an impact at the edges of plate structure / ground support for analysing its dynamic behaviour due to impact excited vertically from the automobile wheels. As for the DAQ hardware system, National Instrument-Universal Serial Bus (NI-USB 9234) signal acquisition module was used in this setup to acquire six (6) dynamic signals (1 impact hammer as input signal and 6 accelerometers as output signals). The DAQ hardware system will send the acquired data to the computer software, LabVIEW 2013 for the post-processing to acquire the FRF. Modal analysis was performed in two conditions, i.e. stationary and in-service conditions using EMA and ISMA methods. To create the in-service condition, an electrodynamic shaker (LDS model V201) with external power amplifier (LDS model PA25E) was used to produce a harmonic excitation at 30Hz.

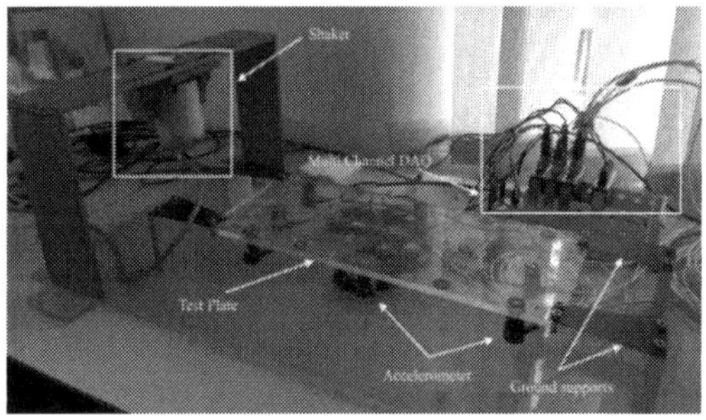

Figure 1. Experimental set-up of test rig

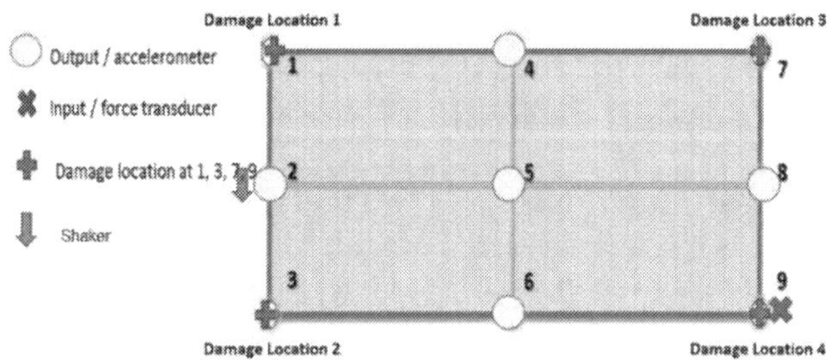

Figure 2. Sensor and damage location

2.2 Damage Identification Scheme

Since the plate simulate a four-wheel car, the location of the damage usually happens at the suspension part where it absorbs the impact done by the tire when hitting a pothole or rough road. To simulate a damage condition, the screws that hold the plate and ground supports together were used as the damage point. The screws will be removed and loosen to an extent of shift in natural peaks were observed to simulate the damage conditions under different severities. Table 1 shows the list of condition along with its description.

Table 1. Damage conditions and descriptions.

Conditions	Description
Undamaged **(UD)**	All four (4) points of the screws were tighten
High Severity Damaged Location 1 **(HD-1)**	Screws at point 1 were removed, as for point 3, 7, and 9 were tighten
High Severity Damaged Location 2 **(HD-2)**	Screws at point 3 were removed, as for point 1, 7, and 9 were tighten
High Severity Damaged Location 3 **(HD-3)**	Screws at point 7 were removed, as for point 1, 3, and 9 were tighten
High Severity Damaged Location 4 **(HD-4)**	Screws at point 9 were removed, as for point 1, 3, and 7 were tighten
Low Severity Damaged Location 1 **(LD-1)**	Screws at point 1 were loosen, as for point 3, 7, and 9 were tighten
Low Severity Damaged Location 2 **(LD-2)**	Screws at point 3 were loosen, as for point 1, 7, and 9 were tighten
Low Severity Damaged Location 3 **(LD-3)**	Screws at point 7 were loosen, as for point 1, 3, and 9 were tighten
Low Severity Damaged Location 4 **(LD-4)**	Screws at point 9 were loosen, as for point 1, 3, and 7 were tighten

Neural network pattern recognition toolbox in MATLAB was used in developing the damage identification scheme in this study. The toolbox solves pattern-recognition classification problems using a two-layer feed-forward pattern recognition neural network. It was trained to classify inputs according to the target classes. Total of 45 sets of FRF data at the frequency range 0Hz to 199.5Hz (9 damage conditions x 5 sensor points) were used and input into pattern recognition neural network for training. The size of input is 400 (0.5Hz frequency resolution gives 2x200 input data). Each set of data was trained with their target data of 1 at respective row to differentiate the condition and detect damage. The setting was set by dividing the input data into 90% of training data, 5% of validation data and 5% of testing data with the number of hidden neurons of 400. The data with confusion matrix approaching 100% indicates the data were well-trained. Any set of untrained data can be input into the trained network to identify the condition in the system.

3 Results and Discussions

3.1 Frequency Response Functions Analysis

From the FRF graph analysis, it is found that the peaks at each FRF for a specific condition is distinct and carrying its own dynamic behaviour. When the structure is damaged, the peaks at FRFs graph are shifted to the left side towards lower frequency range due to the loss of stiffness of the system. In addition, these FRFs obtained for a specific damage condition at their respective sensor point consistent and stable during the real time measurement This shows that directly-measured FRF is easy to obtain, reliable and sensitive to be used as input for damage detection process. In ISMA, ISTA averaging technique is utilized to de-noise the unaccounted forces exerted by the shaker. This can be seen from Figure 3 where no much differences between FRFs graph pattern of EMA and ISMA at the same sensor point under same damage condition is observed. Both the frequency and magnitude at which modes exist in both EMA and ISMA show very high similarity. Hence, it is proven that ISMA is reliable by providing static comparable FRF data through its de-noising method during in-service condition.

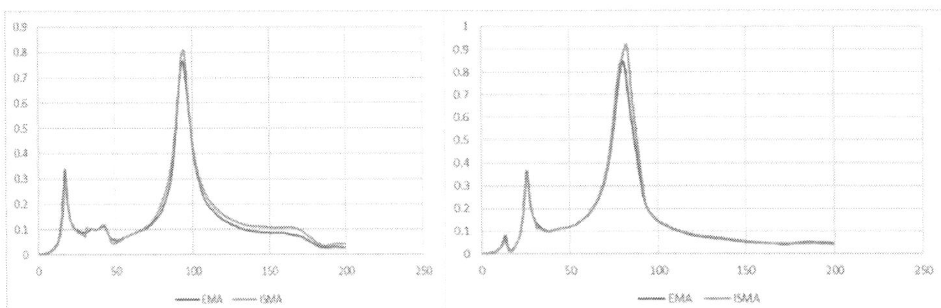

Figure 3. Comparison of EMA and ISMA FRFs for (a) UD condition, (b) HD-1 condition at Point 5

3.2 ANN Model Validation

The output data after training is Tabulated in Table 2, it is obvious that all the data are successfully trained with accuracy approaching 100%. It is observed that only sensor Point 2 and Point 8 which are slightly deviated from the target data of '1' with very small standard deviations of 0.001154 and 0.000767 respectively. Hence, this training network is considered a good damage detection scheme.

Table 2. Output data after training.

Condition / Point	UD	HD-1	HD-2	HD-3	HD-4	LD-1	LD-2	LD-3	LD-4	Standard Deviation
Point 2	0.9966	1.0000	1.0000	0.9988	1.0000	1.0000	1.0000	1.0000	1.0000	0.001154
Point 4	1.0000	1.0000	1.0000	1.0000	1.0000	1.0000	1.0000	1.0000	1.0000	0
Point 5	1.0000	1.0000	1.0000	1.0000	1.0000	1.0000	1.0000	1.0000	1.0000	0
Point 6	1.0000	1.0000	1.0000	1.0000	1.0000	1.0000	1.0000	1.0000	1.0000	0
Point 8	1.0000	1.0000	1.0000	1.0000	0.9977	1.0000	1.0000	1.0000	1.0000	0.000767

It is noted that all the conditions are successfully classified into their respective damage locations and severities for all sensor points with the maximum deviation of only 0.0034 (1.0000-0.9966) which is very approaching the target data (1.0000). Data of sensor Point 4, Point 5 and Point 6 are 100% similar with target data with no deviation. This show the consistency of the FRF data obtained and its suitability to be used as input in damage detection scheme. The damage detection scheme is validated to be successful under stationary condition in which sensor Point 4, Point 5 and Point 6 are recommended to be used for testing and actual implementation of the scheme.

Following this, the damage detection scheme is used to test and validate with other sets of untrained data. ISMA FRF data which collected under in-service condition is input into the damage detection scheme and the testing result is shown in Table 3.

Table 3. Testing of untrained ISMA data.

Condition / Point	UD	HD-1	HD-2	HD-3	HD-4	LD-1	LD-2	LD-3	LD-4	Standard Deviation
Point 2	0.9883	1.0000	0.9984	0.8430	0.9999	0.9969	1.0000	1.0000	0.9999	0.051780
Point 4	0.9997	1.0000	1.0000	0.9071	1.0000	1.0000	0.9969	1.0000	1.0000	0.030842
Point 5	0.9999	1.0000	1.0000	0.9999	0.9996	1.0000	1.0000	0.9998	1.0000	0.000136
Point 6	1.0000	1.0000	1.0000	0.9926	1.0000	0.9998	1.0000	0.9988	1.0000	0.002440
Point 8	0.9998	0.9999	0.9998	0.9967	0.9990	1.0000	0.9980	1.0000	1.0000	0.001164

It is noted that for HD-3 at Point 2, the deviation is more than 10% in which 0.8430 (84.30%) indicates high level of severity at damage location 3 while 0.1535 (15.35%) indicates low level of severity at the same damage location 3. This is due to the non-significance difference in the collected FRF data for these 2 conditions (HD-3 and LD-3) as shown in Figure 4. It is shown that the first peak of FRFs for high severity (HD-3) has minimal down shift in frequency compared with the low severity case (LD-3) whereas the rest of the peaks remain unchanged for both cases. This indicates the effect of damage severity at location 3 on the change of FRF at sensor Point 2 is minimal. It is also noted that the largest standard deviations of result data happened at Point 2 with 0.051780. Therefore, sensor Point 2 is not recommended to be used for testing or actual implementation of the scheme.

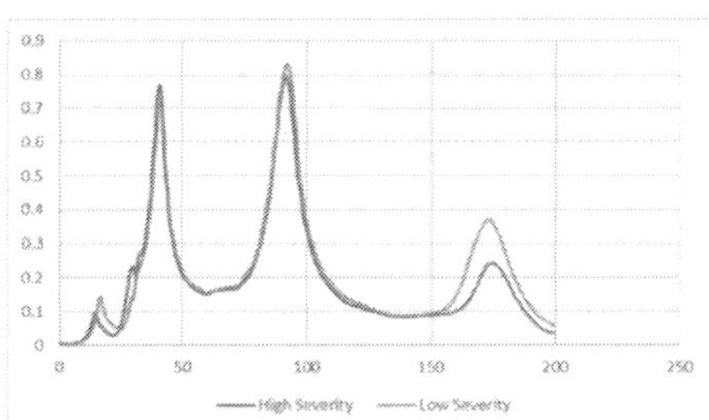

Figure 4. Comparison of FRFs of HD-3 and LD-3 condition at Point 2

Among all five sensor points used, it is observed that sensor Point 5 is the point with smallest standard deviation which is 0.000136. Hence, sensor Point 5 is considered as the best sensor point to be used for testing and actual implementation of the scheme under in-service condition. The developed damage detection scheme proves that the damage detection scheme is successfully developed with 100% accuracy in detecting damage in an in-service system and classifying them into respective locations and level of severities provided the suitable sensor point is selected for the scheme.

4 Conclusion

A damage detection scheme based on FRF is developed under ANN environment. The trained ANN network detected and classified the damages into respective locations and level of severities under both stationary and in-service conditions with high accuracy with the help of de-noising method used in ISMA. This happens due to three contribution factors. First, directly-measured FRF is one of the easiest dynamic responses to obtain in real-time, it is a more reliable and sensitive parameter to be used as input for damage detection process. Second, ISMA allows modal analysis to be performed during in-service condition, it gives static comparable FRF data through its de-noising method, i.e. Impact-Synchronous Time Averaging Technique. Third, the developed damage detection scheme using FRF is easy to use and well trained using the neural network pattern recognition toolbox and hence, able to detect and classify all the damages accurately.

References

1. Rahman, A.G.A., et al., *Study of open crack in rotor shaft using changes in frequency response function phase.* International Journal of Damage Mechanics, 2013. **22**(6): p. 791-807.
2. Ong, Z.C., A.G.A. Rahman, and Z. Ismail, *Determination of Damage Severity on Rotor Shaft Due to Crack Using Damage Index Derived from Experimental Modal Data.* Experimental Techniques, 2014. **38**(5): p. 18-30.
3. Ismail, Z. and Z.C. Ong, *Honeycomb damage detection in a reinforced concrete beam using frequency mode shape regression.* Measurement, 2012. **45**(5): p. 950-959.
4. Ismail, Z., et al., *Approach to Reduce the Limitations of Modal Identification in Damage Detection Using Limited Field Data for Nondestructive Structural Health Monitoring of a Cable-Stayed Concrete Bridge.* Journal of Bridge Engineering, 2012. **17**(6): p. 867-875.
5. Ismail, Z., *Application of residuals from regression of experimental mode shapes to locate multiple crack damage in a simply supported reinforced concrete beam.* Measurement, 2012. **45**(6): p. 1455–1461.
6. Hakim, S.J.S. and H.A. Razak, *Modal parameters based structural damage detection using artificial neural networks - a review.* Smart Structures and Systems, 2014. **14**(2): p. 159-189.
7. Hakim, S.J.S., et al., *Structural Damage Detection Using Soft Computing Method*, in *Structural Health Monitoring, Volume 5: Proceedings of the 32nd IMAC, A Conference and Exposition on Structural Dynamics, 2014*, A. Wicks, Editor. 2014, Springer International Publishing: Cham. p. 143-151.
8. Hakim, S.J.S., H. Abdul Razak, and S.A. Ravanfar, *Fault diagnosis on beam-like structures from modal parameters using artificial neural networks.* Measurement, 2015. **76**(Supplement C): p. 45-61.
9. Ong, Z.C. and C.C. Lee, *Investigation of Impact Profile and Isolation Effect in Automated Impact Device Design and Control for Operational Modal Analysis.* Journal of Dynamic Systems Measurement and Control-Transactions of the Asme, 2015. **137**(9).
10. Cheet, L.H. and O.Z. Chao. *Development of adaptive phase control impact device for enhancement of frequency response function in operational modal testing.* in *Proceedings of ISMA 2016 - International Conference on Noise and Vibration Engineering and USD2016 - International Conference on Uncertainty in Structural Dynamics*. 2016.

Optimal Design of Multi-section Proportional Directional Valve Throttle Grooves with Artificial Neural Networks

Xiaolu Zhang[1], Anlin Wang[1, a] and Jiangwei Tang[1]

[1] *Research Institute of Mechanical and Electronic Engineering, School of mechanical and energy engineering, Tongji University, Shanghai 200029, China*

Abstract. This paper presents a method for design multi-section proportional directional valve Throttle grooves with ANN method, which aims at getting a better flow stability. There exists a coupling matter during the opening and closing process between the throttling notches, so that it's difficult to parameterize the complex flow field characteristics *Cd* and the structure boundary of the spool grooves. However, in this paper, an ANN was built with data from CFD results, while the typical structural parameters (U type, the O-type and C-type), operating parameters was input vectors, the discharge coefficient as output vectors. Meanwhile, all of the needed data is taken from the three-dimensional CFD analysis, which are organized properly and verified by a bench scale test on a rig. Then, with throttling stiffness as optimization objective to evaluate flow stability, an optimal design process is carried out to optimize to optimize the structure of coupling grooves with ANN models and genetic algorithm. Ultimately, the optimized structure is verified better by the physical test on test rig, therefore, the significance of design method is proved.

1 Introduction

As the main power distribution element of the hydraulic excavator system, which is an open loop system, the multi-section proportional directional valve plays a vital role in the stability and reliability of the system. In the opening and closing process, there is multi-notch coupling throttling effect, while the mapping relationship between the topological structure of the throttling groove and its complex working conditions is difficult to parameterize. Due to the interaction of fluid and solid, the solid boundary affects the flow field performance directly.

As for the difficulty to realize the parameterization of flow field characteristics Cd during the design process of the multi-section proportional directional valve, many achievements have been made by researchers. With the help of the development of computer and computational fluid dynamics, JIANG Tao etc. realized the dynamic 3-D CFD simulation of complex structure sliding vale [1], and also tried to apply Neural Network Model into the process of the valve design .Wang Anlin [2-3] used the method of numerical fitting to fit the parameter Cd and reynolds number Re. Cavitation and the flow forces in hydraulic proportional directional valves by the means of experimental and numerical analysis [4-5] have been carried out by AMIRANTE R etc. As for the flow force acting on the spool

[a] Corresponding author : 1410289@tongji.edu.cn

of sliding valve, LISOWSKI E etc. [6-7] have made many efforts on the simulation method and realized the pressure compensation in a multi-section proportional directional control valve.

For the above analysis, this paper has taken a certain type of multi-section proportional directional valve as an example, based on the rig test to verify the validity of the three-dimensional CFD analysis and the means of artificial neural networks for prediction of discharge coefficient, to optimize the coupling groove s structure with the aim of better throttling stability. As a result, an ANN has been built with data from CFD results, while the structural parameters (U type, the O-type and C-type), operating parameters as input, the discharge coefficient as output. The effectiveness of optimization result has been verified.

2 object of research

2.1 The mechanism of the coupling throttling notches

In this paper, a certain type of multi-section proportional directional valve, which has been widely used in hydraulic excavator system to conduct the swing condition, is taken as the research object. One part of the pressure flow passes through the A port to the executing agency, and the other pressure oil passes out port C through the middle passage to the return tank.

Above all, the coupling flow satisfies the following:

$$Q = Q_1 + Q_2 \tag{1}$$

Where Q is the total flow inputting from port P, Q_1 is the flow flowing out port A, Q_2 is the flow flowing out port C.

$$Q_1 = C_{d1}A_1\sqrt{\frac{2(P - P_1)}{\rho}} \tag{2}$$

$$Q_2 = C_{d2}A_2\sqrt{\frac{2(P - P_2)}{\rho}} \tag{3}$$

Where Cd_1 is discharge coefficient of the throttle notch from port P to port A, Cd_1 is discharge coefficient of the throttle notch from port P to port C, A_1 is the throttling area about port A, A_1 is the throttling area about port C, A_1 is the throttling area about port A, P is the pressure in port P, P_1 is the pressure in port A, P_2 is the pressure in port C, ρ is the density of the fluid.

2.2 The design variable space

There are three types grooves used in this paper, which is O, C and U types. The spatial distribution is shown in Figure 1, and the same structure is distributed symmetrically around the circumference of the spool.

According the above prerequisites, the topology variable space **X**=[***Xa***, ***Xc***] are proposed basing the original model, that **Xa** are the design variable space for the port A, and the **Xc** is the design variable space of the port C. As shown in Figure 1, the parameters of port A, the O-type groove can be expressed by the parameter r_1;the U-type can be expressed by the r_2 ,and L_2 should keep a limit because of the continuity of the throttling groove; the C-type can be made by the a Parameter r_3 and r_2; that is, **Xa**=[r_1, r_2, r_3, L_2]; Among the parameters of port C; the U-type on can be expressed by parameter r_4, and L_3 should keep constant because of the total displacement of the spool ; C-type one can be r_5 by parameter, **Xc**=[r_4,r_5] . The design space of the swing valve topology is **X**=[**Xa**, **Xc**]=[r_1, r_2, r_3,L_2 r_4, r_5].

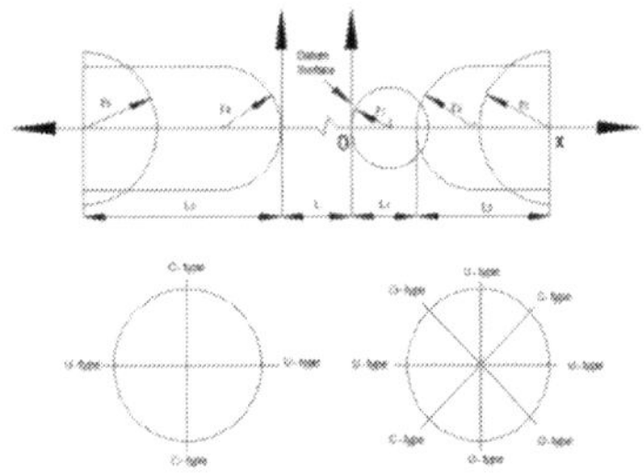

Figure 1. Topology diagram of the swing valve

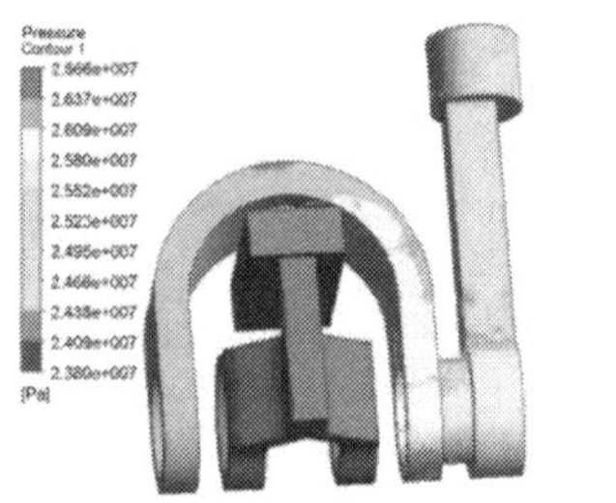

Figure 2. The pressure diagram of CFD result

3 CFD analysis and test

The multi-groove coupling throttling effect of multi-way valve is studied with the three-dimensional CFD technique[3, 6-7] being used to accurately describe the valve cavity flow field. At the same time, in order to verify the effectiveness of simulation methods, it is necessary to design corresponding tests to prove the accuracy and rationality of the three-dimensional CFD simulation.

3.1 CFD simulation models and settings

The three-dimensional model is established with the base of the actual valve port structure which refers to the template of a kind of commercial multi-way valve swing. With this as the object, the boundary conditions of the inlet and outlet are flow and outlet pressure for the fluid domain in the flow field. The fluid state setting for the calculation process is defined as: 1) the fluid is Newtonian fluid and incompressible; 2) the hydraulic oil density ρ is 890 kg/m^3, the dynamic viscosity μ is 0.036 Pa•s, the bulk modulus κ is 700 MPa; 3) the fluid state is turbulence, the standard k-ε model is used; 4) The spool displacement is from the zero opening position, in order of 0.2, 0.5, 1.5, 2, 2.5, 3, 3.5, 4, 4.5, 5.2mm; 5) Setting port A pressure p_{out} to 25Mpa, port C pressure to 6Mpa, inlet flow Qin to 250L/min. The simulation model and the calculation result cloud diagram are shown in Figure.2, and the flow curve of simulation result is shown in Figure. 4.

Figure3. The diagram of test system

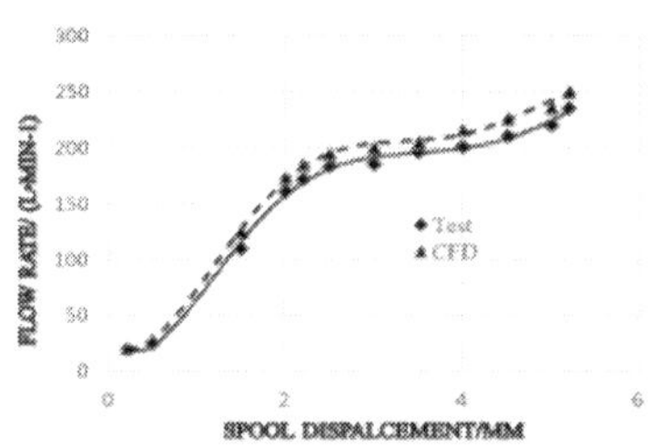

Figure 4. The contrast of the test result and the CFD result

3.2 Experimental verification of CFD simulation

To verify the effectiveness of the CFD simulation throttling at any position of valve opening [1-3], a system is required to accurately control and measure the displacement of the spool and adjust the pressure at the inlet and outlet of the tested valve. As shown in Figure.3, a test system is set up to measure the throttling characteristics of the swing work port. The schematic diagram includes the main pump, the tested multi-way valve swing, the loading valve, the spool micro-displacement control device, and the oil return measurement system.

For the simulation scheme and test system, the following settings (As shown in Table 1) are made for the test verification.

Table 1. The test setup and system parameters.

test setup		system parameters	
Displacement of the Spool /mm	0.2,0.5,1.5,2,2.5 ,3,3.5,4,4.5,5.2	Maximum displacement	112 ml/r
P_{out}	25 MPa	Rated pressure	35Mpa
Q_{in}	250L/min		

The experimental result is compared with the CFD simulation result so that the flow characteristics of the work port are obtained. The research shows that the CFD simulation and the test results of the flow characteristics of the work port have similar trends, so it can be considered that the CFD method can better simulate the actual work of the directional valve within the allowable range of error. The conclusion is consistent with the literature [2], this proves that the three-dimensional CFD simulation technique can be equivalent to bench test to some extent.

4 Mapping model and optimization

On the basis of the bench test to verify the effectiveness of three-dimensional CFD analysis, combined with the orthogonal test method, the structure parameters, working parameters combination and the mapped data of the throttle coefficient of the three types of throttling grooves structures are obtained respectively by the orthogonal test method, and then, the training samples of the ANN model are realized respectively. The neural network expressions of structural parameters, working conditions and throttling coefficients of port A and C are presented.

4.1 orthogonal test scheme and the ANN model

Table 2 is the orthogonal test table of port C, for example, table 3 is the orthogonal test table of port A. During the design process of orthogonal arrays, we need to satisfy the continuity of structure, and there are the following constraints in the orthogonal test table:

Table 2. The orthogonal test table of port C

Sample	r4/mm	r5/mm	level\factor	r4/mm	r5/mm
1	1.2	2	9	2	2
2	1.2	1.6	10	2	1.6
3	1.2	2.4	11	2	2.4
4	1.2	2.8	12	2	2.8
5	1.6	2	13	2.4	2
6	1.6	1.6	14	2.4	1.6
7	1.6	2.4	15	2.4	2.4
8	1.6	2.8	16	2.4	2.8

Table 3. The orthogonal test table of port A

Sample	r1/mm	r2/mm	r3/mm	L2/mm	Sample	r1/mm	r2/mm	r3/mm	L2/mm
1	0.9	1.2	1.6	3.2	9	1.2	2.4	2.4	3.2
2	1.2	1.6	1.6	3.6	10	0.9	2	2.4	3.6
3	1.5	2	1.6	4	11	1.8	1.6	2.4	4
4	1.8	2.4	1.6	4.4	12	1.5	1.2	2.4	4.4
5	1.8	2	2	3.2	13	1.5	1.6	2.8	3.2
6	1.5	2.4	2	3.6	14	1.8	1.2	2.8	3.6
7	1.2	1.2	2	4	15	0.9	2.4	2.8	4
8	0.9	1.6	2	4.4	16	1.2	2	2.8	4.4

According to the combination of each structure in the above orthogonal test table, the simulation model of the opening of the valve is 0.2,0.5,1.5,2,2.5,3,3.5,4,4.5,5.2mm in turn. For each model, the flow rate boundary is set 5, 55,105,155,205,255L/min.

The simulation results are processed through computer programming: the throttling area function A_1, A_2 of each structure of each structure is obtained, and the throttling coefficient of each group of simulation is calculated with the throttling formula $C_{d,i} = \frac{q_i}{A_i}\sqrt{\frac{\rho}{2\Delta p_i}}$.

As shown in Figure 5, A is a neuron network model of port A with [r1, R2, R3, L2, Xv, Q1] as input, [Cd1,A1] as output; Figure 6 shows a neuron network model of port C with [r4, r5, Xv, Q2] , [Cd1,A1] as output. For the total data, 15% of the data points were randomly extracted for use in testing. The remaining data points were used as training patterns. Both of the ANN models were one hidden layer, and the number of neurons in the hidden layer of the ANN model for port A and port C is 9 and 7.

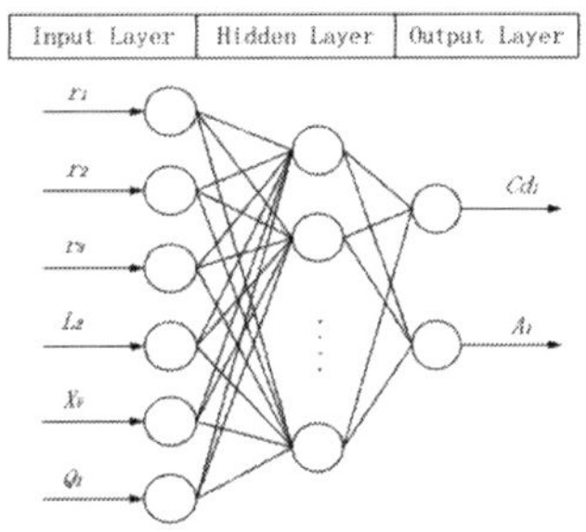

Figure 5. Schematic diagram for the ANN of port A

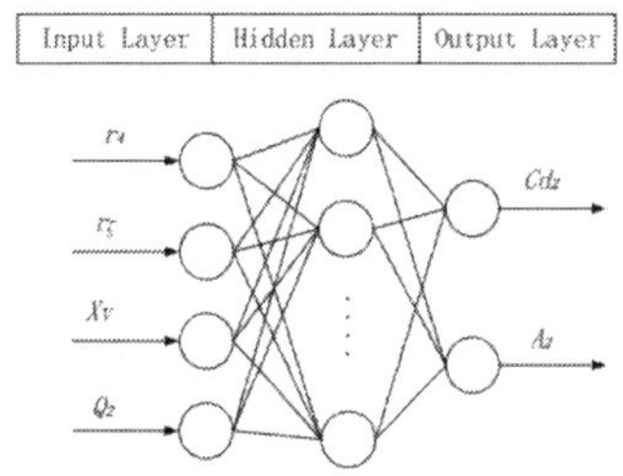

Figure6. Schematic diagram for the ANN of port C

To evaluate the models accuracies, the root square errors (RMSE), mean absolute errors (MAE) were used. The RMSE sizes the goodness of the fit related to high discharge coefficient values whereas the MAE measures a more balanced perspective of the goodness of the fit at moderate discharge coefficients. The RMSE and MAE are defined as:

$$RSME = \sqrt{\frac{1}{N}\sum_{i=1}^{N}\left(Cd_{test,\ i} - Cd_{predict,\ i}\right)^2} \tag{4}$$

$$MAE = \frac{1}{N}\sum_{i=1}^{N}\left|Cd_{test,\ i} - Cd_{predict,\ i}\right| \tag{5}$$

in which N is the number of data set ; $Cd_{test,i}$ is the original data for testing, which number is i ; $Cd_{predict,i}$ is the prediction data, which number is i.

For the total evaluation,the RSME and MAE for the above ANN models is shown as flllowing table 4.

Table 4. The RSME and MAE for each model in validation period

Model	RSME	MAE
ANN for port A	0.156	0.135
ANN for port C	0.246	0.212

4.2 Optimization of the throttling grooves

During the changing process of the multi-section valve of the hydraulic excavator, various indexes need to be satisfied for the system design. The most typical one is to meet the demand of the throttle robustness for the load change, which means that the multi-section valve should be of good flow stability. Therefore, in this paper, throttling stiffness was taken as the evaluation criterion for flow stability, which was defined as following formula 6.

$$f(X) = \frac{\partial Q}{\partial p} \tag{6}$$

In which,the f(X) is the throttling stiffness, the Q is the flow rate through the working port, and the p is the pressure on the working port as the working load.

However, with throttling stiffness as optimization objective to evaluate flow stability, an optimal design process is carried out to optimize to optimize the structure of coupling grooves with ANN models and genetic algorithm, which flow chart is shown as Figure.7, and the significant outcome of the optimization is shown in table 5, in which the F is the optimization objective function.

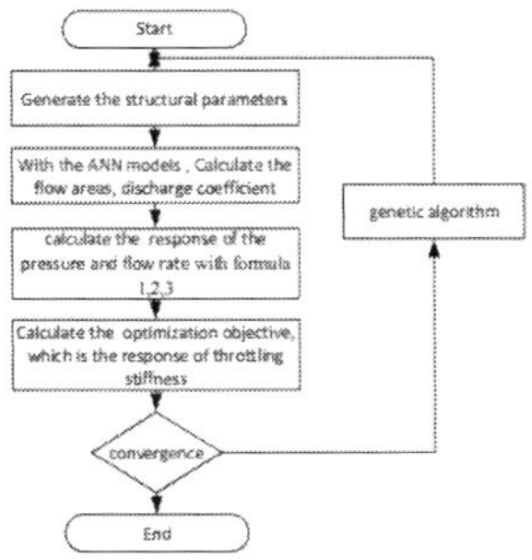

Figure 7. Flow chart of the optimization

Table 5. The Comparison before and after the optimization

	r1	r2	r3	L2	r4	r5	F
Before	0.5	2.0	2.0,	4	2.0	2.0	12.3
After	0.56	1.78	2.38	3.84	1.98	2.34	16.1

Ultimately, the optimized structure is verified better by the physical test on test rig, therefore, the significance of design method is proved.

5 Conclusion of the research

Combining the results of the above research, this paper realized the following conclusions:

1) An ANN model is established for variables of design space, parameters of working condition and throttling coefficient, which accurately expresses the mapping relationship between the structure parameters, parameters of working condition and parameter of the flow field characteristics of the multi-section spool valve, avoiding the large scale computing and CFD analysis of large degree of freedom, which purpose is to realize the optimization of topology structure for the multi-section spool valve under the complex working condition.

2) The optimized structure is verified better by the Physical test on test rig, therefore, the significance of design method is proved.

References

1. WANG Anlin, KUANG Long, ZHANG Xiaolu. *A Study on Flow Coefficient of Combined Throttling Groove in Spool Valves.* Journal of Xi'an Jiaotong University, 52(02): 110-117 (2018).
2. WANG Anlin, KUANG Long, ZHANG Xiaolu. *Optimization research for topology variable space of multiway valve coupling groove structure.* Journal of Xi'an Jiaotong University, 51(06): 103-109 (2017).
3. JIANG Tao, XIA Mingjie, WANG Anlin. *A robust evaluation method of micro characteristics for directional control valve of hydraulic excavators.* Journal of Xi'an Jiaotong University, 50(2): 118-123(2016).
4. AMIRANTE R, DISTASO E, TAMBURRANO P. *Experimental and numerical analysis of cavitation in hydraulic proportional directional valves. Energy Conversion and Management,* 87: 208-219 (2014).
5. AMIRANTE R, MOSCATELLI P G, CATALANO L A. *Evaluation of the flow forces on a direct (single stage) proportional valve by means of a computational fluid dynamic analysis. Energy Conversion and Management,* 48(03): 942-953 (2007).
6. *LISOWSKI E, FILO G, RAJDA J. Pressure compensation using flow forces in a multi-section proportional directional control valve. Energy Conversion and Management,* 103: 1052-1064 (2015).
7. *LISOWSKI E, CZYZYCKI W, RAJDA J. Three dimensional CFD analysis and experimental test of flow force acting on the spool of solenoid operated directional control valve. Energy Conversion and Management,* 70: 220-229 (2013).

Rule Optimization of Self-organization Control of Traffic Signals in Urban Net Based on Hydrodynamic

Fusheng Zhong[1, a] and Anlin Wang[1]

[1]*Mechanical-electronic Engineering, Tongji University, CHINA*

Abstract. Prior researchers indicate that hydrodynamics models of traffic-flow is lack of description of changing mechanism under urban traffic, and self-organization control system can not explain the dynamic characteristics of urban traffic flow clearly. The aim of the paper is to puts forward an optimized method on control rules that make the united application of hydrodynamics and self-organization system in signal control. The parameter sets of control rules are built from parameter sets of road network which are evolution under hydrodynamic equations such as the length of each lane, phase, queue length and so on .With the aim of the maximum traffic volume at each intersection in the road network, the control rules optimize its parameter sets to adapt to the dynamic change. By means of the computer simulation, the application of signal self-organizing control under hydrodynamic is proved effective in urban traffic.[a]

1 Introduction

The urban traffic signal control system is a complex system as the change of traffic flow is complex.The equations are not accurate enough because of some unknown traffic mechanism, on the other hand the higher the degree of discretization is, the easier it is to deviate from the reality.Traffic-flow under Hydrodynamics Models are commonly assumed to be continuous fluid and compressible, what makes it good to describe the collective behavior of vehicles. In fact, the interaction of vehicles, the judgement of the driver and other factors have great influence on the dynamic characteristics of traffic flow to make it discontinuous. Self-organizing control in urban traffic can easily linked the intersection to its interaction downstream and upstream which makes the control rules more reasonable, but if the parameter sets of rules are not sensitive to adapt to the change of traffic then the evolution of self-organizing are not determined.

Lighthill [1] defines the first order continuous module and Nagatani T [2] defines the discrete module with cellular automata(CA),both are research basis in their area. Most researchers focus on the how the traffic flow varies propagating disturbance in continuous mode or take more factors such as driver's behavior into consideration to modify the process during phase transition in discrete mode. XUE [3] proposes a two-time-scale equation to distinguish different relaxation time in speed propagating. GUPTA[4] takes driver's prospect into consideration to find out that reduces traffic jam efficiently. Biham [5] defines the traffic-flow models in two dimensions that explain how traffic flow change from Low density to high density under self-organization. Debashish [6] takes two typical CA

[a] Corresponding author : ttxhway@126.com

module to show that how traffic become jam and point out the progress depends on the linked signal cycle and phase difference. WEI [7] proposes a simple CA module to study the relations between intersections and regarded the intersection as the cell. The aim of this paper is to find more efficiently parameter sets of self-organization signal control, to use Hydrodynamics Models to enhance the description of the dynamics characteristics of the traffic flow. We put forward a new CA module which the intersections are the cell for control equation and lanes are also the cell for traffic flow's transfer. We also build the flow distribution module to modify the parameter sets more reasonable.

2 Road-net and parameters

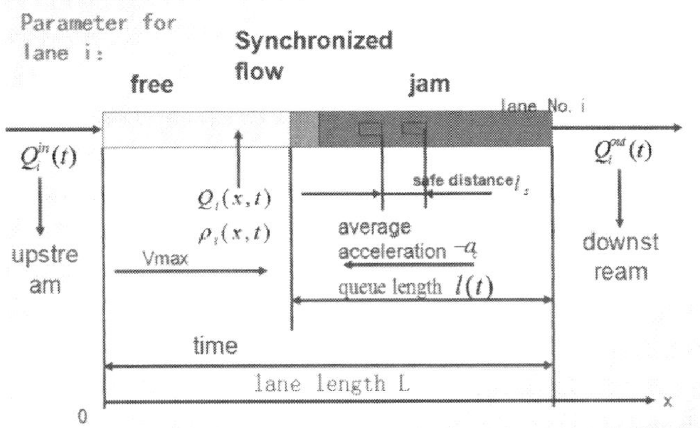

Figure 1. LANE MODULE

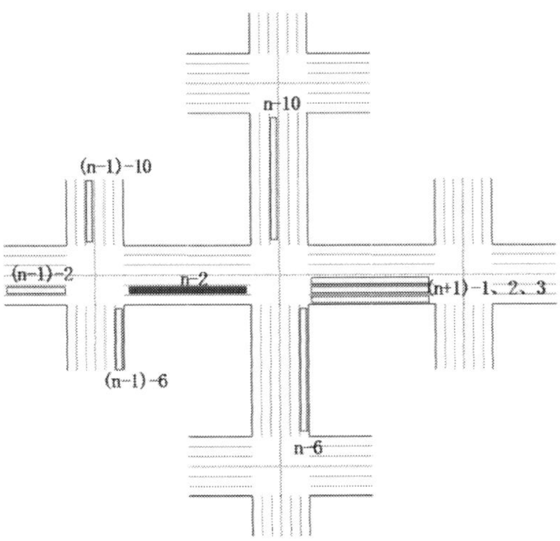

Figure 2. CA(intersection)MODULE

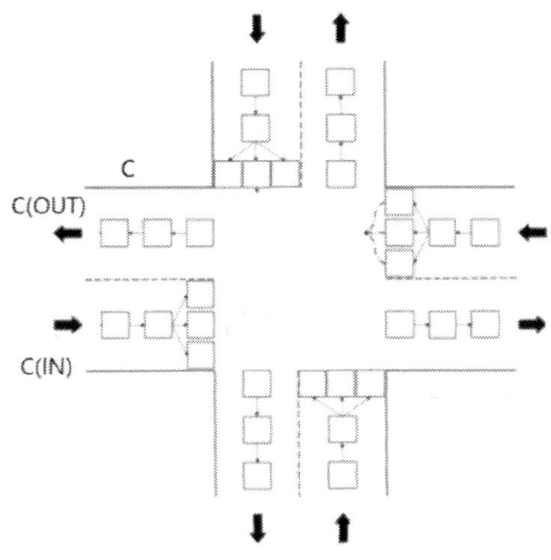

Figure 3. CA(lane)MODULE

Fig. 1 illustrates the basic lane module and parameters for it, we define every single vehicle or some vehicles linked into queue in the equations related to its current position and the time, which make us to calculate the queue with any time t that is wanted, the queue length l(t) is taken into the control equation and the prediction module that we do not study in this paper. Variable Q presents the flow value and variable ρ presents the flow density.

Fig. 2 shows that how CA(intersection) module works, variable n presents intersection number while n-1, n+1 means upstream and downstream. When the dark colour marked lane n-2(intersection n, lane 2) is called to judge whether to turn the current phase signal to next, the information of all those light color marked lane in fig.2 will be collected by the control module. The parameter sets of lane n-2 are construction of all the lanes interacted which are marked.

The basic CA(intersection) module comes from Wei[8] with one lane per road and simple parameter, the basic control equation is built from reconstruction of N-S equation as follow:

$$\frac{du}{dt} = f_b - \frac{1}{\rho}\nabla p + \mu\nabla^2 V \tag{1}$$

$\frac{du}{dt}$ --inertia force, f_b --mass force, $-\frac{1}{\rho}\nabla p$ --pressure gradient force, μ --viscosity coefficient;

To use finite difference method to decompose equation(1) on x axis to get the control module:

$$l(t)_{n,k}^{t+1} = (1-2\lambda_1-\lambda_2)l(t)_{n,k}^{t} + (\lambda_1-\lambda_2)l(t)_{n-1,k}^{t} + (\lambda_1+\lambda_2)l(t)_{n+1,k}^{t} + \frac{\lambda_2}{2} \tag{2}$$

With: $\lambda_1 = \frac{\mu\Delta x}{(\Delta t)^2}$, $\lambda_2 = \rho\mu\Delta t$,k--signal phase

Fig.3 explains that CA(lane) module consist of 3 lanes per road, which enabled left turn, straight and right turn.

Here we define that each cell in the lane can be occupied with 3 vehicles also with each safe distance. The CA(lane) module provide the control system the dynamic characteristics of traffic flow at any moment t. The constraint for the vehicles and cells are reasonable when jam occurs do not talk in this paper.

3 Rule optimization

According to Fig 3, to maximize the traffic volume of each intersection represents that every cell x of CA(lane) in each intersection counts sum of vehicles either put in or put out during interval t.

$$F(x) = MAX \sum_{\forall t \in T} \sum_{\forall i \in C_{IN}} x_i^t \tag{3}$$

3.1 flow distribution module

Based on the flow distribution control rule, when any chosen lane corresponding to any phase is in the process of heading for the downstream lane, it will confluences with other associated with lanes related to same phase at downstream entrance of the local intersection. At the same time, its own traffic flow rate associated with adjacent lanes will determine the traffic allocation of upstream intersections. With different forms of injections at the entrance of each intersection different distribute rate will be chosen [9]:

$$\sum_i M_n Q_n^{out}(t) = \begin{cases} M_{n+1} Q_n^{in}(t, L_n) / M_n \\ M_{n+1} Q_n^{\max} / \mathrm{M}_n \\ \sum [\min(M_{n+1} Q_{n+1}^{out}(t, L_{n+1}))] / \mathrm{M}_n \end{cases} \tag{4}$$

M--distribution coefficient related to l(t)

3.2 optimization

The parameter sets are defined with each initial value under Fig1.,when any thresholds of the lanes in the net are achieved by current pressure, the local control system aims to solve its own demand so that a green phase time g_p is allocated and the start time t_{gp} is arranged. With comparison of prediction superposition and current release of the characteristic pressure, the control system reassigns the matching rule parameters between linked input and output in their own range. α_j --all the parameter sets.

$$F(g_p) = \sum_{j-1}^{n} \alpha_j * P_j$$

$$P_j = F[(t_j^{gp} - t_{j+1}^{gp}), (t_j^{gp} + t_{j-1}^{gp})L] \qquad (5)$$

This solution analyzes the changing process of the traffic flow by hydrodynamic. It basically explains how traffic flows can transfer continuously. In this way we can effectively characterize the instability of traffic states when they are at any of the critical points, and then return the parameters' change area. Hence, this basic rule designs the space for the self-organizing control system, which makes avoiding abrupt system concussion in a sudden.

4 Simulation verification

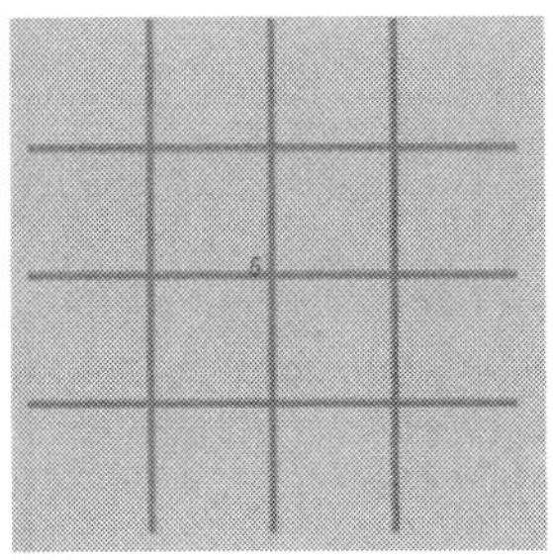

Figure 4. Road net of simulation

Fig.4 shows that the road net of simulation consist of 9 intersections which are linked as CA(intersection) module,8 phases are chosen for all intersections. Maximum and minimum of signal cycle length are designed to 120s and 60s; Maximum and minimum of vehicles' speed are designed to 60KM/H and 8 KM/H. The vehicles input into boundary from 600puc/h to1800pcu/h. The whole simulation lasts for 3600s.Time step is defined as 15s, data analysis will start after vehicles have uniformly distributed in the network. The intersection 5 is chosen for observation.

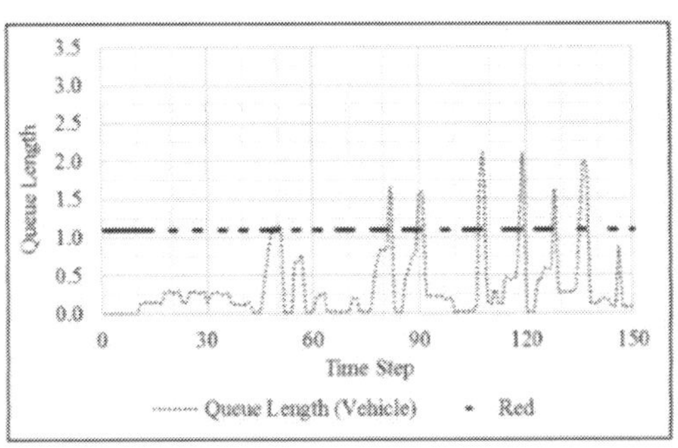

a

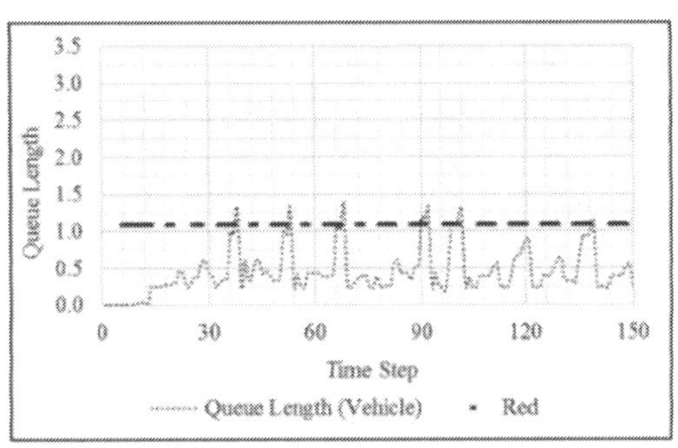

b

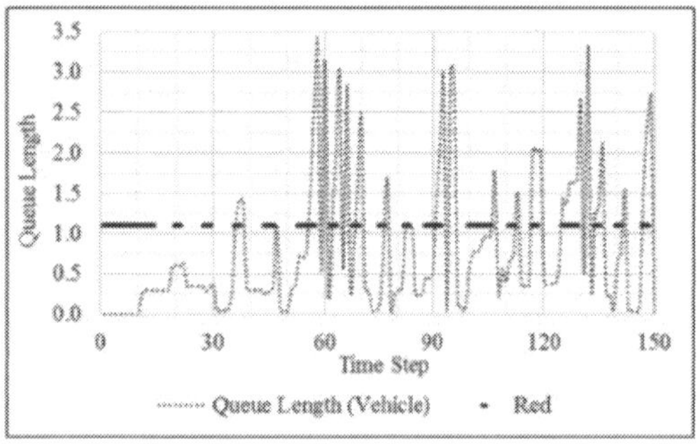

c

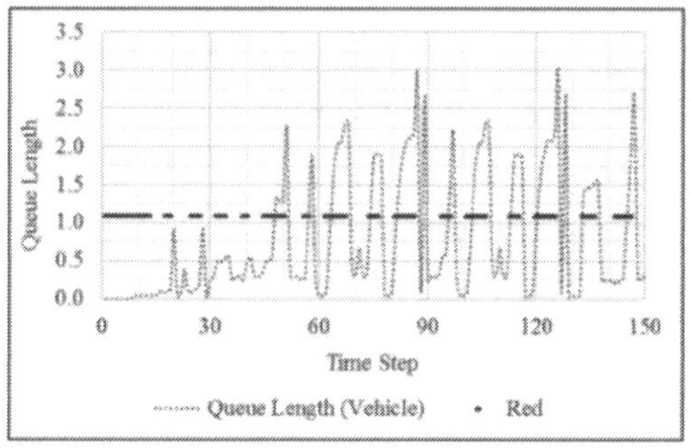

d

Figure 5. a.East to West without optimization b.East to West with optimization
c. North to South without optimization b.North to South with optimization

Comparing with a/b and c/d in Fig.5,it is obviously that the self-organization signal control with rule optimization produce less traffic-jam than that without rule optimization, it also shows that after rule optimization the signal control system try to avoid jam in advance that is what self-organization works. The results confirms that the method really modify the evolution in a fast way to adapt to traffic flow's dynamic change.

5 Conclusions

This research has successfully demonstrated how to find different intersections' parameter sets and optimize them, which to satisfied their own maximum traffic volume as objective function. The new method can describe the dynamic characteristics by using the hydrodynamics models of traffic-flow, but also can create connections between interacted intersections under self-organization. The response speed of the signal control system is strengthened so that the signal control system can start to eliminate traffic-jam much earlier, the result is that each lane increases output vehicles during each phase no matter whether it is main road. By means of the computer simulation, the application of signal self-organizing control under hydrodynamic is proved effective in urban traffic. It provides comparison and reference for further solving such the complex system of traffic signal control.

Acknowledgments

This work is supported by the International Exchange Program for Graduate Students, Tongji University (No. 201801222).

References

1. M.J.Lighthill,G.B.Whithan,*kinematic waves:2*,**229**,317-345(1955)
2. T.Nagatani,Physi A, **261**:599-607(1998)
3. Y.XUE,[D],ShangHai Unversity,21-29(2002)
4. A.K.Gupta,S.Sharma,P.Redhu,*NonlinerDyn*,**80**:1091-1108(2015)
5. O.M.Biham,A.Levine,Physi.E **46**:6124-6127(1992)
6. C.Debashish,S.Andreas,Physi.E,**59**:1311-1314(1999)
7. J.HWei,A.L.Wang,N.C.Du,*IEEE TVT*,**54**:744-747(2005)
8. J.H.Wei,[D],Shanghai JiaoTong University 32-39(2005)
9. D.Helbing,S.Lammer,J.P.Lebacque,*OCDG*,239-274(2009)

Research Method of Assembly Sequence Planning Based on Assembly Constraint Relationship Analysis

Yang Yi[1,2,a] Xiaojun Liu[1,2] Zhonghua Ni[1,2] Xia Tang[3] and Xiaokang Xu[1,2]

[1]*School of Mechanical Engineering, Southeast University, Nanjing 211189, China*
[2]*Jiangsu Key Laboratory for Design and Manufacture of Micro-Nano Biomedical Instruments, Southeast University, Nanjing 211189, China*
[3]*Wuxi vocational college of science and technology, Wuxi 214028, China*

Abstract. In order to effectively reduce the human-computer interaction (HCI) workload before the assembly process simulation in the digital assembly environment, and to enhance the intelligence of assembly sequence planning, we introduce the assembly constraint relationship into the automatic generation of assembly sequences and propose a practical assembly sequence planning method for intelligent assembly. Firstly, starting from the assembly geometry model, an assembly model liaison graph is constructed by combining the analysis of assembly constraint relationship and the definition of disassembly direction rules. Then, according to the liaison graph, disassembly priority constraint matrix based on constraint relationship is constructed, and the automatic generation and dynamic update of the disassembly priority diagram can be realized by the definition of disassembly doubly linked list. According to the diagram, the rationality of the disassembly sequence is determined through the interference check analysis, and then the assembly sequence is automatically generated from the reversed disassembly sequence. Finally, the method in the text is verified by an example case study.

1 Introduction

In recent years, with the wide application of model-based definition (MBD) technology, three-dimensional (3D) digital assembly technology as a further extension and deepening of virtual assembly technology has become a hot topic in the field of manufacturing. The product process design based on the 3D model is the bridge to connect product design and manufacturing based on MBD. The 3D digital assembly is an important part of the product process design. Among them, assembly sequence planning (ASP) is the core of the 3D digital assembly process design, and is also one of the focuses of the current assembly process automation and intelligence.

Currently, many scholars have conducted extensive research on the methods and techniques of ASP. Some typical ASP methods have been proposed, which can be summarized as "graph model + solving & search method" and "product disassembly model + optimization algorithm for disassembly solution" [1]. Combined with the application of ASP in practical engineering, we believe that all the current ASP methods have limitations. The method based on priority relationship [2] is simple and intuitive and easy to operate, but the priority relationships of assembly constraint relationship are

[a] Corresponding author: seu169307@163.com

provided by HCI rather than automatic recognition. The method based on cut-set [3-4], with the sharp increase of the number of parts and the solution space, is easy to encounter the assembly sequence explosion problem. For the proposed knowledge-based ASP method [5], it is difficult to ensure the matching degree between actual assembly models and standard knowledge bases in terms of knowledge acquisition and expression. The method based on GCA [6-7] solves the problem of generating a complete assembly sequence in a 3D environment, however, the further promotion of the method is hampered due to the amount of calculation quantities and results. The methods based on heuristic intelligence algorithm [8-9] usually need to establish complete assembly constraints relationship and are restricted by search space and time, which leads to sacrifice algorithm efficiency largely to obtain optimal or sub-optimal solution.

2 An overview of this paper

An assembly sequence planning method based on assembly constraint relationship analysis for intelligent assembly was proposed in this paper. Firstly, in order to identify and acquire the assembly constraint relationship between parts, and to derive a set of feasible disassembly directions, the geometric information of product assembly model is extracted. Then, the assembly model liaison graph (AMLG) is constructed, and the disassembly priority constraint matrix is generated by the assembly priority constraint relationship. The disassembly priority relationship diagram is inferred by interference check in the feasible disassembly directions. Finally, the assembly sequence can be generated by reversing the generated disassembly order. When constructing the liaison graph of product assembly model, a product hierarchy connection diagram with a hierarchical structure can be established according to product hierarchy tree, and a layered disassembly is performed in a bottom-up order. In addition, according to the principle of "can be removable and can be assembled, can be reassembled after being disassembled", product assembly is disassembled layer-by-layer from the original assembly state of product. In this way, the disassembly sequence of the entire product is generated finally, and the product assembly sequence can be obtained by reversing the acquired disassembly sequence.

The overall flow diagram of the method in this paper is shown in Fig. 1. The diagram takes a given 3D assembly model as input, and finally generates a feasible assembly sequence, in which n denotes the number of product assembly layer.

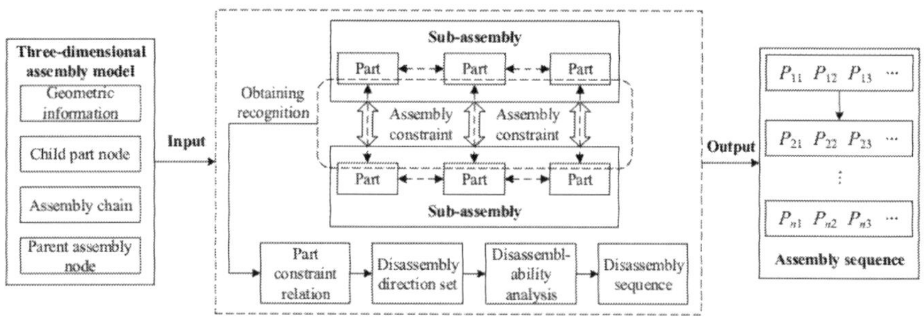

Figure 1. Overall flow diagram of the method in this paper.

3 Analysis of the assembly constraint relationship

3.1 Assembly constraint relationship

Product assembly modeling is the process of creating assembly by defining assembly constraint relationship on several parts. The constraint relationship is mainly a collection of areas with engineering significance on parts that match each other by cooperative actions such as coaxial, coplanar, align, offset, etc., which can be used to fix the position and orientation in the assembly and limit its corresponding the degree of freedom. Since these relationships between parts directly affects the assembly direction, movement form, assembly sequence, etc., they significantly determine the difficulty degree of product assembly. Therefore, the modeling, expression, and identification of assembly constraint relationship are prerequisites for product intelligent assembly process planning and evaluation.

These parts in the product assembly are associated with each other through point, line, and face. The association can form assembly pairs and the corresponding associated point/line/face are termed as assembly auxiliary elements. From a topological viewpoint, there are 6 basic forms of assembly pairs: vertex-vertex, vertex-edge, vertex-face, edge-edge, edge-face, and face-face. However, since most of parts are connected by face-face constraint in the assembly, only the condition of face-face constraint is considered in this paper. The surface type elements are restricted to plane, cylinder, conical surface, sphere and torus. The CAD models with complex shape (e.g. thread surface, involute gear surface, etc.) generally are processed with simple surfaces (e.g. cylinder).

The mating relations of geometric surface of assembly constraint can be classified as against, coplanar, fit and align. The mating relations between geometric surfaces is shown in reference [10].

3.2 Disassembly direction reasoning

In the process of 3D digital assembly sequence planning, a general adopted method to obtain assembly sequence is to analyze the disassembly feasibility with product assembly what have been assembled. Therefore, the initial assembly position and orientation of the parts in the assembly have been determined by assembly constraint relationship. Assume that all parts of the assembly are rigid bodies, and there are not irreversible operations during product disassembly process. The analysis of disassembly feasibility is based on the product assembly, and the assembly sequence is generated by the reverse order of feasible disassembly sequence. Among them, the determination of the disassembly direction is a very key step in the process.

According to the definition of assembly constraint relationship, it can be seen that mating between parts is mainly realized by face-face constraint. The contact mating relation determines the disassembly direction of part, which is called surface contact characteristics. First of all, based on the mating relations, the basic form of surface contact is divided into two types: contact between planes and contact between rotary surfaces. Then, the disassembly direction set (DDS) is inferred and generated automatically using the surface contact characteristics and the automatic identification of the mating relations based on the surface contact. The specific algorithm flow is shown in references [11-12].

4 Assembly sequence planning based on Assembly Constraint Relationship Analysis

4.1 Assembly model liaison graph (AMLG)

The reasonable representation of assembly information (topological information, geometric information, and constraint information) in product assembly is the basis of ASP. The rationality of assembly information expression will seriously affect the efficiency of ASP. Therefore, in the process of ASP, it is necessary to identify and extract the relevant geometric constraint information from the product assembly model, and establish the liaison graph model of product assembly.

Product assembly model can be represented as a liaison graph connected by a group of part nodes in term of geometric constraint relationship. The expression formula is as follows:

$$G=<N, E> \quad (1)$$

where G represents the assembly model, and $N=\{v_1, v_2, \ldots, v_n\}$ represents a collection set of part nodes attribute, n is the number of part node, and E represents a collection set of edges attribute connected part nodes, $E=e_{ij}=\{E_1, E_2, \dots, E_m\}$, $(i, j \in n)$, m is the number of connected edges, and e_{ij} represents connection constraint information between v_i and v_j .

A simplified assembly of pulley mechanism and the corresponding assembly model liaison graph are shown in Fig. 2. Among them, these circles represent the nodes of liaison graph, and these connected edges between nodes represent the constraint relationship between two parts. The dashed line represents the fastening connection relation, and the solid line represents the mating contact relation.

Component	
a_1	Base plate
a_2	Support1
a_3	Support2
a_4	Washer1
a_5	Bolt1
a_6	Washer2
a_7	Bolt2
a_8	Center shaft
a_9	Wheel

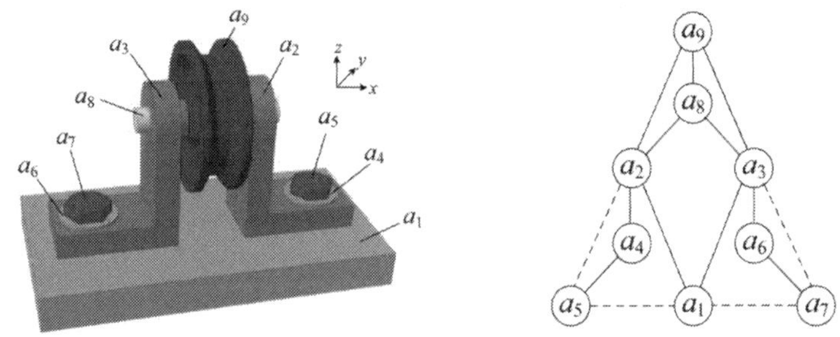

Figure 2. Simplified assembly of pulley mechanism and the corresponding AMLG.

4.2 Generation method of assembly sequence planning

The process of ASP can be seen as the reverse of disassembly sequence. In the 3D digital assembly environment, the main task of interactive disassembly is to generate information about priority constraint relationship, disassembly direction, etc. For the disassembly sequence planning of product assembly, the interference information of parts can be inferred by using the breadth-first search algorithm of AMLG and assembly constraint relationship between parts, and then the disassembly priority constraint matrix can be established to automatically calculate and generate the disassembly priority relationship diagram.

In order to obtain the interference information between parts, the disassembly direction in the initial state of product assembly can be determined according to the assembly constraint relationship. Then the benchmark part is selected by the selection principle of reference part and it is regarded as the starting point of searching. Based on them, the undirected liaison graph of product assembly is traversed by using the breadth-first search algorithm. The searching program can exit until all of nodes in AMLG have been accessed and the interference information between parts is generated. The interference information of disassembly pulley mechanism is shown in Table 1, and the symbol "#" represents no interference information between parts.

Table 1. Part interference information of disassembly pulley mechanism.

disassembly part	interference part	disassembly part	interference part
a_1	#	a_6	a_7
a_2	a_4, a_5	a_7	#
a_3	a_6, a_7, a_8	a_8	#
a_4	a_5	a_9	a_2, a_3, a_8
a_5	#		

The disassembly priority constraint matrix is derived from the interference information of AMLG and the matrix can be represented as a two-dimensional table with n-1 rows and n-1 columns. Excluding the reference part of assembly, the matrix $[a_{ij}]_{n-1,n-1}$ can be built. Each element a_{ij} represents the priority constraint relationship at the row ai and the column a_j , and the element a_{ij} can be defined as follows:

$$a_{ij} = \begin{cases} -1, P_i \Leftarrow P_j \\ 0, P_i \neq P_j \\ 1, P_i \Rightarrow P_j \end{cases} \tag{2}$$

In Eq. (2), $P_i \Leftarrow P_j$ indicates that the part P_i must be disassembled after the part P_j , and $P_i \neq P_j$ indicates that there is no explicit constraint relationship between the parts P_i and P_j , and $P_i \Rightarrow P_j$ indicates that the part P_i must be disassembled before the part P_j . According to the Eq. (2), the disassembly priority constraint matrix of pulley mechanism assembly (Fig. 2) can be inferred as shown in Table 2, and the part a_1 is benchmark part.

Table 2. The disassembly priority constraint matrix of pulley mechanism assembly.

a_i \ a_j	a_2	a_3	a_4	a_5	a_6	a_7	a_8	a_9
a_2	0	0	-1	-1	0	0	0	1
a_3	0	0	0	0	-1	-1	-1	1
a_4	1	0	0	-1	0	0	0	0
a_5	1	0	1	0	0	0	0	0
a_6	0	1	0	0	0	-1	0	0
a_7	0	1	0	0	1	0	0	0
a_8	0	1	0	0	0	0	0	1
a_9	-1	-1	0	0	0	0	-1	0

According to the disassembly priority constraint matrix, it can be seen that multiple parts may be disassembled before one part, or it may appear that multiple parts are disassembled simultaneously after one part. In order to realize the dynamic adjustment of disassembly priority relationship, using the double linked list of disassembly parts can realize the dynamic update of *previous* pointer and *next* pointer in the process of disassembly priority relationship reasoning. The algorithm flow of constructing product disassembly priority relationship diagram is shown in Fig. 3.

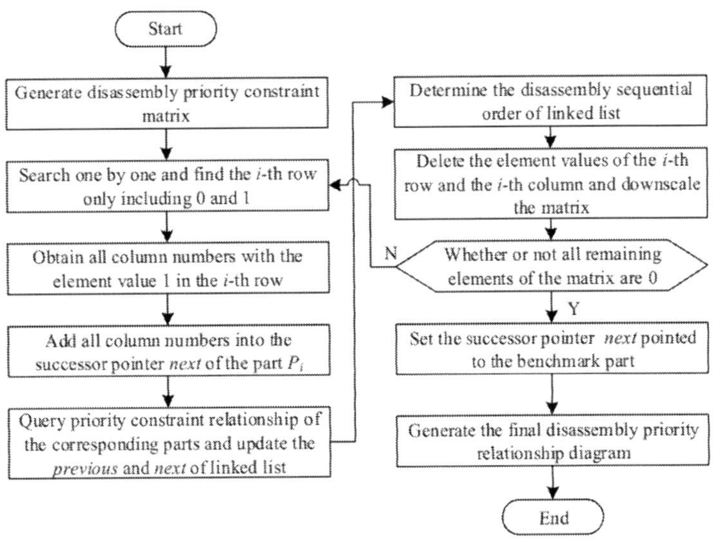

Figure 3. Generation algorithm flow of disassembly priority relationship diagram.

According to the algorithm flow of the above, the priority constraint relationship of pulley mechanism assembly is $a_5 \Rightarrow a_4 \Rightarrow a_2 \Rightarrow a_7 \Rightarrow a_6 \Rightarrow a_3 \Rightarrow a_8 \Rightarrow a_9 \Rightarrow a_1$. The disassembly sequence also accords with the practical engineering application. Finally, the assembly sequence of pulley mechanism is generated in the reverse order of disassembly sequence. It needs to point out that, due to the symmetry of the pulley mechanism assembly, the disassembly sequence of $a_5 \Rightarrow a_4 \Rightarrow a_2$ and $a_7 \Rightarrow a_6 \Rightarrow a_3$ can exchange the sequence or be executed in parallel.

5 Case study

Based on the above-mentioned methods and key techniques, the assembly sequence planning of product assembly is implemented in a prototype computer-aided assembly process planning (CAAPP) system named AMTProcesser in our laboratory [13], which is developed based on ACIS and HOOPS using VC 2008. The following is a detailed description of concrete example case.

The explosion diagram of 3D assembly model of engine body and the liaison graph are shown in Fig. 4. The 3D assembly model of engine body is built in Pro/E and imported into 3D digital assembly process design system environment named AMTProcesser. Using the disassembly method in this paper, the disassembly priority constraint matrix can be generated, which is similar to Table 2. P_{20}-P_{26} represent parts of a piston rod sub-assembly and they can be viewed as a part named P_{link}. At the same time, to be easy for analysis, engine body1 and engine body2 can also be viewed as a part named $P_{8\text{-}9}$, and fasteners (e.g. bolt-washer) can be grouped as three part named $P_{6,7}$, $P_{13,14}$ and $P_{16,17}$, respectively.

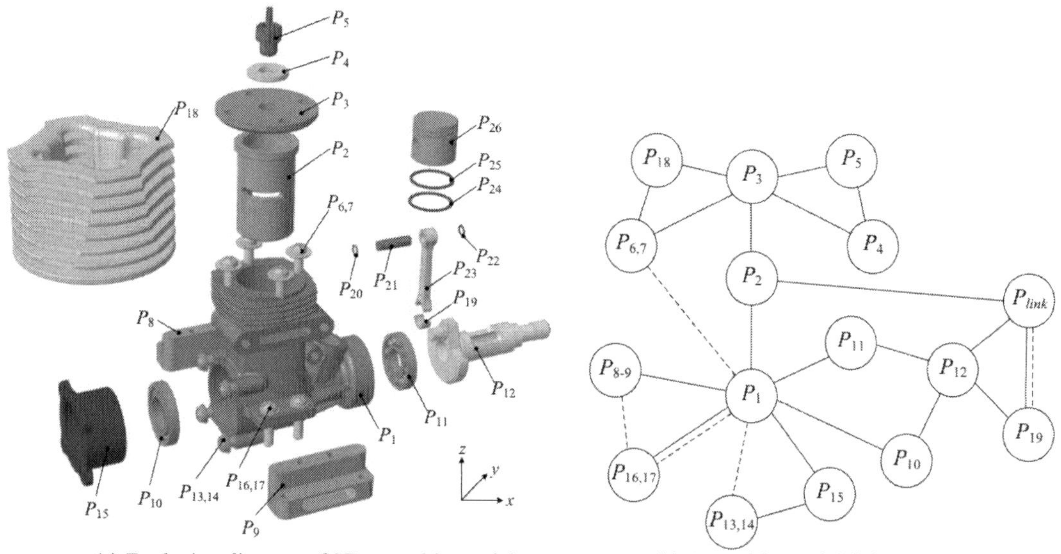

(a) Explosion diagram of 3D assembly model (b) Assembly model liaison graph

P_1-Engine body, P_2-Cylinder sleeve, P_3-Cylinder sleeve, P_4-Copper backing, P_5-Ignition plug, P_6-Washer 4×M3, P_7-Bolt 4×M3×12, P_8-Engine base1, P_9-Engine base2, P_{10}-Bearing1, P_{11}-Bearing2, P_{12}-Crankshaft, P_{13}-Washer 4×M3, P_{14}-Bolt 4×M3×8, P_{15}-End cover, P_{16}-Washer 4×M3, P_{17}-Bolt 4×M3×12, P_{18}-Radiator, P_{19}-Rod head, P_{20}-Ring1, P_{21}-Axis pin, P_{22}-Ring2, P_{23}-Rod body, P_{24}-Piston ring1, P_{25}-Piston ring2, P_{26}-Piston

Figure 4. Explosion diagram of engine body and the corresponding liaison graph.

According to the analysis of the assembly constraint relationship between parts of engine body, the disassembly direction of each part can be obtained. Then, using the generated disassembly priority constraint matrix, the disassembly priority relationship diagram is inferred. Finally, the feasible disassembly sequence is generated by implementing the detection of disassembly feasibility. The generating disassembly priority relationship diagram is shown in Fig. 5.

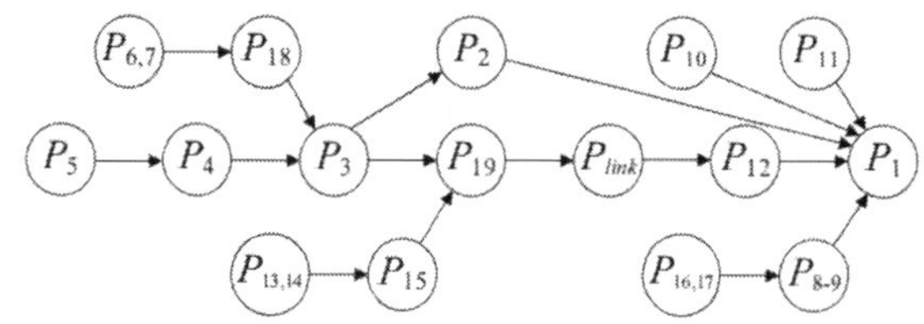

Figure 5. Generating disassembly priority relationship diagram of engine body.

Since dismantling engine body is a more complicated process, other than satisfying the disassembly geometric feasibility, many factors such as disassembly cost, disassembly time, tool operability, disassembly difficulty, etc. should also be taken into consideration. Therefore, a more practical disassembly sequence can be obtained from the above example about the disassembly of engine body and the dismantling scheme is $P_5 \Rightarrow P_4 \Rightarrow P_{6,7} \Rightarrow P_{18} \Rightarrow P_3 \Rightarrow P_{13,14} \Rightarrow P_{15} \Rightarrow P_{19} \Rightarrow P_{link} \Rightarrow P_{12} \Rightarrow (P_2 \cap P_{10} \cap P_{11}) \Rightarrow P_{16,17} \Rightarrow P_{8\text{-}9} \Rightarrow P_1$, where there is no sequential relationship between P_2, P_{10} and P_{11}. Finally, the assembly sequence of engine body is obtained by reversing the above disassembly sequence.

6 Conclusions

Assembly sequence planning has always been a very concerned issue in the field of manufacturing. The efficient, convenient and practical assembly sequence generation strategy is one of the key technologies for improving the efficiency of 3D digital assembly process design. In consideration of the shortages of automatic and human-computer interactive assembly planning in the three-dimensional environment, an assembly sequence planning method based on assembly constraint relationship analysis for intelligent assembly is proposed in this paper. The results of case study clearly show the capability of generating assembly sequence simply and reasonably. The automatic generation method of assembly sequence based on the identification and acquisition of assembly constraint relationships can be used in 3D digital assembly process design for assembly sequence planning in a more convenient and practical way.

However, further research questions arise. Since the complexity and diversity of the product assembly process design, the computational volume and search efficiency will be one of the major issues affecting intelligent assembly or disassembly planning. How to properly mix different methods to obtain the optimal assembly sequence from several geometrical feasible assembly sequences by combining the advantages of precise calculation method and various heuristic intelligence algorithms will be the next step in our study. At the same time, more detailed research is required to validate and apply these ideas using more complex case-studies.

Acknowledgments

This paper is financially supported by Certain Ministry of China under Grant Numbers 41423010402, 41423010203, Qing Lan Project, the Fundamental Research Funds for the Central Universities, and the Six talent peaks project in Jiangsu Province of China.

References

1. YM. Rong, H. Wang, LX. Zhang, et al, Assembly Automation, **33**, 78-85 (2013)
2. J. M. Henrioud, A. Bourjault, Computer-Aided Mechanical Assembly Planning, 191-215 (1991)
3. D. F. Baldwin, T. E. Abell, M. C. M. Lui, et al, IEEE Trans Rob Autom, **7**, 78-94 (1991)
4. YL. Fu, LZ. Tian, L. Xie, et al, Chinese J Mech Eng, **39**,59-62 (2003)
5. TY. Dong, RF. Tong, L. Zhang, et al, Proceedings of the 5th World Congress on Intelligent Control and Automation. IEEE, 2748-2752 (2004)
6. Q. Su, Int J Adv Manuf Technol, **33**, 48-57 (2007)
7. SJ. Lai, Q. Su, Computer Integrated Manufacturing Systems, **14**, 1791-1795 (2008)
8. YF. Xing, YS. Wang, Int J Prod Res, **50**, 7303-7312 (2012)
9. JF. Wang, JH. Liu, YF. Zhong, Int J Adv Manuf Technol, **25**, 1137-1143 (2005)
10. E. B. Hamidullah, M. A. Irfan, Int Conf Emer Technol. IEEE, 617-623 (2007)
11. K. Sambhoos, B. Koc, R. Nagi, J Comput Inf Sci Eng, **37**, 1-15 (2009)
12. TY. Dong, RF. Tong, L. Zhang, et al, Journal of Computer-Aided Design & Computer Graphics, **17**, 782-788 (2005)
13. XJ. Liu, ZH. Ni, JF. Liu, et al, Int J Adv Manuf Technol, **82**, 1-15 (2016)

Study on the frame load spectrum combining mileage and percentile extrapolation

Qingchun Song[a], Zhenghe Song and Yuefeng Du

Beijing Key Laboratory of Optimized Design for Modern Agricultural Equipment, College of Engineering, CAU, Beijing, 10083, China

Abstract. In order to get the load spectrum that reflects the actual operation condition of agricultural machinery, this paper proposes to compile the load spectrum of frame by combining the method of load mileage extrapolation and percentile extrapolation. In this paper, the research object was carried out in the frame of vegetable tool carrier to study load spectrum and fatigue life. The finite element analysis of the frame is carried out, and a set of stress load test system is established based on the analysis results, and the stress time course is obtained under typical working conditions. The time domain signal is converted into frequency domain signal by rainflow counting method, and frequency statistics are carried out. According to the "From-to" matrix, the "Mean based ellipse" kernel function is used for non-parametric rainflow matrix extrapolation, considering the influence of different drivers in the actual operation, the load of three conditions is extrapolated by percentile. According to wave center method, developing eight two-dimensional load spectrum of the vegetable tool carrier frame and one-dimensional load spectrum that reflects of actual loading situation, which provides a reference for the compilation of load spectrum in agricultural machinery.

1 Introduction

In the development stage of agricultural machinery products, the fatigue strength of structural parts should be investigated and the fatigue life should be evaluated. Dynamic loads exist in the use of mechanical structures, which will cause fatigue damage to components. The evaluation of fatigue damage is a very practical research content [1-2]. Load spectrum is the basis of component fatigue analysis, so it is particularly important to compile the load spectrum reflecting the actual working conditions [3]. Based on vegetable field machinery frame as the research object, combined with finite element analysis results to select relevant measuring point, testing load data through the establishment of load testing system and compiling the load spectrum, the fatigue life prediction and damage analysis was carried out on the frame. The structure design life is so long that it is difficult to obtain the load course in the service life through general test, so it is necessary to extrapolate the load course. Rodzewicz[4] integrated the measured data and extrapolated it to accurately predict the long-term load of glider. Socie[5] used the non-parametric estimation of rain flow matrix expansion method and applies it to practice, proving the effectiveness of the method. Drebler et al. [6] proposed a non-parametric extrapolation method to carry out loading process extrapolation by introducing kernel

[a] Corresponding author : cauqingchun@163.com

function and adaptive bandwidth. Fansong Li et al. [7] proposed a modified KDE theory and application method based on the kernel density estimation theory, gave a KDE application method under various loading processes. Nonparametric rain flow extrapolation method is a method to estimate the unknown distribution parameters, it does not depend on the distribution model assumption, avoids the subjectivity in the process of extrapolation, this paper argues that nonparametric rain flow extrapolation is adaptable load extrapolation method on the load mileage extrapolation processing, using this way. At the same time, on the basis of mileage extrapolation, this paper proposes to combine mileage extrapolation and percentile extrapolation in order to reflect the difference in load characteristics.

2 Frame load collection

2.1 Test principle

In this paper, a strain test was used to collect the load time history of three different drivers in the ploughing condition of the dynamic mechanical vehicle frame, the strain condition of the structure was measured by pasting strain gauge, and then collected by data acquisition card. The schematic diagram is shown in figure 1.

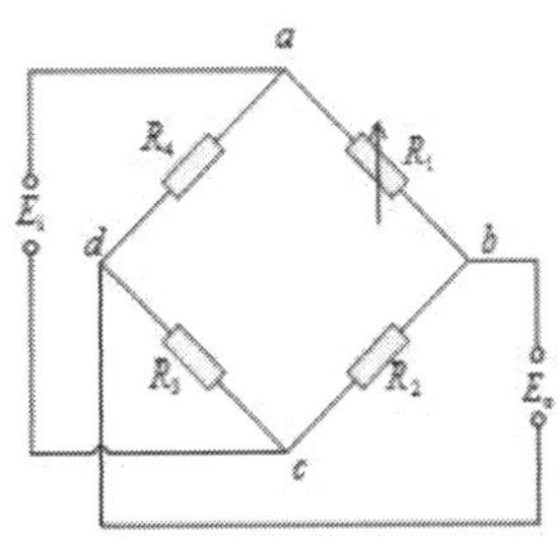

Figure 1. Schematic of wheatstone bridge test.

In the figure, E_i is the excitation voltage of the bridge. E_o is the bridge output voltage signal; R_1 is the measuring strain gauge, the rest is the fixed resistance.

$$E_o = U_{ab} - U_{ad} = \frac{R_1R_3 - R_2R_4}{(R_1+R_2)(R_3+R_4)} E_i \quad (1)$$

When R_1=R_2=R_3=R_4, the bridge is a full-arm bridge and the output voltage E_o=0. When R_1 resistance changes ΔR, bridge output voltage is:

$$E_o = \frac{\Delta R}{4R + 2\Delta R} E_i \quad (2)$$

Because ΔR<<R, ignore the item $2\Delta R$, then :

$$E_o = \frac{1}{4}\frac{\Delta R}{R} E_i = \frac{E_i}{4} K\varepsilon \quad (3)$$

ε is strain value, K is the sensitivity coefficient of strain gauge, then:

$$\varepsilon = \frac{4E_o}{E_i K} \quad (4)$$

Therefore, given the excitation voltage, sensitivity coefficient K, collecting the output signal E_o, the measured point strain data can be obtained.

2.2 Selection of measuring points

Firstly, the location of the measuring point is selected. In this paper, using the Workbench to do finite element analysis on frame , figure 2 is finite element model, and combined with the actual testing requirements, selecting 7 points as shown in figure 3.

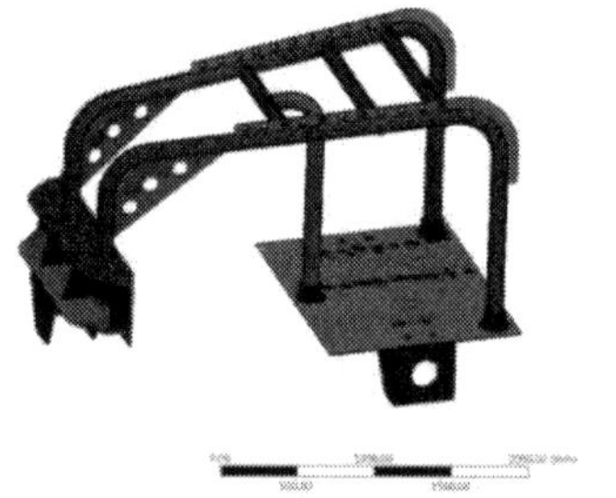

Figure 2. Finite element analysis model of frame.

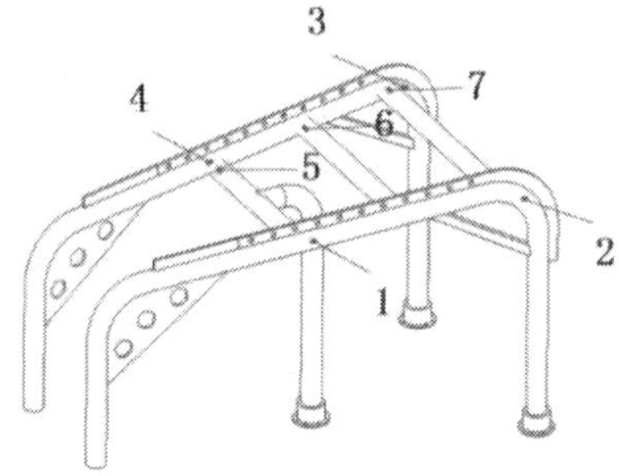

Figure 3. Measuring points.

2.3 Test system construction

Taking on load test in the working condition on the vegetable tool carrier driven by three different drivers. The strain testing system mainly includes BX120-3AA type single-axial strain gauge, and the sensitive grid resistance is 120 Ω , YSV7008 dynamic strain adjuster, YSV8016 network signal acquisition instrument, which consists of 16 data channels with a maximum sampling frequency of 16KHz/ channel and an AD precision of 24 bits. The sampling frequency is 1024Hz, with a total of 7 data channels. The strain signal is collected and archived in ASCII data format, and each group is tested twice for comparison and analysis of data within the group.

3 Signal processing and statistics

3.1 Signal processing

Taking the first measuring point of ploughing condition as an example, the collected signal is filtered in nCode, the trend term is removed, and the singularity is eliminated [8] , the comparison is shown in figure. 4.

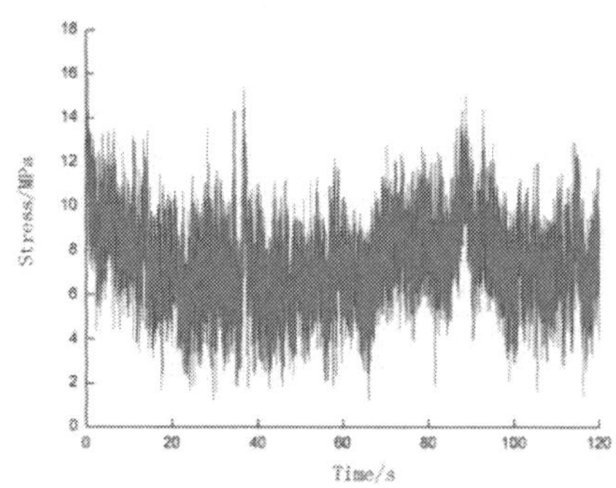

a. Unprocessed signal

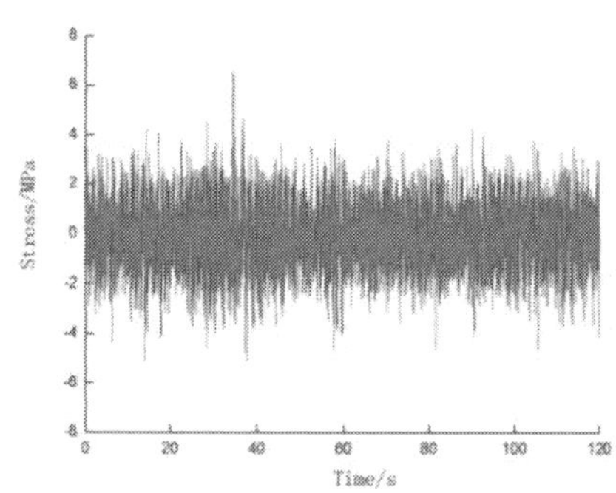

b. Processed signal

Figure 4. Signal of before and after processing.

3.2 Signal statistics

Mechanical components in practical engineering load is random, its characteristics is described by statistical method, the rain flow count method is the most widely used method in counting method [9], after counting post-processing by matrix and amplitude condition is shown in figure 5.

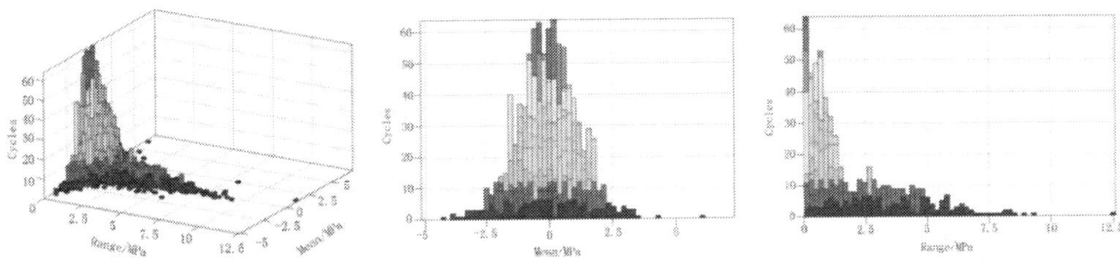

a. Load rainflow histogram b. Load mean distribution c. Load range distribution

Figure 5. Rain flow counting results of frame.

According to the amplitude probability distribution and test [10], the test results, determine the mean load obeys normal distribution, the load amplitude obeys weibull distribution, which is consistent with the characteristics of load data.

4 Load mileage extrapolation

4.1 Kernel density estimation

Based on the "From - to" rain flow matrix of non-parametric estimation method, under the premise that ensure that does not destroy the structure of the hysteresis loop back to load extrapolation, calculate the hysteresis loop back in the whole life cycle possible frequency [11]. $X_1,\ldots,X_n$ are sample of population, then the kernel density of the overall probability density function $f(x)$ at any point is estimated as:

$$\overline{f_h}(x)=\frac{1}{nh}\sum_{i=1}^{n}K(\frac{x-X_i}{h}) \tag{5}$$

$K(\frac{x-X_i}{h})$ is kernel function, h is the bandwidth, and the kernel function is required to satisfy:

$$K(x)\geq 0\,,\int_{-\infty}^{+\infty}K(x)dx=1 \tag{6}$$

In kernel density estimation $\overline{f_h}(x)$, bandwidth h determines the degree of smoothness, therefore, selection of proper bandwidth is particularly important in kernel density estimation, to select the appropriate bandwidth:

$$M(\overline{f_h})=E\{\int[\overline{f_h}(x)-f(x)]^2dx\} \tag{7}$$

Where $f(x)$ is the overall true distribution density and M is a function of bandwidth h [12], the optimal bandwidth estimate can be obtained by finding its minimum point.

4.2 Rain flow matrix processing and extrapolation

The load data of the first measuring point is transformed into "From-to" rain flow matrix, and the mileage extrapolation of the test load is carried out, as shown in figure 6.

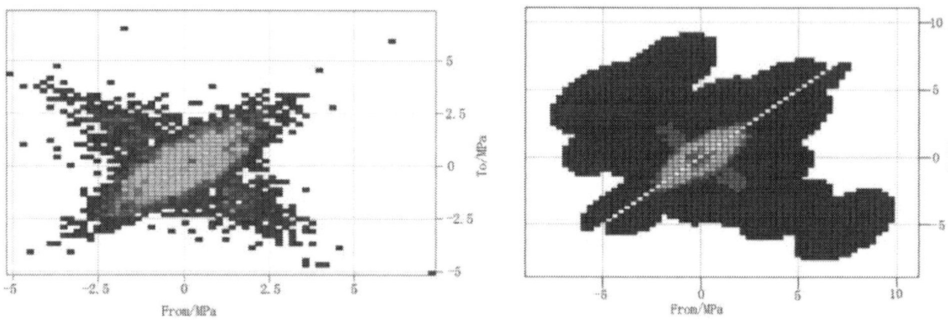

a. Original "From-to" matrix b. Mileage extrapolation matrix

Figure 6. Matrix of before and after mileage extrapolation.

5 Load percentile extrapolation

Collected vegetable tool carrier load of three different drivers, the load data of the three working conditions of mileage extrapolation were quantitatively extrapolated, taking first measuring point as an example. The results are shown in figure 7:

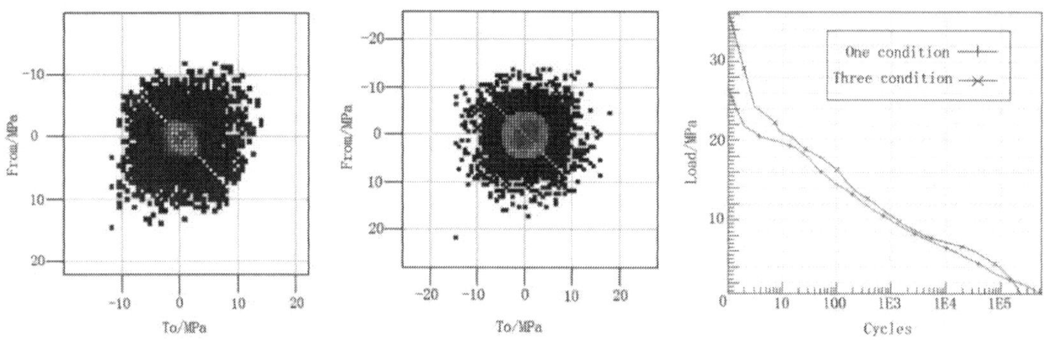

a. One condition b. Three condition c. Comparision

Figure 7. Matrix of before and after percentile extrapolation.

6 Frame load spectrum preparation

In engineering practice, the non-uniform interval method is usually adopted for load classification [13], divided into 8 levels. Rain-flow statistics are carried out on the load data after mileage extrapolation and percentile extrapolation. The mean amplitude is divided into eight levels to form the eight-level two-dimensional load spectrum, as shown in the following table 1:

Table 1. Two-dimensional load spectrum.

Range /MPa Mean /MPa	15.132	14.375	12.862	10.971	8.701	6.431	4.161	1.892
6.566	0	0	0	0	0	0	3	111
4.655	0	0	0	4	23	0	96	2612
2.744	1	24	52	37	106	715	5224	83113
0.833	0	6	19	296	2879	22200	49486	223724
-1.708	0	0	0	2	11	414	1432	46886

-2.989	0	0	0	0	0	0	10	1507
-4.900	0	0	0	0	0	0	0	2
-6.811	0	0	0	0	0	0	0	1

References

1. Li Jianmin, Pei Zhongtao, Wen Buzheng, et al. An Analysis on the Load Spectrum and Fatigue Life of Loader's Driving Axle House[J]. Automotive Engeering, 2012, 34(11): 1020-1023+1032.
2. Lin Mingfang, Zhang Hongxin, Du Zhiqi, et al. Prediction of Fatigue life and Reliability Design in Automotive Semiaxle[J]. Automotive Engeering, 1991(04): 235-242.
3. Gao Yunkai, Xu Chengmin, Fang Jianguang. Study on the Programed Load Spectrum of the Body Fatigue Bench Test[J]. Journal of Mechanical Engeering, 2014, 50(04): 92-98.
4. Rodzewicz M. Determination and extrapolation of the glider load spectra[J]. Aircraft Engineering and Aerospace Technology, 2008, 487–496.
5. Socie D F, Pompetzki M A. Modeling Variability in Service Loading Spectrum[J]. Probabilistic Aspects of Life Predictions, ASTM STP, 2004, 1450: 46-57.
6. Drebler K, Gründer B, M. Hack et al. Extrapolation of rainflow matrices[J]. SAE Technical Paper, 1996.
7. Li Fansong, Wu Pingbo, Zengjing. Extrapolation of Load Histories Based on Kernel Density methods[J]. Journal of the China Railway Society, 2017, 39(07): 25-31.
8. Jia Haibo. Study on the Test and Generation Methods about Load Spectrum of Wheel Loader Driveline[D]. Changchun: Jilin University, 2009.
9. Liu Yongchen, Wang Guolin, Sun Li. Fatigue damage analysis and life prediction for vehicle control arm[J]. Transactions of the Chinese Society of Agricultural Engineering (Transactions of the CSAE), 2013, 29(16): 83-91.
10. Yan Chuliang. Investigation of compilation of fatigue load spectrum by two-parametric counting method[J]. Transactions of the Chinese Society for Agricultural Machinery, 1986(02): 94-102.
11. Gao Tianyu, Li Huanliang, Zheng Zheng, et al. Compiling method of load spectrum of ZLK50 type loader hydraulic cylinder based on rain flow frequency extrapolation[J]. Journal of Mechanical Strength, 2017, 39(04):951-956.
12. Silverman B. Density Estimation for Statistics and Data Analysis[M]. London: Chapman and Hall, 1986.
13. Zhai Xinting, Zhang Xiaochen, Jiang Zhujin, et al. Load spectrum compiling for wheel loader semi-axle based on mixed distribution[J]. Transactions of the Chinese Society of Agricultural Engineering (Transactions of the CSAE), 2018, 34(08): 78-84.

An Electromechanical Co-Simulation Model Based on Lumped Parameter Model of Ball Screw Feed Drive System

Liang Luo[1], Weimin Zhang[1,2,a], Haonan Sui[1] and Jürgen Fleischer[2,3]

[1]*School of Mechanical Engineering, Tongji University, 201804 Shanghai, China.*
[2]*Sino-German School for Postgraduated Studies (CDHK), Tongji University, 201804 Shanghai, China.*
[3]*Institute of Production Science （wbk）, Karlsruhe Institute of Technology, 76131 Karlsruhe, Germany.*

Abstract. The continuous search for efficiency put forward higher requests to the machine tool for high speed and high acceleration, which makes the large-size and lightweight-designed feed drive system more likely to produce vibration during high-speed and high-acceleration feed operation. Ball screw feed system is the most widely used linear drive system in the field of industrial automation. Electromechanical Co-Simulation for ball screw feed drive dynamics is an important technique for solving vibration problems occurs in the feed motion. In view of the shortcomings of the current dynamic simulation model in the study of vibration of ball screw feed drive system, taking a ball screw feed drive system test bench as an example, an electromechanical co-simulation model based on the lumped parameter model of ball screw feed drive system was built up in this paper. Firstly, based on the axial and rotation vibration integrated dynamic modeling method of ball screws, the lumped parameter model of ball screw feed system was established. Secondly, through the integration of the simulation model of semi-closed-loop cascade control system and the lumped parameter model of ball screw feed drive system, an electromechanical co-simulation model was built up. Simulation result shows that, the co-simulation model of ball screw feed drive system can predict the vibration occurs in the feed operation caused by the servo controller, ball screw feed system or the coupling between them.

1 Introduction

Ball screw feed system is the most widely used linear drive system in the field of industrial automation[1]. In order to enhance the velocity and accuracy of present systems further, current research focus on the vibration reduction and avoidance of the feed drive. Additional damping modules or structures are integrated in the feed drive system to achieve this goal, such as semi-active damping system and set point filtering etc. Active damping system only reacts once a vibration is present and set point filtering can lead to path deformation [2-4]. Another way to solve this problem is generate a smoother trajectory. For this purpose, numbers of trajectory algorithms are established and the frequency contents of the trajectory are discussed and compared [5,6]. The vibration caused by the trajectory is difficult to analysis on hardware because of the coupling of variety excitation sources. All these researches needs a simulation method to help the searchers or engineers to study or optimize the design and parameter setting of the feed drive system.

[a] Corresponding author : iamt@tongji.edu.cn

Finite element model of ball screw feed drive system can predict the accurate dynamic characteristics. However, it is difficult to integrate with the simulation model of servo control system. Lumped parameter model of ball screw feed dirve system can simplify the simulation model by reduce the number of degrees of freedom (DOF) of the whole system. More importantly, it can easily integrate with the simulation model of the servo control system. A reasonable simplification of the lumped parameter model is the key to accurately predict the vibration of feed drive system[7-9]. This paper in view of the shortcomings of the current modeling method in the vibration study of ball screw feed drive system, an electromechanical co-simulation method was established, which can be used to study the dynamic characteristics and vibration behavior of the feed drive system. Simulation result shows that, the co-simulation model of ball screw feed drive system can predict the vibration occurs in the feed operation.

2 Lumped parameter model of ball screw feed system

A typical ball screw feed system consists of a servomotor, coupling, ball screw, work table and base (Figure 1). The ball screw is supported by two sets of bearings, which are fixed to the base. The servomotor torque is transmitted through a coupling onto the ball screw shaft to drive the work table. The linear guideways constrain the movement of the work table in axial direction. The base is fixed on the machine bed or place on the ground. The transformation from the rotational movement of the screw shaft into the linear motion of the work table is realized by the ball screw system with its transmission ratio i, which is defined as the distance of travel h during one revolution of the shaft as following :

$$i = \frac{h}{2\pi} \tag{1}$$

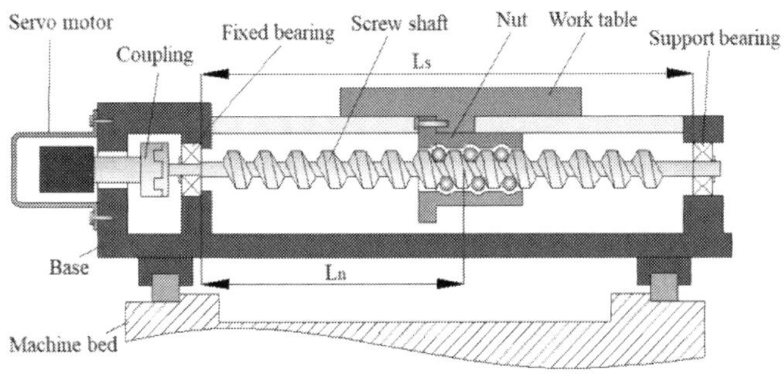

Figure 1. Typical structure of ball screw drive system

Low-order modes are the main factors affecting the dynamic characteristics of the ball screw feed drive system of machine tools. Typically the first axial or rotational mode of the ball screw show a dominant influence on the overall dynamics while the relevance of higher order modes for most technical applications is rather small[10]. The lumped parameter model of ball screw feed dirve system can reasonably reduce the DOF number of the simulation model while reserving the low-order modes of the system to simplify calculations. Figure 2 shows the lumped parameter model of a ball screw feed drive system. The influence of the shaft on the rotational mode and axial mode of the drive system are explicitly included into the lumped parameter model here. Therefore, the shaft is separated into two different branches, an axial and a rotational branch while the coupling once more is realized using constrained equations. Since all components are expressed by discrete springs and dampers, the rigidity values of shaft, coupling and bearing are combined to an overall axial K_{ax} and rotational value K_{rot}[11].

In this model the parameters are defined as following: rotary inertia of servo motor J_M, screw shaft side equivalent rotary inertia J_S, mass of base M_B, screw shaft side equivalent mass M_S, mass of the work table M_T, equivalent torsional rigidity K_{rot}, equivalent axial rigidity K_{ax}, rigidity of ball screw nut

K_n, axial rigidity of base K_B, servo motor torsional damping C_M, equivalent torsional damping C_{rot}, screw shaft side damping C_S, ball screw nut damping C_n, equivalent axial damping C_{ax}, axial damping of the base C_B, axial damping of the guide C_g. The equivalent method of the lumped parameters of this model can refer to reference article [12].

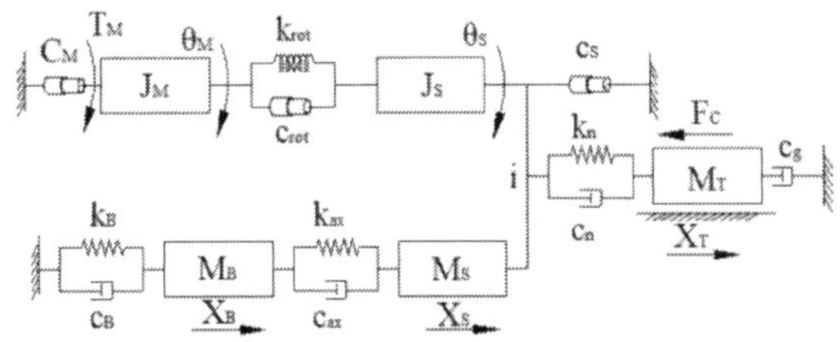

Figure 2. Lumped parameter model of ball screw feed system

According to the Lagrange's equations, with the independent coordinate system **q** and the generalized forces of the system **Q**, the lumped mass model of ball screw feed system can be established as equation(2).

$$\mathbf{m}\ddot{\mathbf{q}} + \mathbf{c}\dot{\mathbf{q}} + \mathbf{k}\mathbf{q} = \mathbf{Q} \tag{2}$$

Where: $\mathbf{q} = (\theta_M \quad \theta_S \quad X_B \quad X_S \quad X_T)^T$

$$\mathbf{Q} = (T_M \quad 0 \quad 0 \quad 0 \quad -F_C)^T$$

$$\mathbf{m} = \begin{bmatrix} J_M & 0 & 0 & 0 & 0 \\ 0 & J_S & 0 & 0 & 0 \\ 0 & 0 & M_B & 0 & 0 \\ 0 & 0 & 0 & M_S & 0 \\ 0 & 0 & 0 & 0 & M_T \end{bmatrix} \quad \mathbf{k} = \begin{bmatrix} k_{rot} & -k_{rot} & 0 & 0 & 0 \\ -k_{rot} & k_{rot} - i^2 k_n & 0 & -ik_n & ik_n \\ 0 & 0 & k_{ax} + k_B & -k_{ax} & 0 \\ 0 & ik_n & -k_{ax} & k_{ax} + k_n & -k_n \\ 0 & -ik_n & 0 & -k_n & k_n \end{bmatrix}$$

$$\mathbf{c} = \begin{bmatrix} c_M + c_{rot} & -c_{rot} & 0 & 0 & 0 \\ -c_{rot} & c_{rot} - c_S - i^2 c_n & 0 & -ic_n & ic_n \\ 0 & 0 & c_{ax} + c_B & -c_{ax} & 0 \\ 0 & ic_n & -c_{ax} & c_{ax} + c_n & -c_n \\ 0 & -ic_n & 0 & -c_n & c_n + c_g \end{bmatrix}$$

3 Co-simulation modeling of ball screw feed drive system

Based on lumped parameter model of ball screw feed system and the simulation model of servo control system, an electromechanical co-simulation model of ball screw feed drive system are constructed. The co-simulation schematic of ball screw feed drive system is shown in Figure 3, while (a) is the simulation model of Semi-closed-loop cascade control system and (b) is the lumped parameter model of ball screw feed system. The cascade control structure is the most widely used in machine tool feed drives. As showning in Figure 3, the control system includes a P-position controller, a PI-velocity controller and a PI-current controller.

In Semi-closed-loop cascade control system, the servo motor feed back speed ω_M and the work table feed back position X_T are calculated by the rotor position detected by the encoder on the servo motor. The input P_{ref} is given by the CNC system according to the feed motion command, the input P_{ref} and work table feed back position X_T are compared and the reference speed $\omega_{M\text{-}ref}$ is given by the position controller. Then the reference speed $\omega_{M\text{-}ref}$ is compared with the feed back speed ω_M and the velocity controller gives the reference current $i_{q\text{-}ref}$ for the q-axis and the reference current $i_{d\text{-}ref} = 0$ for d-axis of the stator. The three-phase current of the servo motor is detected and converts into i_d and i_q in d-q coordinate system through the Clark and Park transformation. $i_{d\text{-}ref}$ and $i_{q\text{-}ref}$ are compared with the feedback i_d and i_q respectively, and the current controller calculates the given voltages U_d and U_q of the d and q axes, then they are converted into U_α and U_β in the α-β coordinate system by Park inverse transformation.

Finally, the SVPWM module generates 6-phase PWM to drive the three-phase inverter.The inverter outputs A-B-C three-phase voltage to servo motor stator, which generates rotating magnetic field and produce magnetic torque on the servo motor rotor. This magnetic torque is the output torque T_M of the servo motor and drive the rotor to rotate under the dynamic relations of ball screw feed system.

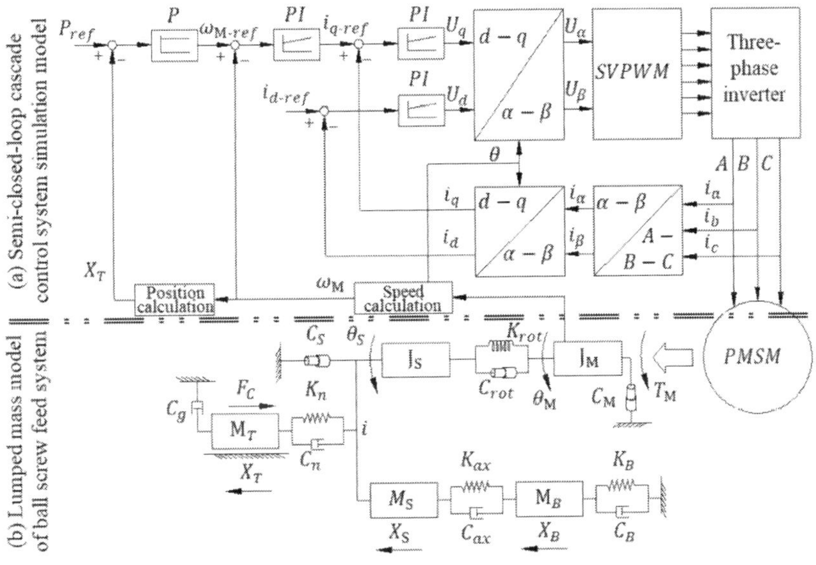

Figure 3. Schematic of ball screw feed drive system electromechanical co-simulation

The electromechanical co-simulation model of the semi-closed-loop controlled ball screw feed drive system is shown in Figure 4. The S_Cal module on the left side generate the trajectory command for the feed drive system according to the acceleration/deceleration strategy. Under the cascade control system the servo motor drive the ball screw accomplish the motion command accordingly.

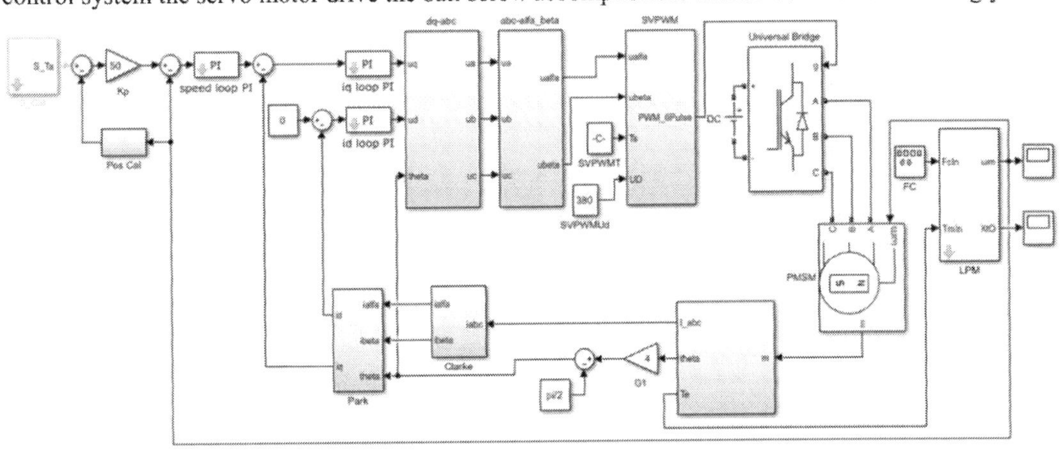

Figure 4. Electromechanical co-simulation model of half-closed ball screw feed system

4 Simulation result of the electromechanical co-simulation model of ball screw feed drive system

Take a ball screw feed drive system test bench of AMTC at Tongji Univesity as an example, an electromechanical co-simulation model as in figure 4 was built up. Table 1 are the simulation parameters used in the lumped parameter model of the test bench, which are calculated according to the specifications of the test bench. The specifications are either obtained from the manufacturers' catalogs, approximated from prior knowledge, or calculated from computer aided design (CAD).

Table 1. Simulation parameter used in the lumped parameter model of test bench

Parameter of the component	Value	Parameter of the component	Value
Mass of work table M_T (kg)	206	Axial rigidity of base K_B (N/m)	1.09×10^8
Screw equivalent mass M_S (kg)	11.28	Screw equivalent rotary inertia J_S $(kg{\cdot}m^2)$	1.7×10^{-3}
Rotary inertia of motor J_M $(kg{\cdot}m^2)$	6.75×10^{-3}	Rotary rigidity of screw K_{rot} $(N{\cdot}m{\cdot}rad^{-1})$	3.14×10^3
Axial rigidity of screw K_{ax} (N/m)	0.743×10^8	Contact rigidity of the screw nut K_n (N/m)	9.8×10^7
Mass of base M_B (kg)	3820		

Taking the servo motor torque as input and the axial acceleration of work table as output, the frequency response characteristics of the lumped parameter model of the test bench are analyzed. The bode diagram is shown in figure 5, simulation result shows that the work table has 4-order natural frequencies, which are 26.2 Hz,76.7 Hz,247 Hz and 633 Hz. Further study shows that, 76.7 Hz is the main axial vibration frequency of the work table, 26.2 Hz is the main axial vibration frequency of the base, 247 Hz and 633 Hz are the rotational vibration frequency.

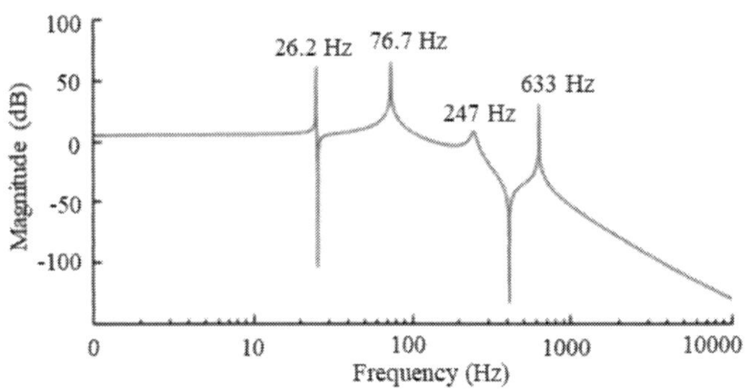

Figure 5. Bode diagram of the lumped parameter model of ball screw feed system

Other imprtant parameters, such as the parameters of the servo motor, the motion command parameters and the control parameters of the servo system, which are used in the electromechanical co-simulation model, are shown in table 2.

Table 2. Parameters of the servo system usd in the simulation model

Parameter name	Value	Parameter name	Value
Number of pole pairs	4	Inductance L_d、L_q (H)	8.15×10^{-3}
Stator resistance per phase (Ω)	1.44	Permanent magnetic flux Ψ_f (wb)	0.21
Feeding displacement (mm)	400	Position loop gain k_p	50
Maximum velocity (mm/s)	400	Velocity loop gain k_v	8
Maximum acceleration (mm/s^2)	2000	Current loop gain k_i	30
Maximum jerk (mm/s^3)	20000		

Figures 6 shows the simulation result of the reference velocity and feedback velocity of the work table at the given operating conditions. (a) and (b) are the partial enlarged detail of constant velocity phase and accelerating phase of the feed motion. Velocity fluctuation can be obviously seen in the feedback velocity.

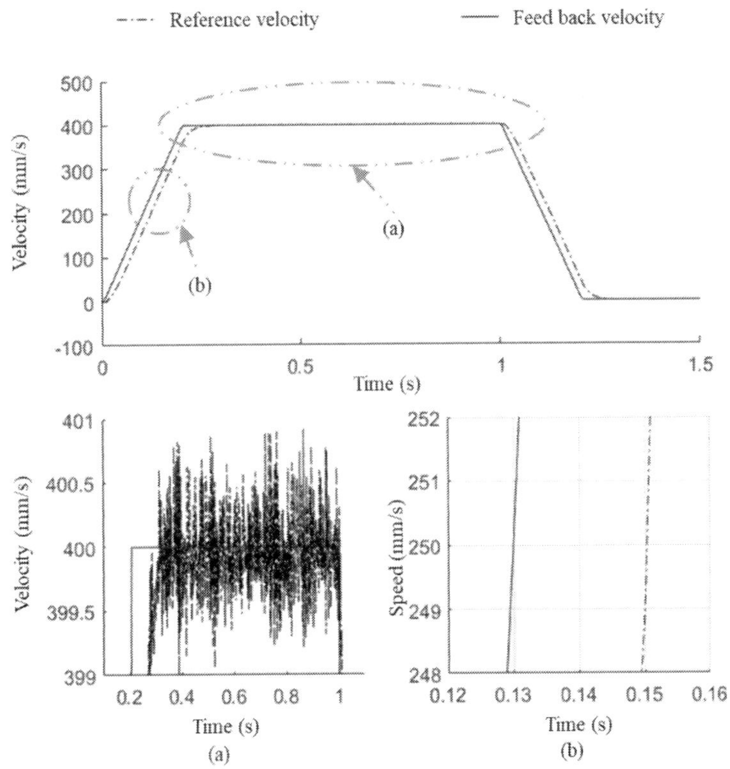

Figure 6. Simulation result of reference velocity and feedback velocity

Use the same operation parameters as seted in the simulaiton model, a feed motion experiment was conduct and the work table position was measured. Figure 7 shows frequency contents of the work table acceleration signals from the experimental and simulation result. Compare the simulation result with the experimental result, the co-simulation model of ball screw feed drive system can predict the vibration occurs in the feed operation. Both result show that in this case the second-order natural frequency (about 75 Hz) but not the first-order natural frequency is the main factors inluencing the performance of feed drive system.

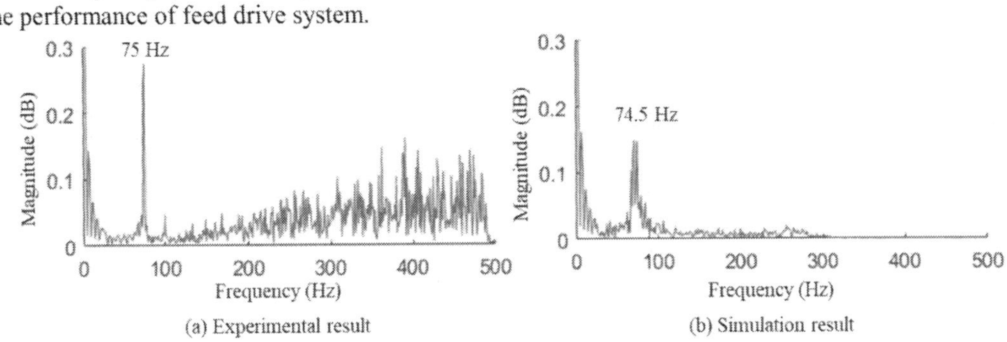

Figure 7. Frequency contents of work table acceleration of the experimental and simulation result

5 Conclusion(s)

In this paper an electromechanical co-simulation model of the ball screw feed drive system was constructed based on lumped parameter model of ball screw feed system, which can be used to study the dynamic characteristics and vibration behavior of the feed drive system. Taking a ball screw feed drive system test bench as an example, an electromechanical co-simulation model was built up. Simulation result shows that, the co-simulation model of ball screw feed drive system can predict the vibration occurs in the feed operation. Because of the integration of lumped parameter model into the detailed modeled cascade control simulation model, the electromechanical co-simulation of ball screw feed drive system could achieve a very good predictability for control performance and vibration behavior study of ball screw feed drive system, which may affact by the servo controller, ball screw feed system or the coupling between them.

Acknowledgments

This work is supported by the major national science and technology projects "high-end CNC machine tools and basic manufacturing equipment" (2012ZX04005031).

References

1. Y. Altintas, A. Verl, C. Brecher, L. Uriarte, G. Pritschow, CIRP ANN-MANUF TECHN, **60** (2011)
2. A. Verl, S. Frey, CIRP ANN-MANUF TECHN, **61** (2012)
3. A. Dietmair, A. Verl, MM Science Journal, (2009)
4. M. Yang, H. Hu, D. Xu, Electric Machines and Control, **16** (2012)
5. Y. Wang , D. Yang, R. Gai, S. Wang, S. Sun, INT J MACH TOOL MANU, **96** (2015)
6. Y. Jin, Y. He, J. Fu, J ZHEJIANG UNIV-SC B, **48** (2014)
7. X. Zhang, J. Zhang, W. Zhang, T. Liang, H. Liu, W. Zhao, MECH SYST SIGNAL PR, **98** (2018)
8. R. Sato, G. Tashiro, K. Shirase, Int. J. of Automation Technology, **9** (2015)
9. R. Sato, KEM, **516** (2012)
10. E C. Okwudire, Y. Altintas, J MECH DESIGN, **131** (2009)
11. S. Frey, A. Dadalau, A. Verl, Prod. Eng. Res. Devel., **6** (2012)
12. L. Luo, W. Zhang, M. Zhou, Mingjian, F. Jürgen, Transactions of the Chinese Society for Agricultural Machinery, **46** (2015)

MATEC Web of Conferences **237**, 03008 (2018) https://doi.org/10.1051/matecconf/201823703008
D2ME 2018

Designing Service Blueprint of Self-service Technology (SST) Based Public Transportation Service in Indonesia using SSTQual, Kano Model, and QFD

Amalia Suzianti[1,a] and Arsila Chairunnisa[2]

[1]*Universitas Indonesia, Industrial Engineering Department, 16424 Depok, Indonesia*
[2]*Universitas Indonesia, Industrial Engineering Department, 16424 Depok, Indonesia*

Abstract. Indonesia as one of developing countries has started to slowly implement self-service technology to its public transportation services which are e-ticketing and e-gate. Regardless of the public's enthusiasm for these self-service technologies implemented, users' complaints regarding the services implies the importance to redesign services that are focused on user satisfaction. This study attempts to result a design recommendation method for self-service technology based public transportation service to improve customer satisfaction by integrating SSTQual method, Kano Model, QFD and service blueprinting. Result shows that there are 12 service attributes belong to attractive category which need to be prioritized in increasing customer satisfaction. 12 attributes are then translated into technical requirements to obtain innovations to be used in the service blueprint. Through this research, there are 10 technical requirements which need to be applied to improve customer satisfaction. The technical requirements are then detailed in operational actions to be mapped in the service blueprint.

1 Introduction

Public service principally is under responsibility of government agencies or state-owned enterprises to fulfil community needs. This study using the case study of public railway station in Jakarta, the capital city of Indonesia. The declaration of Less Cash Society or also known as Gerakan Nasional Non Tunai (GNNT) in August 2014 has influenced the usage of self-service technology in this sector. Self-service technology allows customers to produce services independently. Self-service technology in this case which are e-ticketing and e-gate, has started to be implemented massively in Jakarta. On the other hand, number of train passengers around Jakarta (Jakarta, Bogor, Depok, Tangerang, Bekasi) in the last 7 years has been increasing and can be said to be constantly high [1]. It indicates the importance of service quality related to this kind of public service. A preliminary study is conducted through a questionnaire involving 31 respondents resulting that there are 3 important issues to be followed up which are station's facilities, station's service procedures, and information. These types of complaints can be accommodated with planning of activities, infrastructure, communications, material components, and services or commonly referred to the service design. However, the existing service design still received a lot of complaints which indicate that improvement regarding to its

[a] Corresponding author : suzianti@ie.ui.ac.id

service quality is needed. From the previous explanation, it is necessary to redesign services that are focused on customer satisfaction which leads to the need for service redesign. The goal to be achieved in this study can be divided into three stages sequentially: (1) to analyse the voice of customers, (2) to identify the priority of innovation based on the voice of customers, and (3) to design the service blueprint related to innovation chosen.

2 Theoretical review

2.1 Service design, service redesign, and service blueprint

Service design can be interpreted as the design of services which make the service becomes easier to use [2]. Service design as an essential component in the whole development of the service process focus on basic operational development activities that involve understanding and designing interactions from a variety of physical, technology and human elements [3]. According to ISO 9004-2: 1991, service design is a process that involves changing a service description into specification for service, service delivery and control, as well as looking for options from organizations including goals, policies and costs. A good service design can provide good interaction between human, human with system, and human with environment [4,5]. Meanwhile, service redesign is a process of reconstitution, rearrangement, or replacement of processes in services where innovation in the service process will be formed [6]. This enables an organization to reassess its current service design and provide some alternative ways of delivering more effective services for customers [7]. Redesigned service can add value to customers by increasing the benefits of a service or reducing the burden of the service [8]. There are several methods used in service redesign such as user diaries, safari service, user personas, user journey mapping, design brief, business model canvas, and service blueprinting. In this study, the method used is service blueprinting. A method was found by Shostack in a study of service management regarding the service development tool [9]. This related study was further developed by Kingman-Brundage to visualize the service or service process [10]. The blueprint service is a detailed map of activities and interactions of service designed to describe all the activities required to function, both visible and invisible ones for customers. Therefore, service blueprint can objectively understand and is not influenced by the individual's point of view according to their respective roles in the service [11].

2.2 SSTQual, Kano model, and QFD integration

Self-service Technology Quality or usually known as SSTQual presents the basis of calculation for customer-technology interaction context in the kind of same way as ServQual applies to human interaction. It assumes a linear relationship between customer satisfaction and performance quality. The implication is that low customer satisfaction will result low quality of service. This assumption is not entirely right because of giving extra attention to certain attributes may not always result the higher customer satisfaction, otherwise customer satisfaction can also greatly increase just by giving a small enhancement to other certain attributes with sort of unexpected but pleasant service. The integration of Kano models into SSTQual can answer these limitations. The Kano model can help categorize the attributes of measuring customer satisfaction based on the relationship between performance and customer satisfaction. The integration of Kano and SSTQual models assumes that the relationship between performance and customer satisfaction is not always linear. While Quality Function Deployment (QFD) is used to translate customer requirement into a technical response that can be used to actualize customer desires. This study uses a combination of SSTQual and Kano model which is then integrated into QFD, where the research model will refer to the combined framework of ServQual, Kano model and QFD sourced from Tan & Pawitra (2001) research. Referring to Tan & Pawitra (2001), SSTQual is used to identify gaps between services received with expected services and collect data on the level of importance of service attributes [12]. From these two outputs, value of customer satisfaction score can be calculated from the service attribute. Meanwhile, the Kano model is used to identify the categories based on the kano evaluation tables. The result of customer satisfaction score and weighted kano model is then used to calculate the value of adjusted importance of service attribute in HOQ diagram.

3 Research Methodology

For data collection, a questionnaire of integration of the SSTQual and the kano model is distributed in this stage. Respondents involved in this study amounted to 375 people. After collecting the questionnaire data, the result of voice of customer from the questionnaire is then analysed with QFD to translate customer needs into technical responses that can be used to actualize the needs. Innovation in the form of operational action is then visualized in service blueprint to provide the details of innovation in service.

3.1 SSTQual

According to Lin & Hsieh (2011) SSTQual can be used as a very useful diagnostic method to escort SST service provider from time to time. Furthermore, SSTQual research is a quality method to discover advantages or disadvantages of service and measuring the impact of improvements made. For example, an organization can use SSTQual to compare the performance of its current service with its previous service performance so that it can recognize the repetition actions that need to be applied when one of the following dimensions is below the required level. SSTQual can also be used both quantitatively and qualitatively to categorize the customer segmentation of the service provider. As customer expectations continue to change over time, service providers will gain deeper understanding of customer expectations and pain points in certain areas that have significant gaps (performances – expectations) through SSTQual. The SSTQual used in this study refers to Lin & Hsieh (2011) which consist of 7 dimensions including 20 attributes [13]. Likert scale (1-5) usage in SSTQual questionnaire for service performance and service expectation helps to determine each attribute's gap. The result of SSTQual attributes' gaps are shown in Table 1.

Table 1. SSTQual dimensions and attributes.

No	Dimension	Attribute	Gap
1	Functionality	I can get my service done with the firm's SST in short time (FUN1)	-0.424
		The service process of the firm's SST is clear (FUN2)	-0.515
		Using the firm's SST requires little effort (FUN3)	-0.448
		I can get my service done smoothly with the firm's SST (FUN4)	-0.267
		Each service item/function of the SST is error-free (FUN5)	-0.472
2	Enjoyment	The operation of the firm's SST is interesting (ENJ1)	-0.163
		I feel good being able to use the SSTs (ENJ2)	-0.323
		The firm's SST has interesting additional functions (ENJ3)	-0.075
		The firm's SST provides me with all relevant information (ENJ4)	-0.525
3	Security/Privacy	I feel safe in my transactions with the firm's SST (SEC1)	-0.565

No	Dimension	Attribute	Gap
		A clear privacy policy is stated when I use the firm's SST (SEC2)	-0.493
4	Assurance	The firm providing the SST is well-known (ASU1)	-0.067
		The firm providing the SST has a good reputation (ASU2)	-0.152
5	Design	The layout of the firm's SST is aesthetically appealing (DES1)	-0.523
		The firm's SST appears to use up-to-date technology (DES2)	-0.384
6	Convenience	The SST has operating hours convenient to customers (CON1)	-0.605
		It is easy and convenient to reach the firm's SST (CON2)	-0.488
7	Customization	The firm's SST understands my specific needs (CUS1)	-0.528
		The firm's SST has my best interests at heart (CUS2)	-0.496
		The firm's SST has features that are personalized for me (CUS3)	-0.235

3.2 Kano model

In this study, innovation is only applied for prioritized service attributes. This priority is done so that redesign is focusing on important attributes to generate high satisfaction. In addition, it will certainly be difficult for the organization if they have to make improvements to the overall attributes since it may take longer time and higher cost. Kano model questionnaire is used for prioritizing all 20 attributes by questioning what-ifs for each attribute's functionality and dysfunctionality. Based on 20 SSTQual attributes, there are 12 attributes belong to attractive category. This study focuses on attributes that are included in attractive category because these attributes need to get more attention since they have big impacts in satisfying customer.

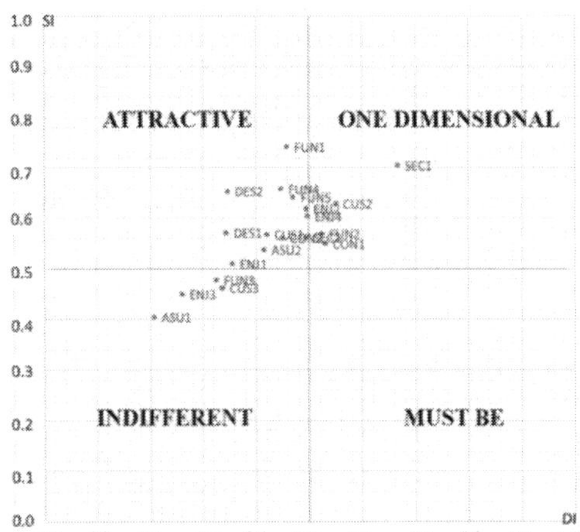

Figure 1. Kano category diagram.

Based on the results which can be seen in Figure 1, FUN1, FUN4, FUN5, ENJ1, ENJ2, ENJ4, SEC2, ASU2, DES1, DES2, CON2 and CUS1 are included in the attractive category. Based on their SI-DI values, these attributes have greatest or significant effect on customer satisfaction and least significant effect on dissatisfaction.

3.3 QFD

Literature review for technical responses is conducted with inputs related to 12 attributes which results 10 corresponding technical responses. SSTQual Integration and Kano model into QFD will provide operational and systematic guidance in improving service quality by obtaining priority of these 10 technical responses. In this stage, experts are needed to determine the relationship between each customer requirement and technical requirement. These relationship scores are then used to calculate the absolute importance and relative importance of each technical requirement. The priority result from QFD where customer requirements are the 12 attributes and technical requirements are the 10 technical responses can be seen in Table 2.

Table 2. Technical requirement priority.

Technical Requirement	Absolute Importance	Relative Importance	Priority
Determination of supporting information facilities in required spots	266,599	16.91%	1
Development of IT service quality support equipment	242.746	15.40%	2
Procurement of information services and 24 hours complaints	196,464	12.46%	3
Use of *social media, websites,* and applications as supporting activities	187.656	11.90%	4
Assessment of employee performance	139.888	8.87%	5
Implementation of *preventive maintenance system*	135,943	8.62%	6
Implementation of *knowledge management* principles in the company	121,379	7.70%	7
Implementation of user satisfaction evaluation system	107,443	6.82%	8
Implementation and management of cooperating with other public transportation organization	95,738	6.07%	9
Standardization of service time	82,537	5.24%	10

3.4 Service blueprint

Based on the technical requirement previously formulated, before visualizing these requirements to the blueprint, there is one stage that need to be done to generate the service innovation needed in the blueprint. In this study, all 10 technical requirement priorities are used to formulate the service innovation by dividing them into three levels: strategic level (level 1), management level (level 2), and operational level (level 3). Examples of a technical requirement at all three levels can be seen in Figure 2.

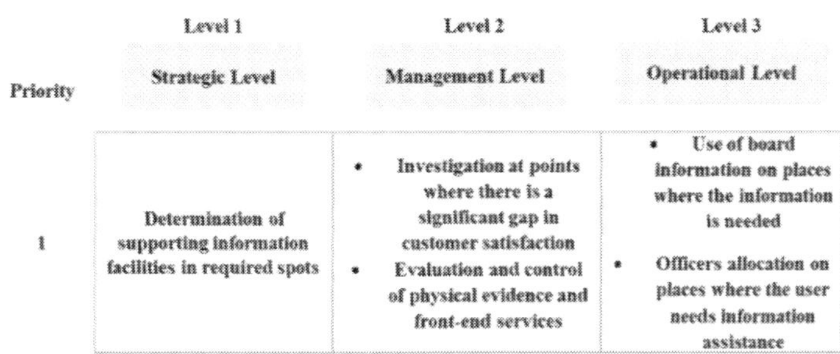

Figure 2. Technical requirement levels.

In result, 10 technical requirements in strategic level are formulated into 16 innovations (operational level technical requirement) for service blueprint. As service design focus on basic operational development activities, the actions that can be continued to be mapped to service blueprint are in level 3 (operational level) based on customer journey. Out of 16 actions, only 14 actions which can be mapped to the service blueprint. Those two operational actions cannot be mapped into blueprint due to their irrelevance to the customer journey of the service which should directly be related to customer.

4 Conclusion

Overall, the integration of SSTQual, kano model, and QFD is proven to support service redesign process specifically service blueprinting. In analysing the voice of customers, SSTQual is a right method to acquire customer expectation related to service performance by presenting the service gap between performance and expectation. In identifying the priority of innovation based on voice of customers, the integration of kano model and SSTQual helps to categorize 20 attributes of SSTQual in each category allowing service provider to choose the priority based on its goals. Since this study aims to reach high customer satisfaction, category chosen is the attractive one. In designing the service blueprint related to innovation chosen, QFD helps to translate those attributes into 10 technical responses which then formulated into three levels of action. This part is important because the flow of service blueprint is determined by customer journey so that operational actions which directly connected to customer can easily be mapped or visualized.

Acknowledgements

Thanks are due to Product Development and Innovation Laboratory - Industrial Engineering Department Universitas Indonesia, which has supported and facilitated this research.

References

1. Badan Pusat Statistik, Statistik Transportasi DKI Jakarta Tahun 2011-2017 (2017)
2. Davies, Ursula, Wilson, & Kelly. Design Methods for Developing Services: An Introduction to Service Design and a Selection of Service Design Tools, United Kingdom: Design Council UK. (2014)
3. B. Edvardsson, B. Thomasson, J. Øvretveit, *Quality of service: Making it really work*, London: McGraw-Hill Book Company (1994)
4. F. I. Stuart, Designing and executing memorable service experiences: Lights, camera, experiment, integrate, action! *Business Horizons*, **49(2)**, 149-159 (2006)
5. L. G. Zomerdijk, C. A. Voss, Service Design for Experience-Centric Services, Journal of Service Research, **13(1)**, 67-82 (2010)
6. L. L. Berry, S. K. Lampo, Teaching an Old Service New Tricks: The Promise of Service Redesign, Journal of Service Research, **II**, 265-275 (2000)
7. L. Shostack, Breaking Free from Product Marketing, Journal of Marketing, 73-80 (1977)
8. V. A Zeithaml, Consumer Perceptions of Price, Quality, and Value: A Means-End Model and Synthesis of Evidence, Journal of Marketing, 2-22 (1988)
9. L. Shostack, Service design in the operating environment, Chicago: AMA (1984)
10. Kingman-Brundage, The ABC's of service system blueprinting, *Designing a winning service strategy* (pp. 30-33). Chicago: AMA (1989)
11. Zeithaml, Bitner, Services Marketing, New York: Irwin McGraw-Hill (2000)
12. K. C. Tan, T. A. Pawitra, Integrating SERVQUAL and Kano's Model into Qfd For Service Excellence Development, Managing Service Quality, **11 (6)**, 418 (2001)
13. J. C. Lin, P. Hsieh, Assessing the Self-service Technology Encounters: Development and Validation of SSTQUAL Scale, Journal of Retailing, **87(2)**, 194-206 (2011)

MATEC Web of Conferences **237**, 03009 (2018) https://doi.org/10.1051/matecconf/201823703009
D2ME 2018

Kinematic Synthesis of Mechanism for System with a Technical Vision

Baurzhan Tultayev, Gani Balbayev, Algazy Zhauyt[a], Aidos Sultan and Aigerim Mussina
Gylym ordasy, Almaty 050013, Kazakhstan

Abstract. A solution to the problem of synthesizing an initial three-dimensional kinematic chain with spherical and rotary kinematic pairs is presented. It is shown that this chain can be used as a structural module for structural-kinematic synthesis of three-dimensional four-link motion generating lever mechanisms by the preset positions of the in-and output links. This paper affects the actual today's problem of optimal synthesis of spatial link mechanisms. In this regard, the task of developing methods for the synthesis of complex spatial link mechanisms with the desired laws of motion of the input and output elements allowing automatizing the implementation of all design phases with the use of computer is quite relevant. The authors develop machine-oriented method of structural and kinematic synthesis of spatial link mechanisms based on the use of spatial initial kinematic chains (IKC) realizing prescribed motions. A new approach to the design of spatial mechanisms is suggested, according to which the design process is based on the kinematic synthesis of four-link initial kinematic chain (IKC) and associable kinematic chains (AKC).

1. Introduction

By necessity of automation in the field of equipment at research optimum engineering of machine and mechanism according to criteria of force transfer are appeared new approaches of solution the task by Pencic et al. [1]. The new approach for drawing up algorithm and program at the optimum complex solution of the task according to criteria of force transfer of four and multilink mechanism is required for the solution of these tasks by Szkodny [2]. The way of the solution of the task at engineering of the initial kinematic chain (IKC) is presented and kinematic synthesis Jalón and Bayo [3] of its modifications of the four-link flat lever mechanism, and also regularity of the movement of multilink mechanism of the high class carrying out incoming and outgoing link on optimum criterion of force transfer. The research and engineering of regularity of the movement of the multilink mechanism carrying out incoming and outgoing link, and also the high-class mechanism is actual and demands comprehensive scientific and creative approach Lioa [4]. These papers demonstrate that four-link initial kinematic chains (IKC) may be used as a structural module with structural and kinematic synthesis of plain linkage mechanisms. Such an approach to the synthesis of plain mechanisms allows reducing the problem of their structural and kinematic synthesis to the solution of problem of IKC synthesis which is very useful for automation of mechanisms engineering. This paper testifies that specified approach may be applied to the problem of structural and kinematic synthesis of spatial

[a] E-mail: ali84jauit@mail.ru

linkage mechanisms Schwerin [5]. The solution of the problem of synthesis of spatial IKC of SSS type (S spherical kinematic pairs) is represented and its use as a structural module with structural and kinematic synthesis of spatial linkage mechanisms as per predetermined positions of input and output links is shown. A method of solution the problem of IKC synthesis of SSS type is based on the introduction of two movable bodies invariably associated with the input and output links Nussibalieva et al. [6]. This papers contains new scientifically grounded results that solve the following problems: mathematical problems of synthesis of spatial four-link initial kinematic chain of general type with spherical pairs are formulated and solved on the basis of motion specification of two movable bodies relative to non-movable body; under different combinations of the given and required parameters of the spatial initial kinematic chain, its various modifications have been obtained Tultayev [7]. The reliability of the theoretical results is confirmed by numerical tests of all found theoretical laws. Numerical results of the kinematic synthesis indicate the reality and authenticity of regularities that were found theoretically, operating capability of the given algorithm of mechanisms synthesis and confirm the expected theoretical provisions Zhauyt et al. [8]. Raising the technical level of machines requires an expansion of technological capabilities of the equipment and devices of the existing structures, as well as their equipping with fundamentally new mechanisms. Creation and implementation of new structures of industrial robots, equipment for light industry, mining, and oil and gas industry requires the use of mechanisms with a complicated motion of actuating devices. In this regard, the task of developing methods for the synthesis of complex spatial link mechanisms with the desired laws of motion of the input and output elements allowing automatizing the implementation of all design phases with the use of computer is quite relevant Zhauyt et al. [9]. Presented article touches on the actual up to date problem of optimal synthesis of spatial link mechanisms. At the theoretical level, it is proposed a new approach to the design of spatial mechanisms, according to which the design process is based on the kinematic synthesis of four-link initial kinematic chain and associable kinematic chains or two-element chains. Article differs by serious problem statement, interesting theoretical study and practical test Zhauyt et al. [10]. The work contains new scientific results having theoretical and applied significance.

2. Materials and Methods

Problem statement: given N of finite distant positions of two solids Q_1 and Q_2:

$$Q_1(X_A, Y_A, Z_A, \theta_i^1, \psi_i^1, \phi_i^1)\; Q_2(X_{Di}, Y_{Di}, Z_{Di}, \theta_i^2, \psi_i^2, \phi_i^2)\; i = \overrightarrow{1, N} \tag{1}$$

where θ_i^j, ψ_i^j, ϕ_i^j are fixed axis Eulerian angles $OXYZ$ and X_{Di}, Y_{Di}, Z_{Di} are coordinates of the point D_i of the solid Q_2.

It is required to find such points in the fixed axis as $A(X_A, Y_A, Z_A)$, $B(x_B, y_B, z_B)$ of the solid Q_1 and $C(x_C, y_C, z_C)$ of the solid Q_2, so that distance between the points B and C in all positions of the solids Q_1 and Q_2 is little different from some constant value R (see Figure 1).

Problem solution: Let us introduce a weighted difference for the i - th position of the solids in form:

$$\Delta_{q_i} = \left|\overrightarrow{B_i C_i}\right|^2 - R^2 = (X_{C_i} - X_{B_i})^2 + (Y_{C_i} - Y_{B_i})^2 + (Z_{C_i} - Z_{B_i})^2 - R^2 \qquad i = \overrightarrow{1, N} \tag{2}$$

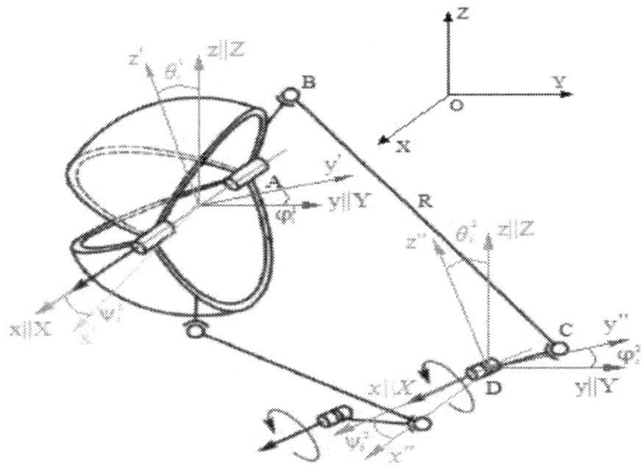

Figure 1. Equivalent RSSR four-link kinematic chain ABCD.

Relative to the fixed plane Q. Toughly connect with planes Q, Q_1 and Q_2 respectively, rectangular coordinate systems *OXYZ*, A_{xyz} and $Dxyz$. Plane Q_1 rotates around the point A with coordinates X_A, Y_A, Z_A, determined by the angle of rotation φ, and the plane Q_2 plane-parallel motion determined by the trajectory of its point D with coordinates X_D, Y_D, Z_D and angle of rotation ψ. To obtain a coordinated motion of two moving planes Q_1 and Q_2. It is necessary to impose geometric constraints on them, which can be done in two ways [1].

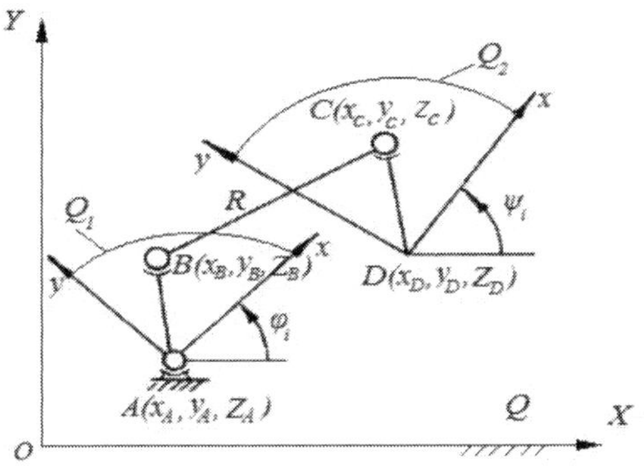

Figure 2. Equivalent SSS kinematic chain ABCD.

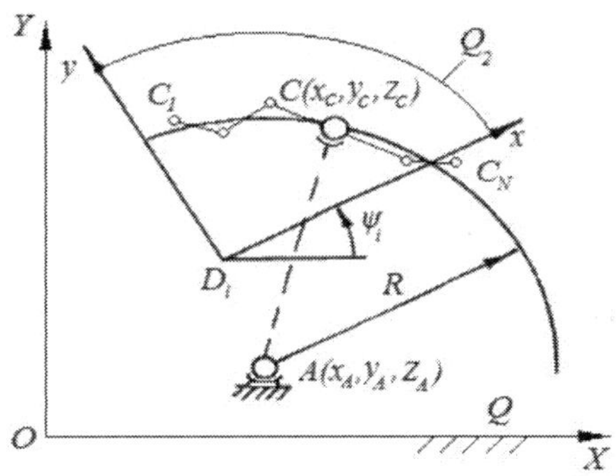

Figure 3. Equivalent SS kinematic chain CD.

It is required to find on the plane Q, Q_1 and Q_2 respectively, such points $A(X_A, Y_A, Z_A)$, $B(x_B, y_B, z_B)$ and $C(x_C, y_C, z_C)$ that the distance between points B or C. In all positions of the planes Q_1 and Q_2 The values determined by the relations differed little from a certain constant value R. Variable coordinates X_{Bi}, Y_{Bi}, Z_{Bi} points B and X_{Ci}, Y_{Ci}, Z_{Ci} points C in the *OXYZ* system are expressed in terms of their constant coordinates x_B, y_B, z_B and x_C, y_C, z_C in systems A_{xyz} and $Dxyz$ known linear transformation formulas (Figure 2) and (Figure 3):

$$\begin{bmatrix} X_{B_i} \\ Y_{B_i} \\ Z_{B_i} \\ 1 \end{bmatrix} = \begin{bmatrix} & [T_{10}^i] & & X_A \\ & & & Y_A \\ & & & Z_A \\ 0 & 0 & 0 & 1 \end{bmatrix} \cdot \begin{bmatrix} x_B \\ y_B \\ z_B \\ 1 \end{bmatrix} \quad \begin{bmatrix} X_{C_i} \\ Y_{C_i} \\ Z_{C_i} \\ 1 \end{bmatrix} = \begin{bmatrix} & [T_{20}^i] & & X_{Di} \\ & & & Y_{Di} \\ & & & Z_{Di} \\ 0 & 0 & 0 & 1 \end{bmatrix} \cdot \begin{bmatrix} x_C \\ y_C \\ z_C \\ 1 \end{bmatrix} \tag{3}$$

$$T_{jo}^i = \begin{bmatrix} e_{i1}^j & e_{i2}^j & e_{i3}^j \\ m_{i1}^j & m_{i2}^j & m_{i3}^j \\ n_{i1}^j & n_{i2}^j & n_{i3}^j \end{bmatrix} \qquad \begin{matrix} j = 1,2 \\ i = 1, N \end{matrix} \tag{4}$$

It is a function of ten parameters: X_A, Y_A, Z_A, x_B, y_B, z_B, R, x_C, y_C, z_C. By grouping these parameters in fours with the common parameter R, let us represent the weighted difference in three different forms [5-7]:

$$\Delta_{q_i}^{(1)} = (\widetilde{X}_{A_i} - X_A)^2 + (\widetilde{Y}_{A_i} - Y_A)^2 + (\widetilde{Z}_{A_i} - Z_A)^2 - R^2 \tag{5}$$

$$\Delta_{q_i}^{(2)} = (\widetilde{x}_{B_i} - x_B)^2 + (\widetilde{y}_{B_i} - y_B)^2 + (\widetilde{z}_{B_i} - Z_B)^2 - R^2 \tag{6}$$

$$\Delta_{q_i}^{(3)} = (\widetilde{x}_{C_i} - x_C)^2 + (\widetilde{y}_{C_i} - y_C)^2 + (\widetilde{z}_{C_i} - Z_C)^2 - R^2 \tag{7}$$

$$\begin{gathered} \begin{bmatrix} \widetilde{X}_{Ai} \\ \widetilde{Y}''_{Ai} \\ \widetilde{Z}_{Ai} \\ 1 \end{bmatrix} = \begin{bmatrix} & [T_{10}^i] & & 0 \\ & & & 0 \\ & & & 0 \\ 0 & 0 & 0 & 1 \end{bmatrix} \cdot \begin{bmatrix} x_B \\ y_B \\ z_B \\ 1 \end{bmatrix} + \begin{bmatrix} & [T_{20}^i] & & X_{D_i} \\ & & & Y_{D_i} \\ & & & Z_{D_i} \\ 0 & 0 & 0 & 1 \end{bmatrix} \cdot \begin{bmatrix} x_C \\ y_C \\ z_C \\ 1 \end{bmatrix} \\ \begin{bmatrix} \widetilde{x}_{B_i} \\ \widetilde{y}_{B_i} \\ \widetilde{z}_{B_i} \\ 1 \end{bmatrix} = \begin{bmatrix} & [T_{01}^i] & & 0 \\ & & & 0 \\ & & & 0 \\ 0 & 0 & 0 & 1 \end{bmatrix} \cdot \begin{bmatrix} X_{D_i} - X_A \\ Y_{D_i} - Y_A \\ Z_{D_i} - Z_A \\ 1 \end{bmatrix} + \begin{bmatrix} & [T_{21}^i] & & 0 \\ & & & 0 \\ & & & 0 \\ 0 & 0 & 0 & 1 \end{bmatrix} \cdot \begin{bmatrix} x_C \\ y_C \\ z_C \\ 1 \end{bmatrix} \\ \begin{bmatrix} \widetilde{x}_{C_i} \\ \widetilde{y}_{C_i} \\ \widetilde{z}_{C_i} \\ 1 \end{bmatrix} = - \begin{bmatrix} & [T_{02}^i] & & 0 \\ & & & 0 \\ & & & 0 \\ 0 & 0 & 0 & 1 \end{bmatrix} \cdot \begin{bmatrix} X_A - X_{D_i} \\ Y_A - Y_{D_i} \\ Z_A - Z_{D_i} \\ 1 \end{bmatrix} + \begin{bmatrix} & [T_{12}^i] & & 0 \\ & & & 0 \\ & & & 0 \\ 0 & 0 & 0 & 1 \end{bmatrix} \cdot \begin{bmatrix} x_B \\ y_B \\ z_B \\ 1 \end{bmatrix} \end{gathered} \tag{8}$$

where T_{kj}^i is the transfer matrix from the k coordinate system to the j system determined as:

$$T_{01}^i = [T_{10}^i]^T \quad T_{02}^i = [T_{20}^i]^T \quad T_{21}^i = T_{01}^i \times T_{20}^i \quad T_{12}^i = T_{02}^i \times T_{10}^i \tag{9}$$

the necessary conditions for minimum of the sum of squares of the weighted difference:

$$S = \sum_{i=1}^{N} [\Delta_{q_i}^{(k)}]^2 \quad (k = 1,2,\ldots,3) \tag{10}$$

may be written as the following system of equations:

$$\frac{\partial S}{\partial X_A}=0 \quad \frac{\partial S}{\partial Y_A}=0 \quad \frac{\partial S}{\partial Z_A}=0 \quad \frac{\partial S}{\partial R}=0 \tag{11}$$

$$\frac{\partial S}{\partial x_B}=0 \quad \frac{\partial S}{\partial y_B}=0 \quad \frac{\partial S}{\partial z_B}=0 \quad \frac{\partial S}{\partial R}=0 \tag{12}$$

$$\frac{\partial S}{\partial x_C}=0 \quad \frac{\partial S}{\partial y_C}=0 \quad \frac{\partial S}{\partial z_C}=0 \quad \frac{\partial S}{\partial R}=0 \tag{13}$$

From (11-13), considering (5-7) and (10), we obtain:

$$\left.\begin{array}{c} \sum_{i=1}^{N}\Delta_{q_i}^{(1)}\cdot\left(\widetilde{X}_{A_i}-X_A\right)=0 \\ \sum_{i=1}^{N}\Delta_{q_i}^{(1)}\cdot\left(\widetilde{Y}_{A_i}-Y_A\right)=0 \\ \sum_{i=1}^{N}\Delta_{q_i}^{(1)}\cdot\left(\widetilde{Z}_{A_i}-Z_A\right)=0 \\ \sum_{i=1}^{N}\Delta_{q_i}^{(1)}\cdot R=0 \end{array}\right\} \tag{14}$$

Assume that $R \neq 0$. Then from the last equality of system (14), it follows that:

$$\sum_{i=1}^{N}\Delta_{q_i}^{(1)}=0 \tag{15}$$

With provision for (15), the system of equations (14), takes the form:

$$\sum_{i=1}^{N}\Delta_{q_i}^{(1)}\widetilde{X}_{Ai}=0 \quad \sum_{i=1}^{N}\Delta_{q_i}^{(1)}\widetilde{Y}_{Ai}=0 \quad \sum_{i=1}^{N}\Delta_{q_i}^{(1)}\widetilde{Z}_{Ai}=0 \quad \sum_{i=1}^{N}\Delta_{q_i}^{(1)}=0 \tag{16}$$

By substituting expressions for $\Delta_{q_i}^{(1)}$ from (5-7) into system (16), we obtain:

$$\left.\begin{array}{c} \sum_{i=1}^{N}\left[\widetilde{X}_{Ai}^2X_A+\widetilde{X}_{Ai}\widetilde{Y}_{Ai}Y_A+\widetilde{Z}_{Ai}\widetilde{X}_{Ai}Z_A+\frac{1}{2}\left(R^2-X_A^2-Y_A^2-Z_A^2\right)\widetilde{X}_{Ai}\right]=\frac{1}{2}\sum_{i=1}^{N}\left(\widetilde{X}_{Ai}^2+\widetilde{Y}_{Ai}^2+\widetilde{Z}_{Ai}^2\right)\widetilde{X}_{Ai} \\ \sum_{i=1}^{N}\left[\widetilde{X}_{Ai}\widetilde{Y}_{Ai}X_A+\widetilde{Y}_{Ai}^2Y_A+\widetilde{Z}_{Ai}\widetilde{Y}_{Ai}Z_A+\frac{1}{2}\left(R^2-X_A^2-Y_A^2-Z_A^2\right)\widetilde{Y}_{Ai}\right]=\frac{1}{2}\sum_{i=1}^{N}\left(\widetilde{X}_{Ai}^2+Y_{Ai}^2+Z_{Ai}^2\right)\widetilde{Y}_{Ai} \\ \sum_{i=1}^{N}\left[\widetilde{Z}_{Ai}\widetilde{X}_{Ai}X_A+\widetilde{Y}_{Ai}\widetilde{Z}_{Ai}Y_A+\widetilde{Z}_{Ai}^2Z_A+\frac{1}{2}\left(R^2-X_A^2-Y_A^2-Z_A^2\right)\widetilde{Z}_{Ai}\right]=\frac{1}{2}\sum_{i=1}^{N}\left(\widetilde{X}_{Ai}^2+\widetilde{Y}_{Ai}^2+\widetilde{Z}_{Ai}^2\right)\widetilde{Z}_{Ai} \\ \sum_{i=1}^{N}\left[\widetilde{X}_{Ai}X_A+\widetilde{Y}_{Ai}Y_A+\widetilde{Z}_{Ai}Z_A+\frac{1}{2}\left(R^2-X_A^2-Y_A^2-Z_A^2\right)\widetilde{X}_{Ai}\right]=\frac{1}{2}\sum_{i=1}^{N}\left(\widetilde{X}_{Ai}^2+\widetilde{Y}_{Ai}^2+\widetilde{Z}_{Ai}^2\right) \end{array}\right\} \tag{17}$$

System (17) is linear with respect to the variables X_A, Y_A, Z_A and $H_1=\frac{1}{2}(R^2-X_A^2-Y_A^2-Z_A^2)$, thus it may be written as:

$$\begin{bmatrix} \sum_{i=1}^{N}\widetilde{X}_{Ai}^{2} & \sum_{i=1}^{N}\widetilde{X}_{Ai}\widetilde{Y}_{Ai} & \sum_{i=1}^{N}\widetilde{X}_{Ai}\widetilde{Z}_{Ai} & \sum_{i=1}^{N}\widetilde{X}_{Ai} \\ \sum_{i=1}^{N}\widetilde{X}_{Ai}\widetilde{Y}_{Ai} & \sum_{i=1}^{N}\widetilde{Y}_{Ai} & \sum_{i=1}^{N}\widetilde{Y}_{Ai}\widetilde{Z}_{Ai} & \sum_{i=1}^{N}\widetilde{Y}_{Ai} \\ \sum_{i=1}^{N}\widetilde{X}_{Ai}\widetilde{Z}_{Ai} & \sum_{i=1}^{N}\widetilde{Y}_{Ai}\widetilde{Z}_{Ai} & \sum_{i=1}^{N}\widetilde{Z}_{Ai}^{2} & \sum_{i=1}^{N}\widetilde{Z}_{Ai} \\ \sum_{i=1}^{N}\widetilde{X}_{Ai} & \sum_{i=1}^{N}\widetilde{Y}_{Ai} & \sum_{i=1}^{N}\widetilde{Z}_{Ai} & N \end{bmatrix} \cdot \begin{bmatrix} X_A \\ Y_A \\ Z_A \\ H_1 \end{bmatrix} = \frac{1}{2} \begin{bmatrix} \sum_{i=1}^{N}\widetilde{R}_{A_i}^{2}\widetilde{X}_{A_i} \\ \sum_{i=1}^{N}\widetilde{R}_{A_i}^{2}\widetilde{Y}_{A_i} \\ \sum_{i=1}^{N}\widetilde{R}_{A_i}^{2}\widetilde{Z}_{A_i} \\ \sum_{i=1}^{N}\widetilde{R}_{A_i}^{2} \end{bmatrix} \tag{18}$$

where $\widetilde{R}_{Ai}^{2} = \widetilde{X}_{Ai}^{2} + \widetilde{Y}_{Ai}^{2} + \widetilde{Z}_{Ai}^{2}$. The solution to this system by Cramer's rule is as follows:

$$(X_A, Y_A, Z_A, H_1) = \frac{1}{D_1}(D_{X_A}, D_{Y_A}, D_{Z_A}, D_{H_1}) \qquad D_1 \neq 0 \tag{19}$$

Similarly, from (11-13), considering (5-7) and (10), we obtain a system of linear equations in the unknowns x_B, y_B, z_B, H_2:

$$\begin{bmatrix} \sum_{i-1}^{N}\widetilde{x}_{Bi}^{2} & \sum_{i-1}^{N}\widetilde{x}_{Bi}y_{Bi} & \sum_{i-1}^{N}\widetilde{x}_{Bi}\widetilde{z}_{Bi} & \sum_{i-1}^{N}\widetilde{x}_{Bi} \\ \sum_{i-1}^{N}\widetilde{x}_{Bi}y_{Bi} & \sum_{i-1}^{N}\widetilde{y}_{Bi}^{2} & \sum_{i-1}^{N}\widetilde{y}_{Bi}\widetilde{z}_{Bi} & \sum_{i-1}^{N}\widetilde{y}_{Bi} \\ \sum_{i-1}^{N}\widetilde{x}_{Bi}\widetilde{z}_{Bi} & \sum_{i-1}^{N}\widetilde{y}_{Bi}\widetilde{z}_{Bi} & \sum_{i-1}^{N}\widetilde{z}_{Bi}^{2} & \sum_{i-1}^{N}\widetilde{z}_{Bi} \\ \sum_{i-1}^{N}\widetilde{x}_{Bi} & \sum_{i-1}^{N}\widetilde{y}_{Bi} & \sum_{i-1}^{N}\widetilde{z}_{Bi} & N \end{bmatrix} \cdot \begin{bmatrix} x_B \\ y_B \\ z_B \\ H_2 \end{bmatrix} = \frac{1}{2} \begin{bmatrix} \sum_{i=1}^{N}\widetilde{R}_{Bi}^{2}\widetilde{x}_{Bi} \\ \sum_{i=1}^{N}\widetilde{R}_{Bi}^{2}\widetilde{y}_{Bi} \\ \sum_{i=1}^{N}\widetilde{R}_{Bi}^{2}\widetilde{z}_{Bi} \\ \sum_{i=1}^{N}\widetilde{R}_{Bi}^{2} \end{bmatrix} \tag{20}$$

By solving this system by Cramer's rule [5-7], we obtain:

$$(x_B, y_B, z_B, H_2) = \frac{1}{D_2}(D_{xB}, D_{yB}, D_{zB}, D_{H_2}) \qquad D_2 \neq 0 \tag{21}$$

From (11-13), considering (5-7) and (11-13), we obtain a system of linear equations in the unknowns x_C, y_C, z_C, H_3:

$$\begin{bmatrix} \sum_{i-1}^{N}\widetilde{x}_{Ci}^{2} & \sum_{i-1}^{N}\widetilde{x}_{Ci}\widetilde{y}_{Ci} & \sum_{i-1}^{N}\widetilde{x}_{Ci}\widetilde{z}_{Ci} & \sum_{i-1}^{N}\widetilde{x}_{Ci} \\ \sum_{i-1}^{N}\widetilde{x}_{Ci}\widetilde{y}_{Ci} & \sum_{i-1}^{N}\widetilde{y}_{Ci}^{2} & \sum_{i-1}^{N}\widetilde{y}_{Ci}\widetilde{z}_{Ci} & \sum_{i-1}^{N}\widetilde{y}_{Ci} \\ \sum_{i-1}^{N}\widetilde{x}_{Ci}\widetilde{z}_{Ci} & \sum_{i-1}^{N}\widetilde{y}_{Ci}\widetilde{z}_{Ci} & \sum_{i-1}^{N}\widetilde{z}_{Ci}^{2} & \sum_{i-1}^{N}\widetilde{z}_{Ci} \\ \sum_{i-1}^{N}\widetilde{x}_{Ci} & \sum_{i-1}^{N}\widetilde{y}_{Ci} & \sum_{i-1}^{N}\widetilde{z}_{Ci} & N \end{bmatrix} \cdot \begin{bmatrix} x_C \\ y_C \\ z_C \\ H_3 \end{bmatrix} = \frac{1}{2} \begin{bmatrix} \sum_{i=1}^{N}\widetilde{R}_{Ci}^{2}\widetilde{x}_{Ci} \\ \sum_{i=1}^{N}\widetilde{R}_{Ci}^{2}\widetilde{y}_{Ci} \\ \sum_{i=1}^{N}\widetilde{R}_{Ci}^{2}\widetilde{z}_{Ci} \\ \sum_{i=1}^{N}\widetilde{R}_{Ci}^{2} \end{bmatrix} \tag{22}$$

From which we obtain x_C, y_C, z_C, H_3:

$$(x_C, y_C, z_C, H_3) = \frac{1}{D_3}(D_{xC}, D_{yC}, D_{zC}, D_{H_3}) \qquad D_3 \neq 0 \tag{23}$$

Eliminating the first four unknowns X_A, Y_A, Z_A, R, based on formula (18), it is possible to bring system (11-13) to a system of six equations with six unknowns x_B, y_B, z_B, x_C, y_C, z_C, which is convenient to be given as:

$$\begin{aligned} &\sum_{i=1}^{N}\Delta_{q_i}^{(1)}\frac{\partial\Delta_{q_i}^{(2)}}{\partial x_B}=0 \qquad \sum_{i=1}^{N}\Delta_{q_i}^{(1)}\frac{\partial\Delta_{i}^{(3)}}{\partial x_C}=0\\ &\sum_{i=1}^{N}\Delta_{q_i}^{(1)}\frac{\partial\Delta_{q_i}^{(2)}}{\partial y_B}=0 \qquad \sum_{i=1}^{N}\Delta_{q_i}^{(1)}\frac{\partial\Delta_{i}^{(3)}}{\partial y_C}=0\\ &\sum_{i=1}^{N}\Delta_{q_i}^{(1)}\frac{\partial\Delta_{q_i}^{(2)}}{\partial z_B}=0 \qquad \sum_{i=1}^{N}\Delta_{q_i}^{(1)}\frac{\partial\Delta_{i}^{(3)}}{\partial z_C}=0 \end{aligned} \tag{24}$$

Apparently, equations of this system are the same as the three equations of the thirteen degree in the three unknown functions given in the work by Zhauyt et al. [8-10], though in this case we have a system of six equations in six unknown functions. Solution of system (24) is labor-intensive task, so it is more effective to apply a search algorithm for the minimum of the function S stated below.

Theorem: if $S_1^{(1)}, S_2^{(1)}, S_3^{(1)}$ minimum values of sum S, determined for N position corresponding to the pairs of points:

$$\begin{aligned} &\left\{B^{(0)}\left(x_B'^{(0)}, y_B'^{(0)}, z_B'^{(0)}\right)\in Q_1,\ C^{(0)}\left(x_C''^{(0)}, y_C''^{(0)}, z_C''^{(0)}\right)\in Q_2\right\}\\ &\left\{A^{(1)}\left(x_A^{(1)}, y_A^{(1)}, z_A^{(1)}\right)\in Q,\ C^{(0)}\left(x_C''^{(0)}, y_C''^{(0)}, z_C''^{(0)}\right)\in Q_2\right\}\\ &\left\{A^{(1)}\left(x_A^{(1)}, y_A^{(1)}, z_A^{(1)}\right)\in Q,\ B^{(1)}\left(x_B'^{(1)}, y_B'^{(1)}, z_B'^{(1)}\right)\in Q_1\right\} \end{aligned}$$

such, that

$$\begin{aligned} S_1^{(1)} &= \min_{X_A,Y_A,Z_A,R} S\left[X_A, Y_A, Z_A, R, x_B'^{(0)}, y_B'^{(0)}, z_B'^{(0)}, x_C''^{(0)}, y_C''^{(0)}, z_C''^{(0)}\right]=\\ &= S\left[X_A^{(1)}, Y_A^{(1)}, Z_A^{(1)}, R_1^{(1)}, x_B'^{(0)}, y_B'^{(0)}, z_B'^{(0)}, x_C''^{(0)}, y_C''^{(0)}, z_C''^{(0)}\right]\\ S_2^{(1)} &= \min_{x_B', y_B', z_B', R} S\left[X_A^{(1)}, Y_A^{(1)}, Z_A^{(1)}, R, x_B', y_B', z_B', x_C''^{(0)}, y_C''^{(0)}, z_C''^{(0)}\right]=\\ &= S\left[X_A^{(1)}, Y_A^{(1)}, Z_A^{(1)}, R_2^{(1)}, x_B'^{(1)}, y_B'^{(1)}, z_B'^{(1)}, x_C''^{(0)}, y_C''^{(0)}, z_C''^{(0)}\right]\\ S_3^{(1)} &= \min_{x_C'', y_C'', z_C'', R} S\left[X_A^{(1)}, Y_A^{(1)}, Z_A^{(1)}, R, x_B'^{(1)}, y_B'^{(1)}, z_B'^{(1)}, x_C'', y_C'', z_C''\right] \end{aligned}$$

then $S_3^{(1)} \le S_2^{(1)} \le S_1^{(1)}$.

Proof: The system 20 when $D_1 \neq 0$ defines for any points $B^{(0)} \in Q_1$ and $C^{(0)} \in Q_2$ only one solution $\left(X_A^{(1)}, Y_A^{(1)}, Z_A^{(1)}, R_1^{(1)}\right)$ corresponding to minimum sum S of variables X_A, Y_A, Z_A and R. Therefore, the following equality can be written for a given pair $(B^{(0)}, C^{(0)})$

$$\begin{aligned} S_1^{(1)} &= \min_{X_A,Y_A,Z_A,R} S\left[X_A, Y_A, Z_A, R, x_B'^{(0)}, y_B'^{(0)}, z_B'^{(0)}, x_C''^{(0)}, y_C''^{(0)}, z_C''^{(0)}\right]=\\ &= S\left[X_A^{(1)}, Y_A^{(1)}, Z_A^{(1)}, R_1^{(1)}, x_B'^{(0)}, y_B'^{(0)}, z_B'^{(0)}, x_C''^{(0)}, y_C''^{(0)}, z_C''^{(0)}\right] \end{aligned}$$

Similarly, the system 22 when $D_2 \neq 0$ defines for the points $A^{(1)} \in Q$ and $C^{(0)} \in Q_2$ only one solution $\left(x_B'^{(1)}, y_B'^{(1)}, z_B'^{(1)}, R_2^{(1)}\right)$ corresponding to minimum sum S of variables x_B', y_B', z_B' and R:

$$S_2^{(1)} = S\left[X_A^{(1)}, Y_A^{(1)}, Z_A^{(1)}, R_2^{(1)}, x_B'^{(1)}, y_B'^{(1)}, z_B'^{(1)}, x_C''^{(0)}, y_C''^{(0)}, z_C''^{(0)}\right] =$$
$$= \min_{x_B', y_B', z_B', R} S\left[X_A^{(1)}, Y_A^{(1)}, Z_A^{(1)}, R, x_B', y_B', z_B', x_C''^{(0)}, y_C''^{(0)}, z_C''^{(0)}\right] \leq$$
$$\leq \min_{x_B', y_B', z_B', R} S_1^{(1)}\left[X_A^{(1)}, Y_A^{(1)}, Z_A^{(1)}, R_1^{(1)}, x_B', y_B', z_B', x_C''^{(0)}, y_C''^{(0)}, z_C''^{(0)}\right] \leq S_1^{(1)}$$

The system 24 when $D_3 \neq 0$ defines for the points $A^{(1)} \in Q$ and $B^{(1)} \in Q_1$ only one solution $\left(x_C''^{(1)}, y_C''^{(1)}, z_C''^{(1)}, R_3^{(1)}\right)$ corresponding to minimum sum S of variables x_C'', y_C'', z_C'' and R:

$$S_3^{(1)} = S\left[X_A^{(1)}, Y_A^{(1)}, Z_A^{(1)}, R_3^{(1)}, x_B'^{(1)}, y_B'^{(1)}, z_B'^{(1)}, x_C''^{(1)}, y_C''^{(1)}, z_C''^{(1)}\right] =$$
$$= \min_{x_C'', y_C'', z_C'', R} S\left[X_A^{(1)}, Y_A^{(1)}, Z_A^{(1)}, R, x_B'^{(1)}, y_B'^{(1)}, z_B'^{(1)}, x_C'', y_C'', z_C''\right] \leq$$
$$\leq \min_{x_C'', y_C'', z_C'', R} S_2^{(1)}\left[X_A^{(1)}, Y_A^{(1)}, Z_A^{(1)}, R_2^{(1)}, x_B'^{(1)}, y_B'^{(1)}, z_B'^{(1)}, x_C'', y_C'', z_C''\right] \leq S_2^{(2)}.$$

Consequently, $S_3^{(1)} \leq S_2^{(1)} \leq S_1^{(1)}$.

Algorithm for minimizing the function S:

i. Give arbitrarily initial points $B^{(0)} \in Q_1$, $C^{(0)} \in Q_2$ and check the condition $D_1 \neq 0$.

ii. Solve the system of linear equations (19) and determine $X_A^{(1)}$, $Y_A^{(1)}$, $Z_A^{(1)}$, $R_1^{(1)}$.

iii. Give points $A^{(1)} \in Q$, $C^{(0)} \in Q_2$ and check the condition $D_2 \neq 0$.

iv. Solve the system of equations (21) and determine $x_B^{(1)}$, $y_B^{(1)}$, $z_B^{(1)}$, $R_2^{(1)}$.

v. Give points $A^{(1)} \in Q$, $B^{(1)} \in Q_1$ and check the condition $D_3 \neq 0$.

vi. Solve the system of equations (23) and determine $x_C'^{(1)}$, $y_C'^{(1)}$, $z_C'^{(1)}$, $R_3^{(1)}$.

vii. Cyclically repeat steps i-vi, taking into account $A^{(i)} = A^{(i+1)}$, $B^{(i)} = B^{(i+1)}$, $C^{(i)} = C^{(i+1)}$, $i = 0,1,2...$

By applying the algorithm, we obtain a decreasing sequence of values of the objective function $S_1^{(1)}, S_2^{(1)}, S_3^{(1)}, S_1^{(2)}, S_2^{(2)}, S_3^{(2)}, S_1^{(3)}, S_2^{(3)}, S_3^{(3)}$ which has a limit equal to the value of the function S at the point of local minimum. When satisfying the inequality

$$\max\left(\left|R^{(i)} - R^{(i-1)}\right|,\ \left|X_A^{(i)} - X_A^{(i-1)}\right|,\ \left|Y_A^{(i)} - Y_A^{(i-1)}\right|,\ \left|Z_A^{(i)} - Z_A^{(i-1)}\right|\right) \leq \varepsilon$$

3. Results and discussion

Judging from deviations, the points C_1, C_2, C_3 are the exact spherical points (found with the given measure of inaccuracy), and C_4 and C_5 are approximate points. To solve this problem, a Visual Basic program has been written. The results of the problem solution are given in Table 1. The calculations have been made within the range:

$-1.2 < MJ < 0.6$; $-1.2 < NJ < 0.6$; $-1.2 < KJ < 0.6$. With a pitch $dm_j = 0.2$. In the entire range of calculations, the process of calculation is concurrent. The global minimum is equal to $S_{\min} = 0.0042921$ and it is achieved at: LL=3, KK=2, II=2. In Table 1, only a part of results in the neighborhood of the global minimum is shown. Its value is highlighted in bold frame in Table 1. Below, the results of calculation of the objective function in form of carpet plots and 3D plots in Figure 4, Figure 5, Figure 6, and Figures 7 are given.

Table 1. Assigned positions of the body for synthesis of a single movable mechanism with $N = 7$.

Position № (i)	x_{Oi}	y_{Oi}	z_{Oi}	**Euler angles [deg]**		
				θ_i^j	ψ_i^j	ϕ_i^j
1	0.30	0.12	0.01	0	0	0
2	0.35	0.17	0.24	28	35	17
3	0.44	0.21	0.25	34	38	5
4	0.51	0.15	0.32	17	24	12
5	0.50	0.30	0.45	50	50	21
6	0.60	0.25	0.41	45	33	24
7	0.55	0.32	0.35	0	0	0

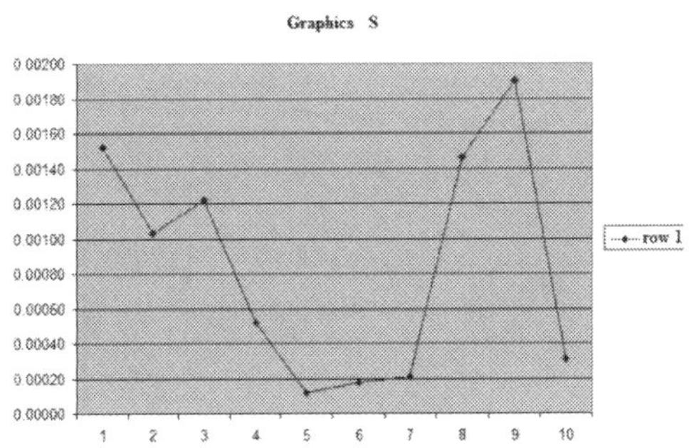

Figure 4. Computed plot of points S.

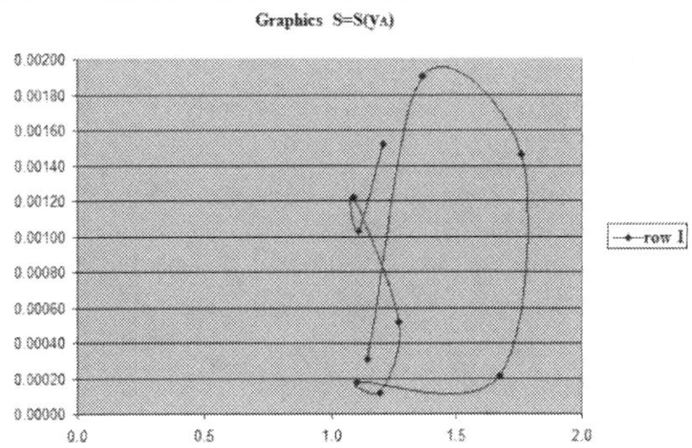

Figure 5. Computed plot of function points $S = S(Y_A)$.

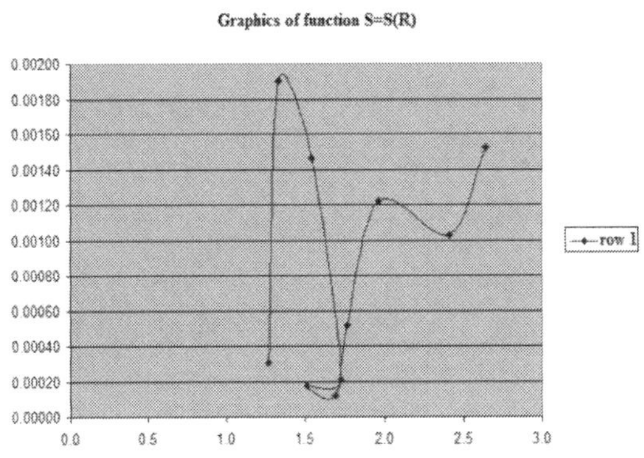

Figure 6. Computed plot of function points $S = S(R)$.

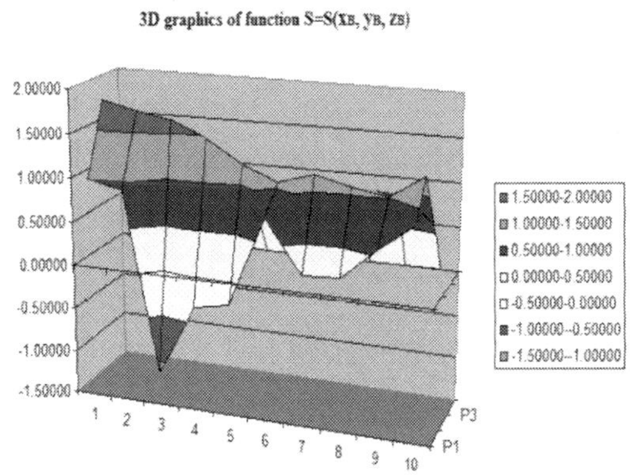

Figure 7. 3D graphics of function $S = S(x_B, y_B, z_B)$.

According to algorithm the value of the global minimum S_{min}=0.004292 has been determined as well as the value of coordinates of the global minimum:

x_B=-0.227631,	x_C=0.652906,	$X_A=Z_A$=0,
y_B=1.21314,	y_C=0.639147,	Y_A=1.189776,
z_B=0.609801,	z_C=0.469694,	R = =1.69127.

4. Conclusions

As can be concluded, the use of one and the same objective function being generated for synthesis of IKC and its modification allows automating of the process of synthesis of spatial linkage mechanisms through predetermined positions of the input and output links of the mechanism. In summary, in the synthesis of IKC with spherical kinematic pairs through predetermined positions of the input and output links of a mechanism when two adjacent links of IKC tend to infinity, it is necessary to replace the spherical kinematic pair for a plain or cylindrical one. In such a case, the synthesized mechanism takes form of a spatial link mechanism after determining the required parameters.

References

1. Pencic M., Cavic M., Rackov M., Borovac B. and Lu Z. Drive system of the robot eyeballs and eyelids with 8 DOFs. Mechanisms and Machine Science, **57**, pp. 47-56 (2018)
2. T. Szkodny Dynamics of industrial robot manipulators. Mechanism and Machine Theory, **30**, pp. 1057–1072 (1995)
3. J.G. de Jalón, E. Bayo Kinematic and Dynamic Simulation of Multibody Sytems: The Real-Time Challenge. Springer-Verlag, NewYork, (1994)
4. M.D. Lioa, V. Cossalter, R. Lot On the use of natural coordinates in optimal synthesis of mechanisms. Mechanism and Machine Theory, **35**, pp. 1367–1389 (2000)
5. R.V. Schwerin Multibody System Simulation Numerical Methods, Algorithms, and Software. Springer-Verlag, Berlin Heidelberg, (1999)
6. Nussibalieva A., Mussina A., Carbone G., Tultayev B., Balbayev G. Neurorobotic Investigation by using a basic vestibulo-ocular reflex system algorithm. New Trends in Medical and Service Robotics: Advances in Theory and Practice, pp. 126-134 (2018)
7. Baurzhan Tultayev, Gani Balbayev, Algazy Zhauyt A kinematic analysis of flat leverage mechanism of the fourth class for manipulators. Materials Science and Engineering, **230,** 2017, pp. 1-7 (2017)
8. Algazy Zhauyt, Kuanyshkali Alipov, Aizhan Sakenova, Adilet Zhankeldi, Raushan Abdirova, Zhastalap Abilkaiyr The synthesis of four-bar mechanism. Vibroengineering PROCEDIA, **10**, pp. 486-491 (2016)
9. Algazy Zhauyt, Gulnar Mamatova, Kuanysh Alipov, Aizhan Sakenova, Raushan Abdirova The kinematic analysis of flat lever mechanisms with application of vector calculation. Vibroengineering PROCEDIA, **8**, pp. 1-5 (2016)
10. Algazy Zhauyt, Indira Zhahanova, Gulbarshyn Smailova, Ulbala Murzakhmetova, Lazzat Kurmangalieva, Akmaral Kurenbay The kinematic analysis of the third class mechanism. Vibroengineering PROCEDIA, **12**, pp. 208-212 (2017)

MATEC Web of Conferences **237**, 03010 (2018) https://doi.org/10.1051/matecconf/201823703010
D2ME 2018

Optimization of Tooth Root Profile Using Bezier Curve with G2 Continuity to Reduce Bending Stress of Asymmetric Spur Gear Tooth

Priyakant Vaghela[1,a] and Jagdish Prajapati[2]

[1]*Mechanical Engineering Department, R C Technical Institute, Gujarat Technological University, Gujarat, India*
[2]*Faculty of technology and Engineering, The Maharaja Sayajirao University of Baroda, Vadodara, Gujarat, India*

Abstract. This research describes simple and innovative approach to reduce bending stress at tooth root of asymmetric spur gear tooth which is desire for improve high load carrying capacity. In gear design at root of tooth circular-filleted is widely used. Blending of the involute profile of tooth and circular fillet creates discontinuity at root of tooth causes stress concentration occurs. In order to minimize stress concentration, geometric continuity of order 2 at the blending of gear tooth plays very important role. Bezier curve is used with geometric continuity of order 2 at tooth root of asymmetric spur gear to reduce bending stress.

1 Introduction

Gears are very important part in all of machinery used in current times. Gear design is an extremely difficult art. High load-carrying capacity, less expensive, quieter running, lighter weight, and longer life are the primary aim of present industry for making gears. Current focus of gear design in industry is on failure occurs at critical section/root of tooth. One major cause of gear failure is fracture at root of the gear tooth due to bending fatigue [1]. The bending strength is influenced by: load location at HPSTC, load value, number teeth on the gear, fillet radius at root of tooth etc.

Failure at root of tooth can be reduced by optimizing geometrical shape of standard symmetric gear tooth [1-2]. In order to achieve this goal, pressure angle on drive side profile (as compare to coast side) can be increases within constraint [3-4]. Due to different pressure angle on both sides of gear tooth gear becomes asymmetric gear. In asymmetric spur gear bending stress at tooth root is reduced as compare to standard symmetric spur gear [5-6].

For further reduction in bending stress, it is necessary to optimize root profile of asymmetric spur gear tooth. Blending of the involute profile of tooth and circular fillet creates discontinuity at root of tooth causes stress concentration occurs [7-8]. In order to minimize stress concentration, geometric continuity of order 2 at the blending of gear tooth plays very important role [9-10]. In this paper a bezier curve is used with geometric continuity of order 2 at tooth root to reduce bending stress of asymmetric spur gear tooth.

2 Mathematical model of gear tooth

[a] Corresponding author : pavaghela@yahoo.co.in

Drive side pressure angle increases, base circle radius for drive side decreases compare to coast side. Hence long involute profile is available on drive side from tip (point A) to base circle radius (point C).

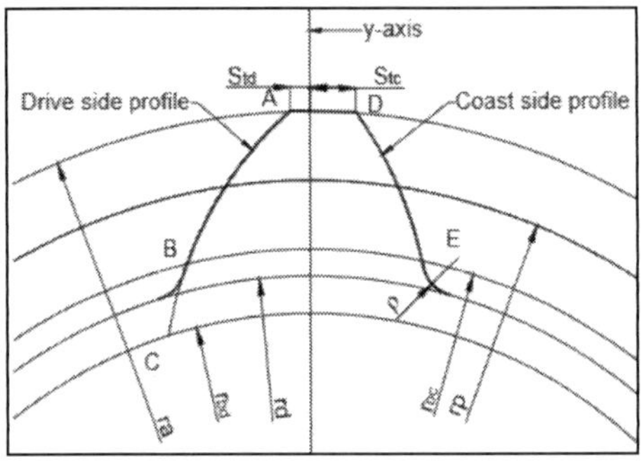

Figure 1. Asymmetric involute spur gear tooth

Whereas on coast side available profile is from tip (point D) to base circle radius (point E) as shown in figure 1.

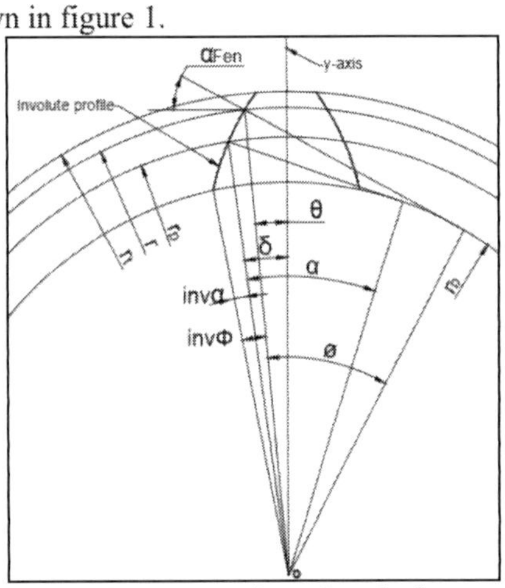

Figure 2(a). Representation of angle θ

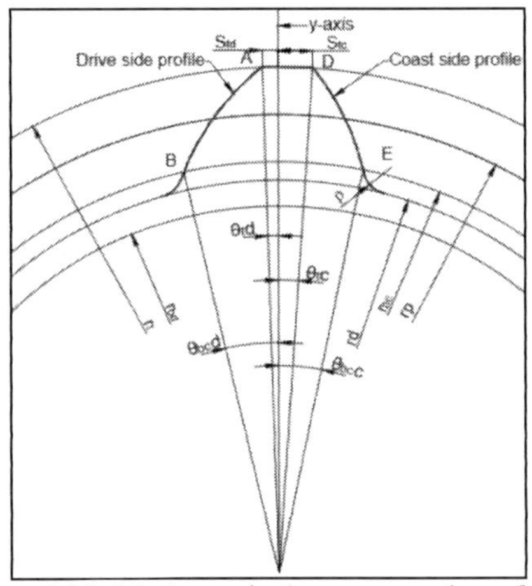

Figure 2(b). Mathematical representation of involute profile/gear tooth

Consider centre of gear as origin and vertical line passes through origin is a y axis. As shown in figure 2(a), θ is an angle at radius r w.r.t y-axis. Angle θ is a variable, which is required to develop involute profile of asymmetric spur gear tooth. Angle θ and radius r define location of points on the involute profile which is given by:

$$\theta = \frac{\pi}{2 \cdot z} + \tan\alpha - \alpha - \left(\tan\left(\cos^{-1}\left(\frac{r_p}{r} \cdot \cos\alpha \right) \right) - \left(\cos^{-1}\left(\frac{r_p}{r} \cdot \cos\alpha \right) \right) \right) \tag{1}$$

To develop drive and coast side profile, θ is varied between $\theta_{bc}d \leq \theta \leq \theta_f d$ and $\theta_{bc}c \leq \theta \leq \theta_f c$ respectively as shown in figure 2(b).

Involute profile coordinate (x_i, y_i) of an arbitrary point i for drive and coast side given by:

$$x_i = r_i \cdot \cos\theta_i$$

$$y_i = r_i \cdot \sin\theta_i$$

$$\theta_i = \frac{\pi}{2.z} + \tan\alpha - \alpha - \left(\tan\left(\cos^{-1}\left(\frac{r_p}{r_i}.\cos\alpha \right) \right) - \left(\cos^{-1}\left(\frac{r_p}{r_i}.\cos\alpha \right) \right) \right) \tag{2}$$

Where design parameter θ_i is varied

$$\theta_{bc}d \leq \theta_i \leq \theta_f d \text{ and } \alpha = \alpha_d \text{ for drive side profile}$$
$$-\theta_{bc}c \leq \theta_i \leq -\theta_f c \text{ and } \alpha = \alpha_c \text{ for coast side profile}$$

3 Optimization of tooth root profile

3.1 G2 Continuity

G2 continuity means geometric continuity of order 2. It represents position of end point, equal tangent and equal curvature continuity between two geometric curves [10]. It is necessary to reduce stress concentration at tooth root to reduce bending stress.

3.2 Bezier curve

Bezier curve mimics the shape of its control polygon and always passes through its ends points [11]. Bezier curve is defined by

$$P(t) = \sum_{i=0}^{n} B_i J_{n,i}(t) \quad , 0 \leq t \leq 1 \tag{3}$$
$$J_{n,i}(t) = \binom{n}{i} \cdot t^i (1-t)^{n-i} \quad , \binom{n}{i} = \frac{n!}{i!(n-1)!}$$

3.3 Optimize Bezier curve fillet with G2 continuity at the blending /critical section

In gear design at root of tooth circular-filleted is used [12] as shown in figure 3(a). Due to this at root, stress concentration occurs. To reduce stress concentration at the common end point of involute tooth profile and circular fillet, Bezier curve (at root of tooth) with G2 continuity has been developed as shown in figure 3(b).

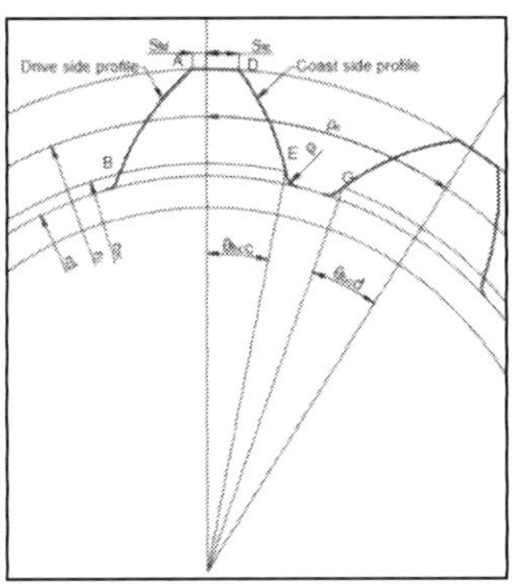

Figure 3(a). Circular fillet at gear tooth root

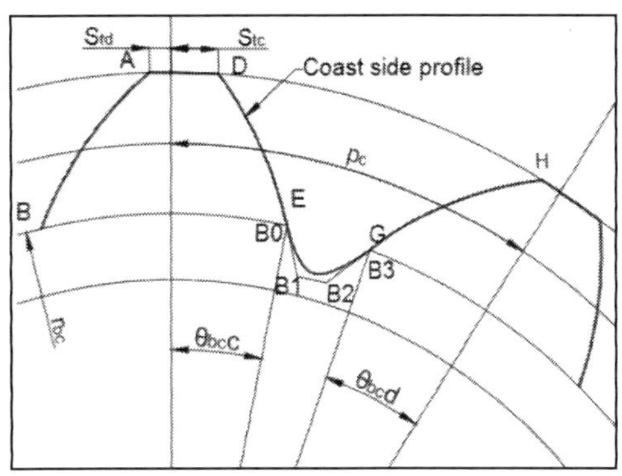

Figure 3(b). Developed fillet at tooth root with G2 continuity

If the radius of curvature is same at common end point, curves are known as curvature continuous. Tangent and curvature continuous is possible only when both first and second derivations of the quations are equal at common end point [10-11]. First and second derivations of bezier curve at end points are [11].

$$P'(0) = n(B_1 - B_0) \tag{4}$$

$$P'(1) = n(B_n - B_{n-1}) \tag{5}$$

$$P''(0) = n(n-1)(B_0 - 2B_1 + B_2) \tag{6}$$

$$P''(1) = n(n-1)(B_n - 2B_{n-1} + B_{n-2}) \tag{7}$$

First and second order derivative of the curve DE and GH at end points gives value of P`(0), P``(0), P`(1) and P``(1) respectively. B_0 and B_3 are end points of curve DE and GH respectively. B_0 and B_3 are known points. Based on equation 4 to 7 a code has been developed to obtain required points B1 and B2 at root of tooth.

4 Geometric model of asymmetric spur gear tooth

Following parameters and material used for geometric model and analysis. It is presented in table 1 & 2 [13].

Table 1. Design parameters [13]

Design Parameters	Value	Unit
Pressure angle, Drive / Coast side	38^0 / 20^0	Degree
Total tooth on pinion and gear	25 and 25	-
Module	4	mm
Power	18	KW
Rotation	1600	RPM

Table 2. Material- (Used for ANSYS Simulation)

Parameters	Value	Unit
Young Modulus	2×10^{11}	N/m^2
Poisson ratio	0.266	-
Density	7860	Kg/m^3
Yield strength	2.5×10^8	N/m^2

As per above parameter and developed mathematical model, involute profile of asymmetric spur gear tooth is generated shown in figure 4(a) and (b).

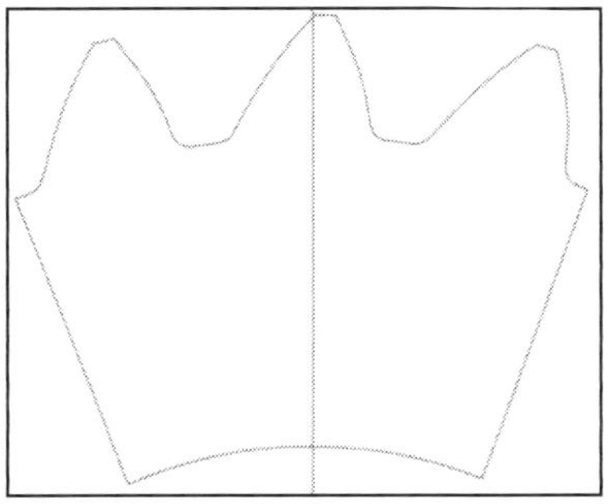

Figure 4(a). Asymmetric involute spur gear (α_d=38^0, α_c=20^0) with normal fillet (ρ=1.2 mm) at root of tooth

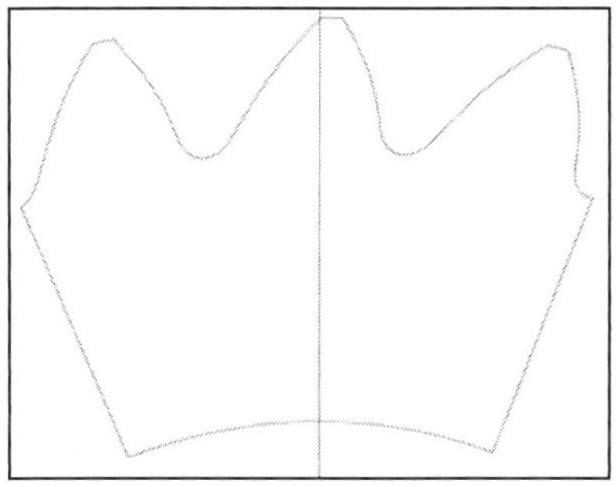

Figure 4(b). Asymmetric involute spur gear (α_d=38^0, α_c=20^0) with G2 continuity at root of tooth

5 Analysis of tooth in ANSYS software

Finite element analyses have been carried out for developed tooth shown in fig. 5(a) and (b). In this simulation tooth is considered as cantilever beam and loaded on drive side. The material used for simulation is presented in table 2. For analysis, following assumptions are made: 1) There is no misalignment in gear transmission, 2) Load is static, 3) Load is uniformly distributed and 4) Gear tooth material is homogenous.

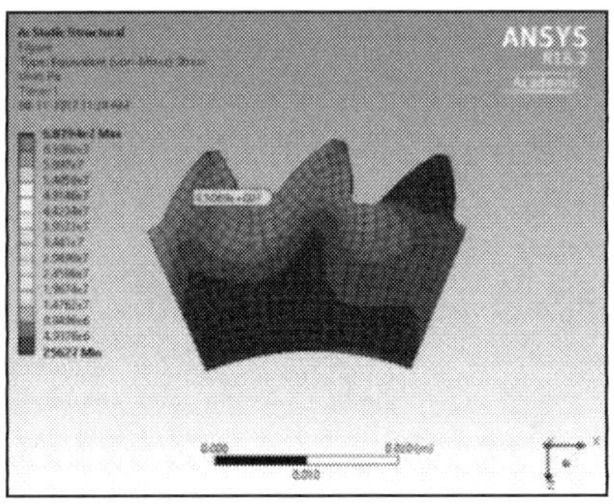

Figure 5(a). ANSYS result for asymmetric involute spur gear (α_d=38^0, α_c=20^0) with normal fillet (ρ=1.2 mm) at root of tooth

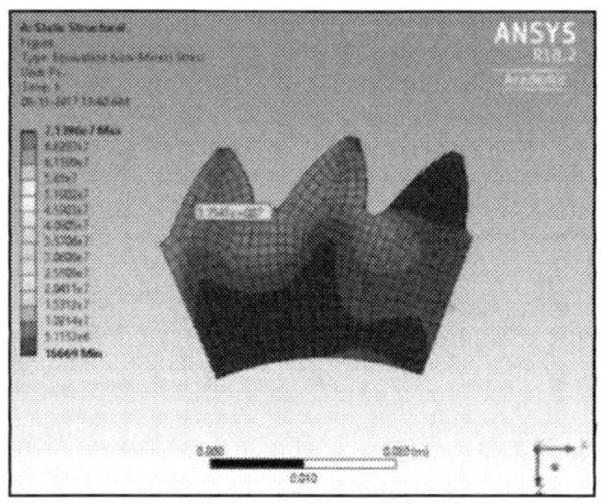

Figure 5(b). Asymmetric involute spur gear (α_d=38^0, α_c=20^0) with G2 continuity at root of tooth.

6 Results

Comparison has been made between normal fillet asymmetric spur gear tooth with modified gear tooth. FEA simulation show that Von mises stress at root of symmetric gear (α_d = 38^0 and α_c = 20^0) with normal fillet is 45.06 Mpa. Von mises stress at root of tooth of asymmetric gear (α_d = 38^0 and α_c = 20^0) with G2 continuity is 37.54 Mpa. Von mises stress is reduced by 16.68% compared with normal fillet asymmetric spur gear. A result has been presented in table 3.

Table 3. Result table

Parameters	Value	Unit
Bending stress (asymmetric gear /20^0-38^0) with normal fillet radius	45.06	Mpa
Bending stress (asymmetric gear/20^0-38^0) with G2 continuity	37.54	Mpa

7 Conclusions

Root profile is optimized to reduce bending stress of asymmetric involute spur gear tooth. Von mises stress of optimized root profile is analyzed and compared with normal fillet profile of asymmetric gear. Von mises stress is reduced by 16.68 % by optimization of root profile. Stress concentration of modified asymmetric spur gear has been significantly reduced.

References

1. Dudley Darle, *Handbook of Practical Gear Design*, (Ist Edition, CRC Press London,(2009)
2. A. Kapelevich,"Geometry and design of involute spur gears with asymmetric teeth" Mechanism and Machine Theory **35**,117-130, (2000)
3. F. Litvin,,Q. Lian , A. Kapelevich,"Asymmetric modified spur gear drives: reduction of noise, localization of contact, simulation of meshing and stress analysis" Computer Methods Applied Mechanics and Engineering. **188**,363-390, (2000)
4. Th. Costopoulos, V. Spitas "Reduction of gear fillet stresses by using one-sided involute asymmetric teeth", Mechanism and Machine Theory **44**,1524-1534, (2009)
5. N. Pedersen, "Improving bending stress in spur gears using asymmetric gears and Shape optimization" Mechanism and Machine Theory **45**,1707-1720, (2010)
6. V. Senthil Kumar, D.Muni, G. Muthuveerappan "Optimization of asymmetric spur gear drives to improve the bending load capacity" Mechanism and Machine Theory **43** ,829-858,(2008)
7. A. Kawalec, J. Wiktor, D. Ceglarek," Comparative Analysis of Tooth-Root Strength Using ISO and AGMA Standards in Spur and Helical Gears With FEM-based Verification", Journal of Mechanical Design, **128** (2006)
8. V. Spitas, Th. Costopoulos and C. Spitas, "Increasing the Strength of Standard Involute Gear Teeth with Novel Circular Root Fillet Design", American Journal of Applied Sciences 2,**6**,1058-1064, (2005)
9. ISO 6336 Standard, *Calculation of Load Capacity of Spur and Helical Gears*
10. Lizheng Lu, "Optimal multi-degree reduction of Bezier curve with G2 continuty" Computer aided geometric design, **23** ,673-683,(2006)
11. D.Rogers, J. Adams, *Mathematical element for computer graphics*, (1990)
12. T. Zou , "Optimization of tooth root profile of spur gears for maximum load-carrying capacity", *Proceedings of ASME International design engineering technical conference and computers and information in engineering conference*,(2014)
13. P.Vaghela, Master degree thesis "*Effect of rim thickness on asymmetric spur gear tooth bending stress", Gujarat technological university*, (2011)

Design of Street Furniture for Famous Historical and Cultural Cities in the View of Cultural and Technological Integration - A Case Study on the Design of Bus Shelter in Bozhou

Yaqin Xia[1,a], Changchun Yang[1] and Xiang Wu[1]

[1]*Fashion and Art Design Institute, Donghua University, 1882 West Yan'an Road, China*

Abstract. It is necessary and urgent to integrate culture and technology into the street furniture of famous cultural and historical cities. A design model of innovative cultural and technological integration is constructed by combining its internal mechanism and analysing the status of street furniture of famous historical and cultural cities in China. Meanwhile, principles and methods of design street furniture for famous historical and cultural cities in the view of cultural and technological integration are proposed, and strategies of applying the design of street furniture for famous historical and cultural cities under the cultural and technological integration are elaborated in detail based on the design case of urban street furniture in Bozhou.

1 Introduction

The famous historical and cultural city is a cultural relic preservation system confirmed and announced by the State Council of the People's Republic of China. As of July 2017, 133 cities had been listed as famous historical and cultural cities in China, as the carriers of protecting historical feature and inheriting characteristic regional culture [1]. With the development of the Internet technology and gradual progress of smart city, it is necessary for many cities including famous historical and cultural cities to conduct urban renewal in order to meet the increasingly diversified demands of people. Street furniture, as the most representative entity sign of urban intelligence, is now drawing increasing attention of urban planners and urban managers.

With respect to the design of urban street furniture, domestic studies mainly focus on the following perspectives. 1). Aiming at the relationship between urban regional culture and street furniture and its application, for instance, Chen Chao interpreted the close relationship between regional culture and public art and mutual influence, and analyzed the application form of regional cultural element in public art with specific examples [2]. 2). Aiming at the application of technology in street furniture, for instance, Zhou Xinhai proposed the design and planning of intelligent street furniture in the background of smart city by aiming at the existing problems in the design of urban street furniture[3]. 3). Aiming at the planning of street furniture, for instance, Li Wenjia et al. sorted out and inspected the relationship between single street furniture and the entire city from the overall spatial structure of city [4]. Liu Quan et al. studied the size of urban communities, spatial planning and layout of street furniture, and re-intepreted the principles of urban street furniture [5]. In conclusion, studies on urban street furniture mainly focus on the culture, technology and planning. Though there are various studies

[a] Corresponding author : summer@dhu.edu.cn

on the cultural and technological aspects of urban street furniture, most separate the two aspects, and there is few theoretical research or case analysis of the integrated innovation of culture and technology in street furniture.

In the trend of Internet and smart city, street furniture of famous historical and culture cities should be equipped with perfect functions and characteristic cultural connotations, in order to create good user experience and meet the diversified demands of people. Therefore, how to integrate modern science and regional culture in the design of urban street furniture to promote urban quality and realize the "harmonious coexistence" with the overall urban environment has already become a pressing problem of the construction of famous historical and cultural cities in the current trend of smart city.

2 Overview of street furniture in famous historical and cultural cities in China

2.1 Development of urban street furniture and its connotation

"Street furniture", originated from Britain, is literally translated as "furniture of street", and it is usually called as urban element in Europe. In China, it is generally referred to as "urban public facilities", "urban environmental facilities", "urban furniture", etc. Street furniture is public goods provided by the government to meet people's demands for outdoor interactions, activities and travelling, etc. In the meantime, it creates the atmosphere of public space, as an indispensable element of urban space. Therefore, its design is especially important, which is a significant approach of improving the urban quality.

2.2 Design status of street furniture of famous historical and cultural cities in China

Famous historical and cultural city refers to the city with profound historical and cultural deposits, which would shoulder important responsibilities in the context of cultural revitalization. With the development of the times, famous historical and cultural cities should also advance with the times and keep updating constantly. Street furniture, as the "first name card" of famous historical and cultural cities, may impact the first impression of residents and tourists. Good street furniture can not only serve urban residents and foreign tourists, but also enhance the spatial connotation and quality of famous historical and cultural cities, and shape the cultural features of the city. Through the field research and collection, classification, settlement and analysis of second-hand information, it is found that most less developed famous historical and cultural cities are evidently disadvantageous in the street furniture, mainly for the following two reasons [6]: firstly, improper planning and lack of cultural connotation. Domestic famous historical and cultural cities were built a long time ago, and the original urban planning failed to reserve space for street furniture. Besides, it was not adjusted timely later, and accordingly, most street furniture was located in corners, of poor regional position. Currently, most cities mainly launch the urban construction by relying on the mass-produced street furniture. As a result, completely the same street furniture may occur in different cities, resulting in the loss of cultural genes, and "the same appearance of a thousand cities". Secondly, lack of the application of new technological achievements, incomplete functions, poor experience and tolerance. At present, street furniture of famous historical and cultural cities in China ignores the application of new technological achievements, such as interaction technique, energy-saving materials, big data, etc. In the meantime, currently, street furniture of cities is greatly defected in the man-machine size, information cue, etc. and of poor tolerance, and it might be dangerous for some weak or disabled users to use the street furniture.

3 Research analysis of street furniture design for famous historical and cultural cities in the background of cultural and technological integration

3.1 Necessity of cultural and technological integration in the design of street furniture for famous historical and cultural cities

Culture is a broad concept, and till now, there have already been hundreds definitions of culture, but all fail to summarize it comprehensively and accurately. Culture is mainly composed of material culture and intangible culture, which can be simply summarized as the general term of living factors in a region, including clothing, food, accommodation, travelling, culture and materials, etc. Owing to the lifestyle and folk culture of different regions, the folk culture and scene of buildings are quite different. Therefore, street furniture should respect the regional differences in the design and reflect characteristics. Technology is a general term of science and technology, which directly promotes the changes of the production mode and lifestyle. The technological development completely changes people's living conditions. The internet, artificial intelligence, auto drive, etc. have already become the hottest technical terms, and people are enjoying the convenient services brought by technological products like the mobile phone, computer, subway, vehicles, etc.

Culture and technology are two core elements promoting the social development. Huge progress has been made in science and technology in recent years. "Intelligence" is one of the important attributes of all cities including famous historical and cultural cities in the future. Smart city, smart community and smart traffic will become an irresistible development trend. Currently, there have already been almost 200 cities launching the smart city strategy. Specifically, in 2017, top 20 cities of smart city construction included 14 famous historical and cultural cities, such as Beijing, Shanghai, Hangzhou, Suzhou, Nanjing, etc. The integration of technology and culture will help to explore the characteristic history and cultural resources of famous historical and cultural cities, and boost the sustainable cultural development ability.

Nowadays, function is no longer the only determinant of product evaluation. What's more, it emphasizes more on the interaction between human and products and people's emotional dependence on products. The integration of culture and technology can create a good spiritual experience and functional experience for users, and satisfy the increasingly diversified demands of people. In such a context, street furniture, as an indispensable element in the construction of famous historical and cultural cities and lives of residents, should be equipped with characteristic cultural connotation, sound functions and good experience in the near future. Therefore, it is necessary, urgent and inevitable to integrate the integration of technology and culture in the design of street furniture.

3.2 Innovative design model of cultural and technological integration

Integration refers to a process of combining two or more different things. Technology and culture are in a co-existence and co-fusion relationship, and design plays a vital role in the integration process. Technology mainly selects and determines the goal, while design substantiates the realization of goal and realizes one of the standards, namely emotional. During the design process, only by making the technology useful and attaining the unity with the social environment system, can it be called as innovation. Culture is the endogenous power promoting the technological development, while technology is an important approach of inheriting the cultural innovation. Design coordinates in the relation between technology and culture (as shown in Fig. 1), promotes the integration and co-existence between the two, and changes people's lifestyle, as the main power of boosting the transformation of technology to be the actual productivity.

Emotional experience is determined by the cultural connotation, Emotion and culture supplement each other, since emotion is the outward manifestation of culture in products, while culture is the internal source of product emotion. Good products must be pleasant, communicable and sustainable.

Functional value is mainly promoted by technological innovation, as the expression carrier and communication tool of culture. Technology is the most direct way of realizing product functions; availability, usability and efficiency are three-step evaluation indexes of the action of technology on products to realize the functions. The application of new technology, new materials and new technology expands the boundary and space of design (as shown in Fig. 1). During the design, designers should apply the latest technological achievements, and combine cultural and aesthetic elements, to realize the development of new products, so that new products can be equipped with complete functions and abundant emotions to meet the diversified needs of users [7].

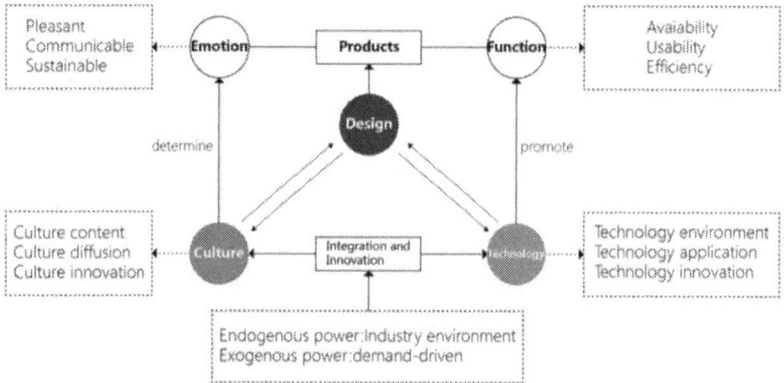

Figure 1. Innovative design model of cultural and technological integration

3.3 Design principles of street furniture for famous historical and cultural cities in the context of cultural and technological integration

Street furniture is mainly designed to meet the practical needs of the public, and such practicability not only requires good technology and processing property of street furniture, but also reflects designers' consideration of the using process and environment, and future development of the society. Therefore, the street furniture design for famous historical and cultural cities in the context of cultural and technological integration must abide by the following principles. Above all, to be "human-oriented" and "demand-oriented"; secondly, to reflect the regional characteristic culture, follow the cultural difference; to improve the functions of facilities with technological means, and optimize the experience, and to abide by the sustainable development design principle.

1) To be "human-oriented" and "demand-oriented"

Design mainly resolves the problems of human and things, and human needs essentially, the needs for all kinds of materials and equipment [8]. To be "human-oriented" is to take human's physiological needs, emotional needs and psychological needs into full account, so as to make full use of things. While designing street furniture, its structure and size should be considered, and its functions and safety should be guaranteed, to meet the physiological needs of people. In the meantime, the street furniture should be endowed with characteristic cultural properties, like the shape, color, texture, etc. to meet the different emotional and psychological needs of people, and enhance people's sense of belonging and involvement. The "human-oriented" design principle attempts to turn the relation between human and facility to be a similar communicable harmonious relation between people, which is the fundamental principle of street furniture design.

2) To reflect the regional characteristic culture, follow the cultural difference

Culture is developed through long-term accumulation and evolution in a region, and it is integrated with the surrounding environment. So culture has regional characteristics. The culture of different famous historical and cultural cities is distinctive, and difference is the unique factor of a city, as a sign of a city differing from other cities. Street furniture is an important media inheriting the regional history and culture. When the cultural value gathers in the street furniture, urban residents or tourists may feel the underlying connotation consciously, reflect the cultural characteristics, be culturally influenced silently, and reach a higher balance between human and the environment [9].

3) To improve the functions of facilities with technological means, and optimize the experience

A huge progress has been made in the technology, and a great batch of new materials and new technologies emerge endlessly, such as carbon fiber, solar energy, internet, big data, interactive technology, etc. In such a context, urban residents have new demands on the life quality and environment quality. Technicalization and intelligence would certainly become the future development trend of urban street furniture. Therefore, the latest technological achievements should be introduced to the design of urban street furniture. The deep integration of street furniture and urban scenes through technological means can enhance the user experience and meet the public demands. [10]

4) To abide by the sustainable development design principle

The sustainable design principle emphasizes on the important role of environment quality and environment input in improving people's life quality, and it puts forward long-term planning for the development of cities and urban street furniture. The design of urban street furniture should fully consider the economic and environmental sustainability, which requires the designers to use energy-saving, environmentally-friendly and recyclable materials in the design process and reduce the environmental burden consciously.

4 Study on the application of innovative cultural and technological integration in the street furniture design of Bozhou

4.1 Introduction of the project

Bozhou, located in the northwest of Anhui Province, is a state-level famous historical and cultural city, one of the first tourist cities in China, and head of the four cities of medicine, known as the "pearl of the north of Anhui Province". With more than 35000 years of history, Bozhou displays profound historical and cultural deposits, and it is also the representative of culture in the north of Anhui Province. Moreover, it has numerous historical sites and cultural heritage, including prehistoric culture, imperial and capital culture, Laozi & Zhuangzi culture, Cao Wei culture, traditional Chinese medicine culture, Gujing wine culture, etc. (as shown in Fig. 2).

Figure 2. Overview of Bozhou culture

4.2 Design of research goal and field research analysis

This research aims to design the bus shelter for Bozhou, and it mainly focuses on the reflection of regional cultural characteristics, application of technology, experience of facility functions and pain point analysis, etc. As a practical design project, the design team would conduct the field research on Bozhou first, record existing problems of bus shelters, crowd behaviors, traffic status, and spatial element through observation, photograph diary, etc. and conduct further analysis. According to the results, bus shelters in Bozhou are old, of incomplete functions, improper size, poor experience, inappropriate planning and lack of regional cultural characteristics (As shown in Fig.3).

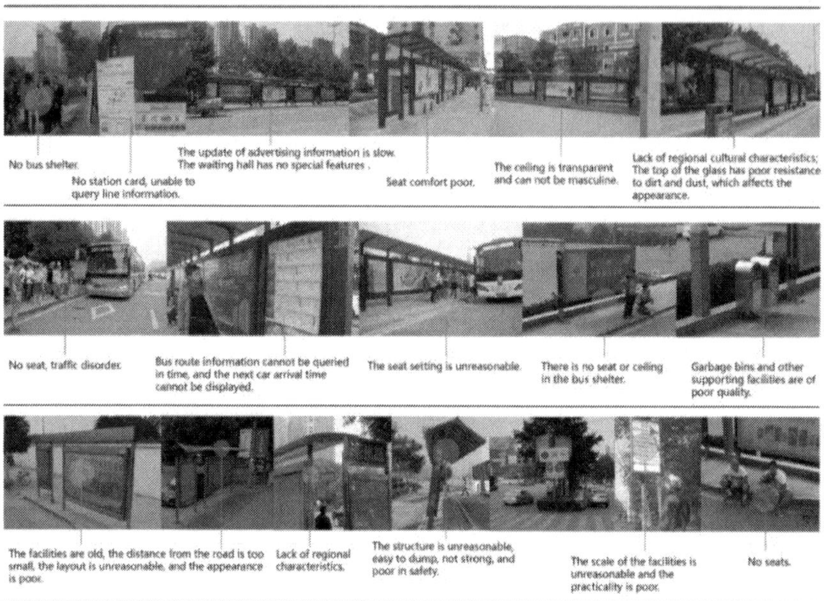

Figure 3. Research analysis of the status of bus shelter in Bozhou

4.3 Analysis of the functional requirements of urban bus shelter

Urban bus shelter is a small building that mainly serves the urban traffic, as the connection point of human and traffic means and playing a vital role in the urban traffic system. The design team conducts the functional requirements analysis for urban facilities like the bus shelter, and clarifies the specific functional requirements of bus shelter from three levels, namely usability, accessibility, and experience of products (as shown in Fig. 4). Meanwhile, functions of culture and technology on the basic functions, convenient functions and emotional experience of bus shelters are analyzed respectively, for guiding the subsequent project design.

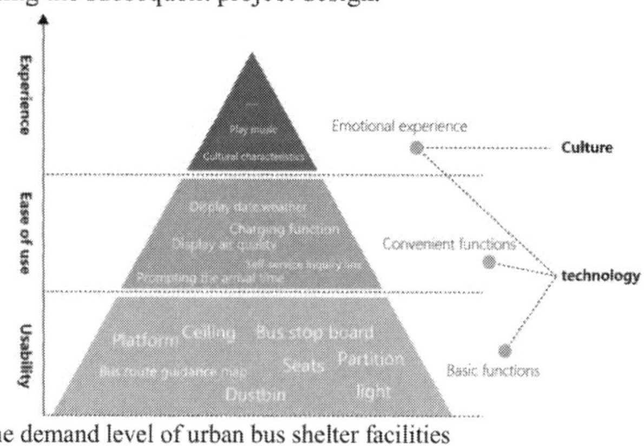

Figure 4. Analysis of the demand level of urban bus shelter facilities

4.4 Design of solutions and innovative points

In accordance with the above stated research conclusions, the design team clarifies the design goals and principles of Bozhou bus shelters in the context of cultural and technological integration. In the meantime, based on the analysis results of problems in existing bus shelters, a set of bus shelter design plans is constructed by integrating the regional culture, technical functions and public demands (as shown in Fig. 5).

Figure 5. Design plan of bus shelter in Bozhou

The design focuses on users and makes innovations from the application of cultural and technological integration on the basis of user demands. Above all, philosophical and physiological demands of different users are taken into full account in the size of bus shelter. The height of station board and display of route information meet people's visual height, featuring high readability. From the cultural level, the design team analyzes the historic culture of Bozhou and architectural characteristics, extracts cultural elements and signs from Cao Cao Park, Wei Wu Temple, Traditional Chinese Medicine City, etc., applies them in the design of bus shelter, expresses the forms and colors through modern design methods, to convey the profoundness of history, and form an organic whole with the urban environment. From the technological level, the design team analyzes the development trend of urban bus shelter design through the smart city strategy, and applies LED to display propaganda posters of the city. Moreover, interactive information display screen is added on the basis of the Internet and big data, to realize the point-to-point service: for instance, display and inquiry of bus routes and real-time route information, tracking and positioning of bus, forecasting the bus arrival time, real-time display of time and weather, etc. which can avoid such problems as the information design, difficulty of identification, etc. In the meantime, based on the consideration of sustainable design, the top of bus shelter adopts new energy materials, which can provide electricity for the shelter. Generally speaking, the application of regional cultural elements in the design of bus shelter in Bozhou enhances the visual recognition of the city. Moreover, functions of the bus shelter are improved through technological methods, and it optimizes the service experience.

5 Conclusion

With the gradual development of the society, the gradual integration of culture and technology is indispensable. Based on this research background, this project applies the innovative cultural and technological integration theory to explore the design strategies of street furniture from the view of cultural and technological integration, and construct an innovative design model of cultural and technological integration. Meanwhile, corresponding street furniture design principles are proposed and applied to the design of street furniture for Bozhou, which provides the theoretical model and practice of street furniture for famous historical and cultural cities in China from the perspective of methofology.

Urban street furniture specifically reflects the spiritual culture of the city, as well as the degrgee of civilization and spiritual quality of the city. Each city is in need of certain amount of street furniture to reach the mutual integration with the space environment. The penetration and integration of culture and technology in urban street furniture will help boost the sense of belonging and sense of place of urban space, and increase the charm of the city.

Acknowledgement

This paper is a staged achievement of the major project of 2017 national social science fund *Research on Design Morphology* (Project No.: 17ZDA020) and innovative fund project of 2017-2018 graduates in Donghua University *Systematic Planning and Design of Urban Image and Environmental Street Furniture in Shiqiu Town of Nanjing* (Project No.: EG2018012). It is funded by the special fund for the basic scientific research in central colleges and universities.

REFFERENCES

1. General Office of the State Council. Order of the State Council of the People's Republic of China, No. 524th. (2008).
2. C. Chao, Art and Design. **05**, 98 (2015).
3. Z. Xinhai, C. Xing, Packaging Engineering. **38**, 207 (2017).
4. L. Wenjia, K. Xiaoyan, G. Jin, Packaging Engineering. **20,** 42(2014).
5. L. Quan, Z. Zhenyu, CITY PLANNING REVIEW. **39,** 46(2015).
6. Xu X, Cao X, Jie L, etal, CAIDCD. **11,** 1002(2011).
7. L. Lin, C. Junyan, W. Hui, Science Technology and Industry. **13,** 13(2013).
8. H. Fei, *Focus on users: UCD concept and Practice* (China Architecture & Building Press) (2009).
9. A. Boting, *Research on urban culture and urban characteristic* (China Architecture & Building Press) (2010).
10. Hopper Leonard J, *Landscape Architectural Graphic Standards* (John Wiley & Sons. Inc) p257 (2007).

Innovation Design and Empirical Research Based on Prevention of Lumbar Muscle Strain and Sanitation Products

Ran Chen[1], Ming Chen[1,2] and Jian-hua Lyu[1,2,a]

[1]*Forestry College, Sichuan Agricultural University, Chengdu 611130, China*
[2]*Key Laboratory of Wood Industry and Furniture Engineering, Sichuan Agricultural University, Chengdu 611130, China*

Abstract: This article takes the lumbar muscles of degeneration status of Chinese sanitation workers as the starting point, analyzed the waist health status of sanitation workers in China. The design conception of sanitation products for the lumbar muscles of degeneration of sanitation workers was obtained by analysis, and the design of today's sanitation products are rethought, use relevant research results to demonstrate that. With sanitation products as the design center, it integrates many research contents such as ergonomics, appearance, function, process and cost. To this article has designed tools that sanitation workers can use to protect their waists while working, It also provides a new idea for the wide design of sanitation products. It is conducive to maintaining the basic rights and interests of sanitation workers and promoting the urbanization process and the harmonious development of society.

1 INTRODUCTION

Urban sanitation workers are very typical vulnerable groups in modern society, They are the main force of developing the urban environmental sanitation, and they have made a great contribution to the civilization, harmony and sustainable development of cities. More than 1000 employees mainly middle-aged and elder people are in sanitation related industries in China. According to incomplete statistics, sanitation workers aged 30-40 account for about 36%, 42% of 40-50 years old, the percentage of people over the age of 50 is 24%. Sanitation workers are mostly women, accounting for 74% of the total, and male workers only account for 26%. Under the condition of long-term fatigue, the lumbar muscles of the middle-aged and elder people are prone to strain. Globally, especially in developing countries, perennially for sanitation workers use the sanitation products which are lacking of humanized design and less maneuverable, which causes that sanitation workers are generally in poor health, one of the most prominent is that the sanitation workers often use the waist as a support point in their working time. Therefore, a in-depth analysis of the health and working situation of sanitation workers in our country and propose corresponding solutions, to help protect the basic rights and interests of environmental sanitation workers, meantime, it is more conducive to urbanization and harmonious social development[1].

2 ANALYSIS OF THE PRESENT SITUATION OF SANITATION PRODUCTS

[a] E-mail: ljh@sicau.edu.cn

2.1 Definition of sanitation products

"Sanitation product" refers to a kind of product which is helpful to keep the environmental hygiene when it is necessary to clean the work to a certain extent, It can reduce the workload of people's cleaning, and can achieve a thorough cleaning effect through some special functions. Sanitation products are the equipment with heavily companionship with sanitation workers, after all the sanitation products affect the worker's health. More than 90% of sanitation operations need to be completed directly or indirectly through sanitation products, unscientific sanitation product design still is one of the causes of human health damage. Therefore, the design of good sanitation products can greatly reduce the damage of operation to human lumbar muscles.

2.2 Analysis of the status of sanitation workers' health

To investigate the health status of sanitation workers, and aurvey results show. Of the 429 sanitation workers inspected, only 32 were completely normal, 7.46% of the total number of persons examined.397 people were found to have abnormal indicators and suffer from various diseases(There is at least one index anomaly),92.54% of the total number of persons examined[2].At the same time, many sanitation workers have expressed a desire to obtain sanitation products that can reduce physical damage. The sanitation products for lumbar muscle strain have played a greater role in protecting the waist of workers. When cleaning the work at ordinary times, they can carry out the work more conveniently and quickly, reduce working hours and workload. The design of such products gives the mission and vitality of new environmental sanitation products to make sanitation workers less injurious to the body while working, and avoid mechanical injury such as lumbar muscle strain.

2.3 Status Analysis of Environmental Sanitation Product Function

With the rapid development of economic development and urbanization, the quantity of urban waste generated is also increasing year by year. The role of urban sanitation facilities in the process of urban modernization is particularly prominent[3]. The design of sanitation products has become an important part of sanitation engineering. At the current stage, China's environmental sanitation products are at a low level of mechanicality, intelligence, and portability. As far as the current situation is concerned, this type of domestic intelligent sanitation products is almost in a blank period. Therefore, the use of ergonomic scientific research methods to improve the design of sanitation products to improve the efficiency of sanitation workers to ensure the quality of life of sanitation workers and promote the harmonious development of the city[4].And effectively strengthen our attention to the construction of sanitation facilities.

3 THE SANITATION PRODUCTS DESIGN PRINCIPLES AND PROOFS FOR AVOIDING LUMBAR MUSCLE STRAIN

3.1 Design process, systems, structure

The design is center on environmental sanitation products and incorporates research contents such as appearance, function, process and cost(Figure 1). The appearance is mainly designed for color and shape to enhance recognition degree. In terms of the function, it mainly considers the operational requirements such as cleaning and portability. As for the process, it mainly aim at the design of structure and processing, and the utilization ratio of reinforced materials. In the cost, mainly considers the raw materials and production processes to enhance product cost performance.

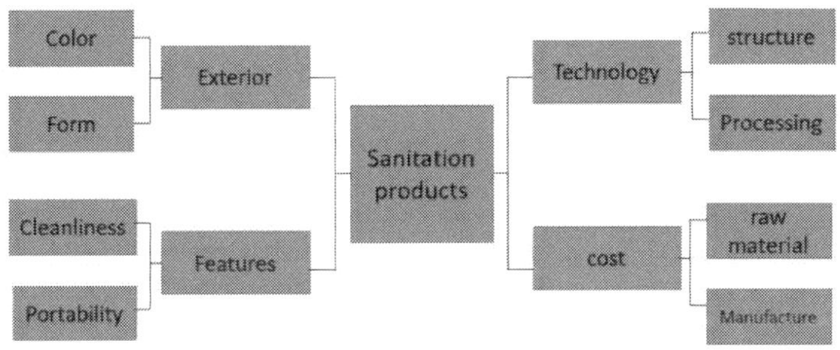

Figure 1. Product function design process, systems, structure diagram

3.2 Design Principles

When sanitation workers work, they often bend over and repeat this mechanical movements. After a period of time, the body parts, such as the waist, can be easily damaged. At the same time, because of the large amount of labor of the sanitation workers and poor product functionality, the design of the structure, selection of materials, and design of functions should be considered in the design of environmental sanitation products for avoiding lumbar muscle strain. Besides, in addition to sanitation products used by sanitation workers, some functional sanitation products for home use should also be designed. Therefore, besides considering the functionality in design, it should also combine the problems of portability, use of space, and user crowds, after that the design of sanitation products can solve problems and meet people's daily needs.

3.2.1 Principle of structural stability

Structure refers to the composition and combination of the elements of a product or object. Structural design is an important part of product design[5]. In the product design, the structure of the product is a necessary condition for achieving the function of the product, and it will directly affect the structural characteristics and appearance of the product. So when design a sanitation product, it is necessary not only to ensure the safety of sanitation workers during operations, but also to avoid mechanical repetitive actions to the greatest extent possible. Make full use of mechanical principles and rationally construct the structural system so that it can be stable and safe.

3.2.2 Material Selection Principles

Tool material is the material basis for product function and it is also the carrier of product design. The material properties and material types of product materials not only affect the appearance of product design, but also have functional restrictions. Functional sanitation products should be portable and easy to be cleaned during operation. Therefore, in the selection of materials, it is necessary to satisfy the variability in structure through the deformation of the material to achieve the desired function. At the same time, the pros and cons of materials can also affect the service life circle and clean-ability. So in the selection of materials, it is necessary to choose a material that is easy to clean on the surface.

3.2.3 Principle of Cleanliness

Sanitation products refer to the products that perform clean operations on facilities. Not only are they required to be able to clean the dirt, but they also need to ensure that the products themselves do not become moldy, smelly or contaminated. During the sanitation working period, where pollution is heavy and air circulation is poor for a long period of time, in order to improve its operating efficiency and satisfaction with work, the first condition is to ensure that it uses clean products and is in a clean environment. Therefore, in the process of product design, in addition to the choice of cleanable materials, it is also possible to use vents, activated carbon and other components, and clean itself to meet the requirements of cleaning products that can clean themselves.

3.2.4 Functional Diversity Principle

In recent years, more and more equipments are needed for cleaning. Due to the lag in product design, many sanitation workers are still using traditional sanitation products, which have a single function and are inconvenient to carry. Therefore, we have added new features such as portability and intelligent technology in product design, so that sanitation workers can play a role in new functional environmental sanitation products during operations, which can improve the practical value of sanitation products, and adapt to the design principles of social development.

3.3 Design evidence

According to the innovative design ideas and approaches discussed in the study, the design verification was carried out. Three design proposals were obtained and utility model patents were declared.

3.3.1 Sanitation tools designed to clean the surface of cylindrical structures

A sanitation tool dedicated to cleaning the surface of cylindrical structures (Figure 2), The rod body is provided with an ultrasonic generator and an ultrasonic transmission cover, the ultrasonic generator is connected with the ultrasonic transmission cover, and the ultrasonic transmission cover is communicated with the open end of the rod body, and the ultrasonic generator is arranged under the sealing plug. The ultrasonic transmission cover is disposed between the ultrasonic generator and the sealing plug. With the principle of clamping mechanism, after deformation, the object to be cleaned is squeezed by force, and the cavitation effect of ultrasonic waves occurs. The ultrasonic wave is transmitted to clean liquid with high frequency transformation mode of alternating force and pressure reduction more than 20,000 times per second. When the pressure is reduced, the phenomenon of vacuum nuclear group bubbles is generated. When the compressible force is applied, the vacuum nuclear group bubble produces a strong impact force when crushed by pressure. This removes the dirt on the surface of the object to be cleaned, thereby achieving the purpose of precise washing. The structure of the tool is simple, the operator can directly contact with the isolation fence without bending, the single contact area is large, and the wiping efficiency is improved.

3.3.2 Equipment for deodorization and effusion elimination based on garbage.

Based on equipment for deodorization and liquid elimination of waste (Figure 3). The top surface of the base is recessed to form a groove, a mounting platform is installed in the groove, a through hole is arranged on the mounting platform, a collecting barrel is arranged in the through hole, and a filter mesh is arranged in the collecting barrel. An exhaust pipe is arranged on the installation platform, an adsorption tank is connected to the bottom of the exhaust pipe, an adsorption layer is arranged in the adsorption tank, and an intake pipe is arranged on the sidewall of the adsorption tank. A suction pump is provided on the installation table, and the suction pump is connected to the intake pipe. One end of the air inlet tube is communicated with the inside of the adsorption tank, and the communication between the air inlet tube and the adsorption tank is located below the adsorption layer. The other end of the air inlet tube is provided with an adsorption hood, and a concentration detector is arranged in the adsorption hood. The side wall of the base is con-caved to form a treatment chamber, a liquid collection tray is disposed in the treatment chamber, and a heating plate is disposed in the treatment chamber. The device can timely process the liquid produced in the garbage can, prevent the liquid from accumulating and damaging the trash can, and also prevent the odor from being emitted into the air. The device reduced the sanitation workers' cleaning process against the trash and greatly reduced the workload of sanitation workers.

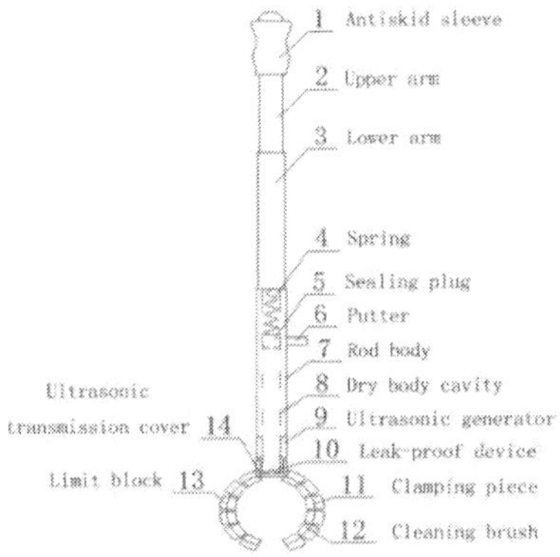

Figure 2.A sanitation tool that can be exclusively used in cleaning the surface of cylindrcal structure

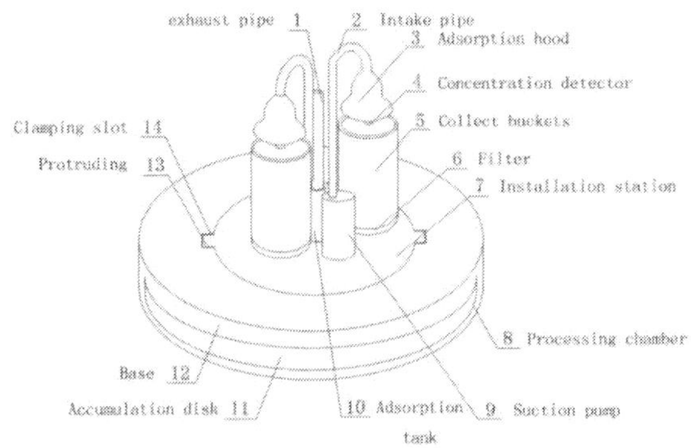

Figure 3. The equipment is based on deodorization and effusion of waste

3.3.3 multifunctional sanitation cleaning tool cart

A multifunctional sanitation cleaning tool vehicle (Figure 4). The bottom of the vehicle body is provided with wheels, a garbage adsorber is installed on the body, a garbage collection bin is arranged in the vehicle body, and a barrier net is arranged in the garbage collection bin. A suction pump is provided in the vehicle body, and the suction pump is connected to a garbage adsorber. A solar panel is arranged on the vehicle body, a steering rod is arranged on the vehicle body, a steering rod is provided with an operating rod, an operating button is arranged on the operating rod, an insertion umbrella frame is fitted on the steering rod, and an umbrella stand is arranged on the insertion umbrella An umbrella hole is inserted, a battery box is arranged on the vehicle body, an extension frame is arranged at the tail part of the vehicle body, and a hinge is arranged between the extension frame and the tail part of the vehicle body. A seat cushion is arranged on the top of the vehicle body, a fan is arranged on the side wall of the vehicle body, and an air outlet is arranged at the top of the vehicle body near the seat cushion. The utility model has the advantages of simple structure, capable of automatically picking up the small garbage, avoiding the bending work of the sanitation workers, reducing the labor intensity and the lumbar muscle strain rate of the sanitation workers, improving the picking efficiency, and improving the degree of cleaning, and achieved versatility

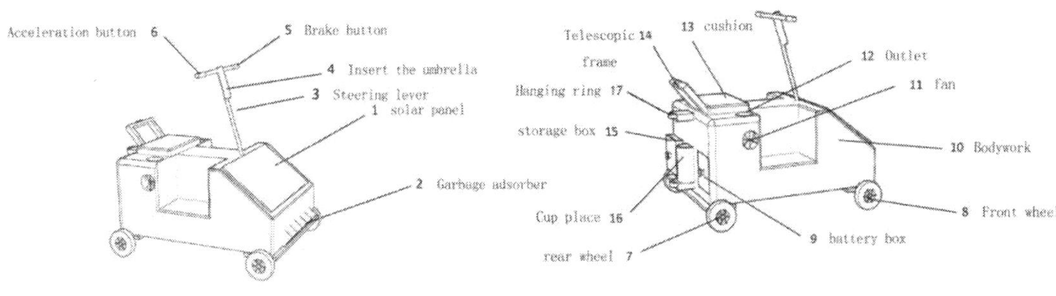

Figure 4. A multi-funcion sanitation vehicle

4 SANITATION PRODUCT DESIGN EVALUATION

The construction of the sanitation product index system will follow the design principles and will proceed from the three dimensions of function, technology and economy, and it also combine other dimensions to establish an evaluation structure for sanitation products. The selection of sanitation products for prevention of lumbar muscle strain must consider the three dimensions of the general evaluation criteria, and at the same time, it should also pay attention to selecting the appropriate evaluation angle, and establish multi-rate evaluations from users to production companies to society, from multiple perspectives of vision, touch, psychology, practicality, protection, function, material, craftsmanship, and cost, benefit, a multi-level comprehensive evaluation system (table 1).

Table 1. Multi-dimensional product evaluation

Evaluation Dimension	Evaluation method	effect
Features	Qualitative and quantitative	Calculate user needs and product operability and satisfaction for product function optimization evaluation
technology	Qualitative and quantitative	Analyze product structure, process, etc., priorities of select technical
economic	Quantitative evaluation	Evaluation of product estimates through product cost estimates and sales statistics

5 CONCLUSION

At present, the demand for products of sanitation workers is not satisfied with the sanitation product when they are working, the needs have already evolved to the stage where higher quality and more convenient products are required. On the one hand, the increase in the number of manufacturers of environmental sanitation machinery has made the manufacturers compete with each other more and more fierce. On the other hand, the market segmentation has become more sophisticated, and a number of more specialized sanitation machinery companies have emerged[6]. The design of environmental sanitation machinery products can guide the design completion with innovative methods.

Through the idea of strategic innovation, reasonable use of different innovative methods to guide product design. When determining the overall plan, through the innovative method of FBS function mapping, starting from the main function of the product, using the scientific effect library to find a number of appropriate principles and programs, and then the optimal program is selected [7]. There are many areas involved in sanitation products. Besides the product design theory itself, it also has a great deal of connection with the fields of dynamics, mechanical engineering, and electronic engineering. With the progress of the society, environmental sanitation has become more and more demanding for intelligent products. To meet the demand for multi-purpose use of sanitation workers, it is necessary to recreate and redesign existing sanitation products at the current stage. Relative research and application are still lagging behind. There is still a great deal of room for research and development in the theoretical foundation, implementation approaches, and technical means.

ACKNOWLEDGMENTS

This research is supported by Research Project of Double Support Program for Art Special Project in Sichuan Agricultural University(2018).This research was carried out at the Sichuan Provincial Colleges and Universities Wood Industry and Furniture Engineering Key Laboratory, Sichuan Agricultural University.

References

1. Liu Hongsheng, Yan Yifan. Research on the Safeguarding of the Operation of Urban Sanitation Workers[J]. Journal of Shaanxi Administration Institute, 2007, 11:21-4.
2. Xiong Chengbiao, Xue Genshan. Health check-ups and health-related factors in Haikou sanitation workers[J]. China Tropical Medicine, 2013,13(8)994-996
3. Gao Chunmei. Research on the practicality of "3R" principle in product design of urban sanitation facilities [D]. Nanchang, Nanchang University, 2012.
4. Bai Rui, Xue Yanmin, Qin Song. Innovative Design of Environmental Sanitation Products Based on Survey[J]. Art and Design, 2010(4):213-215.
5. Sun Delin, Sun Debin. Discussion on the Teaching Mode of "Furniture Structure Design"[J]. Journal of Central South Forestry Science and Technology, 2009(6):127-129.
6. Lin Qi. Research on Sanitation Machinery Industry[D]. Wuhan, Central China Normal University, 2015.
7. Chen Lin, Zhao Wu. Design of Environmental Sanitation Machinery Based on Innovative Methods[J]. Mechanical Design & Manufacture, 2010(12):253-254.

Ontology-based Experimental Support of Conceptual Actions in the Process Production of Aviation Parts Templates

Oleg Chorakaev[1, a], Maxim Grishin[2]

[1]*Ulyanovsk State Technical University, SevernyVenwetc, 32, 432027 Ulyanovsk, Russia*
[2]*JSC "Ulyanovsk Instrument Manufacturing Design Bureau", 10a, Krymov st., 432071, Ulyanovsk, Russia*

Abstract: The paper presents a set of decisions for improving the processes of designing the template tooling in the aviation industry. Specificity of the offered approach to designing this kind of equipment is to obtaining positive effects of reflecting the life cycle of any template on semantic memory, such as a question-answer. The reflection of the life cycle on the semantic memory makes it possible to develop ontological means of support, to organize structural designing the configured templates of aviation details and also use of means of conceptual experimenting. Such type of experimenting suggests the structure and attributes of the semantic memory cells are focused on the coding of models of questions and answers from the viewpoint of their use in solving the creative tasks. In such modeling, any conceptual experiment is conducted as an automated mental experiment the structure and content of which are operatively used in designing of templates. The ontological support helps to reduce the design complexity, improve the initial design quality, get a shorter pre-production term, as well as to accumulate and transfer the experience of the previously implemented design solutions.

1 Introduction

In aircraft manufacturing, template equipment is used for the manufacture, control and assembly of fuselage, wings and aileron parts, including details of their skin, and includes tens of thousands of templates of varying complexity and purpose. Configured templates as specialized equipment for manufacturing and checking of aviation details have a set of features the main of which are the following: complicated geometrical forms, small rigidity, greater scales, and strict requirements to accuracy of manufacturing. In addition, in order to link the details included in each plane section of the aircraft's design, a rigid carrier system is required that fixes the contours of the internal parts that make up this section. So, for example, large parts have to be linked on sheet metal control and contour templates.

The template tooling consisting of tens of thousands of templates is widely adopted in the aircraft construction to produce, check and assemble fuselage, wings, and ailerons, including skin covering parts. Such features of parts determine this fact as the complexity of their geometric shape, low stiffness, large-size dimensions, the strict requirement of production and coupling accuracy. Furthermore, you need to have a system of rigid media fixing inner sides for coupling parts of every aircraft construction plane section where these parts are present.

In general case, a template is not only the carrier of the part geometry and shape, but it also consists of design and engineering data, inner part profiles and axes which are present in this section,

[a] E-mail: olegchorakaev@yandex.ru

different design, and engineering holes. In addition, the templates have different information put on them: template name, drawing code and number, make and thickness of the material, fold line and bevel, hole profiles, hole labeling, etc.

In the general case of designing metal sheet templates, you need to be creative to solve tasks of engineering configuration taking into account engineering conditions to employ templates in production, checking and assembling aircraft parts. For these reasons, any geometry of a configured template is more complex than the geometry of a part developed to work with it.

The interconnected complication of the structure leads to the appearance of a set of points on the life cycle in which to make conceptual actions (in Figure 1). At these points, it is impossible to exclude the operations of checking, correcting and repeating the actions, that is, there is a prolonged experimentation.

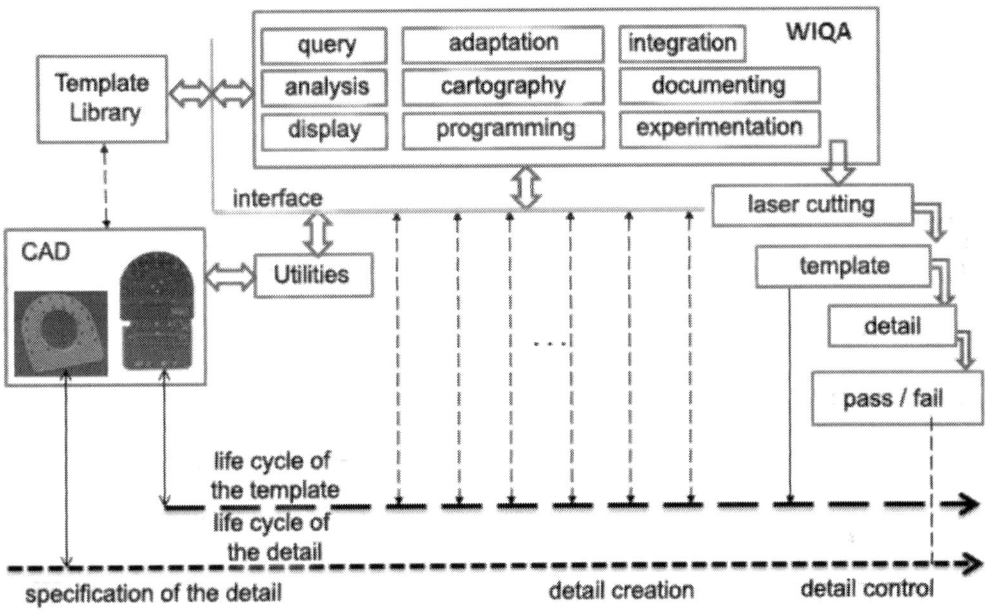

Figure 1. The life cycle of the configurable template

Figure 1 highlights an area responsible for the reuse of the template design experience. In the suggested approach, the experience is represented as an «Ontology» component of the toolset within the framework of which the template model library is used.

Even in cases where the best sample library with «rational» nature proven in large-lot aircraft production is employed to design templates, you anyway need to re-create the template tooling when launching new aircraft production, certainly, using mastered templates, improving them and developing the new ones. It means that the creative nature of template design and interconnection remains.

It is reasonable, in this connection, to include experiments (not with templates, but with their models) into the life cycle design of the template (most complex and requiring configuration).

The article suggests including ontological support means to the tool-and-engineering maintenance in design processes of templates. It will ensure the controllable accumulation of template development experience in the form of precedent models ready for the reuse; systematization of template models based on the interactive classification and coupling using ontological systematization principles, controllable use of vocabulary, including concepts, in documents that are developed during work performance. The important special feature of the suggested means is that they rely on the experimental results of semantic models of the template and CNC software models used in the production of templates. Templates are implemented based on the precedent model [1]. The content of the article concerns notably this component and its use in a template design process.

2 Methods and Instrument of templates support design

The authors use the mapping of aircraft parts with semantic memory (Figure 2 outlines this memory), to create a library of experience and to conduct conceptual experiments. The means and specifics of the conceptual design, oriented to its mapping to the semantic memory of the WIQA toolkit, are detailed in the publications [2-4]. Features of the semantic memory of the WIQA toolkit is the structure and content of the memory cell, which is focused on loading question-and-answer units (question or answer or their composition).

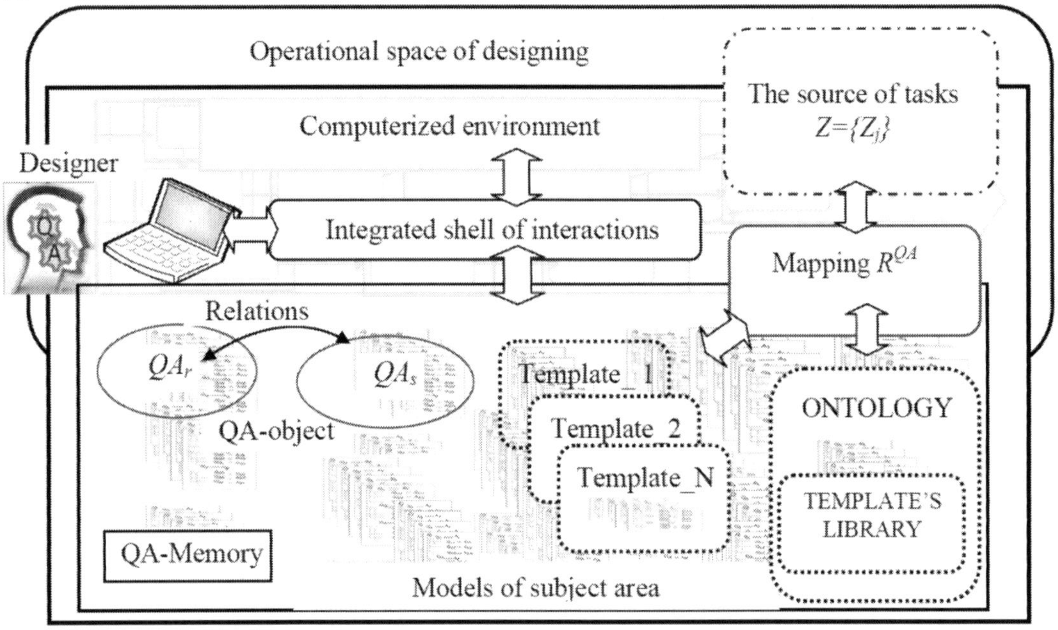

Figure 2. Process of Mapping to QA-Memory

But especially important for QA-memory is that it is defined by the expandable (if necessary) pseudo-code programming language LWIQA. The programs in this language can be executed with the help of the interpreter, the interpreter with the compilation of the selected groups of operators and the compiler, and the designer in this performance acts as an «intelligent processor» [3] interacting with the processor of the computer.

For ontological support of the process, the template model in memory is conveniently represented as a precedent. Such a view accumulates and integrates everything that should be useful for the re-use of such a model.

Moreover, during the creation of each template, it must copy not only the geometrical forms of the corresponding detail but also all necessary elements of utility must be included in the template. Such additional components of the templates are oriented on conditions of their use, for example, in assembly operations with other details. The necessity to take into account the different points of view on a template leads to creative tasks, which are better solved during preliminary experiments (before manufacturing a template) with models of their decisions.

Notice that it includes process results and outlines task names in the life cycle diagram the content of which is specified in detail in source [1,5-7]. Authors draw your attention that the design activities required to optimize the suggested life cycle shall comply with standard. The only difference is that the design activity is guided by a (pseudo-code) program within WIQA environment, at this, mappings of a part and the template original are the inputs to perform actions directly within the modeling environment.

To represent every template, we have selected a model M(Zj) for a task Zj of its reuse by the designer found himself in a particular task situation. The first reaction of any person being in a task situation is to «draw himself on the experience and find a suitable precedent for it» [7].

Regardless the fact that precedents are various by their structure and contents, the precedent template has similar logic of access that in common case leads to a precedent logical model P^L. The model is linked to the precedent (template) construction life cycle, as well as to its mastering during which the following models are created:

- Text model $\mathbf{P^T}$ representing task setting Z (P_i) which solution leads to the creation of a precedent (template) sample as a result of intellectual mastering of the real precedent. Let us give an example: model $\mathbf{P^T}$ can be represented as a query for template design where basic initial data is: «design a template for cutting and tooling on Z-profile with three legs and holes HO and CO, without undercutting and taking into account an additional material».
- Logical model $\mathbf{P^L}$ instantiating models in the form of an equation for predicate logics recorded in the task setting language $\mathbf{P^T}$.
- Let us examine the above-described model $\mathbf{P^T}$ in the form of predicate logic. Here, the task Z (P_i) has the following format including a set of activities as follows:

$$Z(P_i) = P^{RabK} \cup P^{Dob} \cup P^{Risk} \cup P^{Otv} \cup P^{Inf} \cup P^{DopE}, \quad (1)$$

where,

P^{RabK} –working profile including:

$$P^{RabK} = P^{Plk1} \cup P^{Plk2} \cup P^{Plk3} \cup P^{Plk3} \cup P^{OtvSO} \cup P^{OtvNO} \cup P^{RiskDet} \cup P^{InfSO}, \quad (2)$$

where,

P^{Plk1} – designing of the 1st leg;

P^{Plk2} – designing of the 2nd leg;

P^{Plk3} – designing of the 3rd leg;

...

- Precedent (template) graphical model $\mathbf{P^G}$, that outlines the precedent (template) using different means.
- QA precedent model $\mathbf{P^{QA}}$ corresponding to task Z (P_i).

Let us consider an example of a QA model taking into account the developed classification of the above described «template for cutting and tooling» design for task Z (P_i).

Let us determine a list questions and answers:

Q1. What template do we need?

A1. Template for cutting and tooling.

Q2. Is the template single-piece?

A2. Yes //depends on part dimensions.

Q3. What profile to design on?

A3. Z-profile.

......

Q14. Do I need information on CO?

A14. Yes

- Model P^I representing a behavior included in the template (precedent) in the form of the source code.
- Model P^E derived on the executable code and implementing template (precedent) sample.

Note that the special-purpose model P^I allows its representation as the source code of that programming language which is used when developing software for UG NX. But, to extend functionalities of reuse of P^I-type models, it is rational to use pseudo-code language LWIQA.

Precedent-oriented on the design of template equipment ontology model with an extended structure of sections, which include sections that additionally provide effective ontological support in solving problems of pattern search, as well as manufacture, control and alignment of the components of the airframe powerplant set.

Let's consider the technological methods and structural interrelationships of the construction of the template rigging, and also present the practical implementation of the elements of the ontological support complex for the design of the template tooling.

Authors use of the toolkit WIQA (Working In Questions and Answers) that helps to create the ontology and its use in conceptual experimenting in designing. The WIQA tools have been tailored to the engineering support of the template lifecycle with the extension of the LWIQA language based on the description in the Extended Backus-Naur Form (EBNF) [5].

The choice and implementation of a part of the template structure that supplements the shape (and dimensions) included in the template of a particular airplane.

Designing a template is inherently associated with a tool such as CAD. This process begins after the part specifications have found their geometric representation in the CAD system.

Presently, the considered subject (template tooling) is fully modelled within UG NX system in a so-called «manual» mode which increases its labor intensiveness and error number, extend the time for engineering preproduction as a whole.

Construction of a mathematical model for the template tooling comes down to profile modeling. Figure 3 shows a common modelling diagram. All types of template tooling are modeled based on the given diagram.

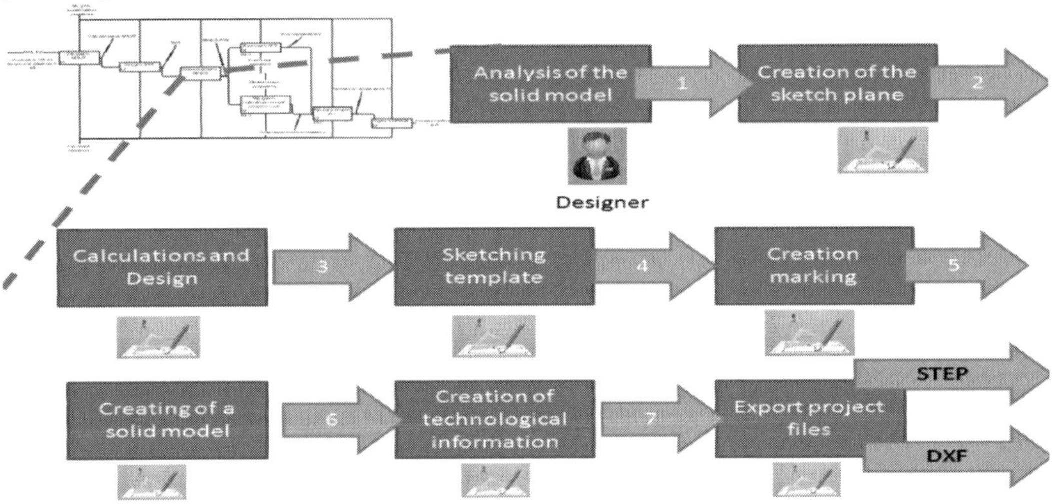

Figure 3. E-Model of Template Tooling Design Process

The diagram shows the following main stages of creating a mathematical model of the template: Analysis of the geometry of EMD.The stage of theoretical comprehension of the future appearance of the template, as well as the choice of the optimal location of the part in the modeling space for the further convenience of designing the mathematical model of the tooling.

Construction of the sketch plane. The template sketch plane is the first item in the design of the template. As a rule, it is built on a flat or non-working surface of the EMD.

Performing the necessary design and planning operations. Number and types of design and design operations, as well as their options, depending on the features of the equipment and experience of the designer.

Design a sketch template. The sketch is created on the plane, and based on the geometry of the part, the complexity and number of operations performed during its design can vary greatly.

Construction of marks. The number and types of required marks directly depend on the type and group-sets of templates. However, in any type of template, there will be at least one mark - this is the mark of the contour of the trimmed part. All marks of the contour in the production of the template are performed by laser engraving, and therefore they are placed on a separate modeling part layer.

When the template snap-in is fully ready, and the contour is closed, the "Extruding" operation creates a solid pattern model. Its main purpose is to link the equipment and control of the manufactured rig to the Control measuring machine (CMM).

Marking technological information on the model of the template. All technological information is applied to the solid body of the template and placed on the modeling reference layer.

Creation output files. With the finished electronic template model, two output files are generated:

- File DXF - serving for cutting a template from a metal sheet with a laser.
- STEP file - for monitoring the contour of the manufactured template on the CMM.

The information about the part includes the geometry plane of the part for which the template is designed. This geometry element is transferred into the WIQA tool-modeling environment with the help of a special utility to display details on the semantic memory of this system. The second purpose of these actions is to analyse this information and to form a query to the library of templates.

All these actions and subsequent (in the course of the life cycle of the template) are performed under the control of the (pseudo-code) program, in the WIQA environment, where the parts mapping and the original version of the template serve as input data for performing actions directly in the modeling environment.

The analysis of the UG NX system and the modeling environment of WIQA allows concluding that the data of the system for a sufficient number of possibilities for creating on their basis a set of ontological support for the process of designing the template rigging.

3 Ontological support for the process of designing aviation template

The central part of tooling design ontology or any other ontology generated in WIQA environment is taken by «Vocabulary». The vocabulary structure has a section for presenting the basic types of templates based on the use of the model templates described above. The vocabulary functions not only contain template definitions but also search keys that allow you to search based on special queries. The WIQA environment is aimed at conceptual modeling [6] based on the pseudo-code programming of designers» activity [7]. It also concerns the routing of accessing the ontology components, including accessing references to the conceptual template models.

The WIQA toolset makes it possible to use the requisite tooling in the vocabulary ontology. With the template, you can create links to formalize templates based on attributes: «whole and parts» because it consists of elements defining the template; «Inheritance», because in most cases the template has a parent or an ancestor. After that, the attributes are populated.

Note that you can both assign names to relations to derive their semantics and comment them. This option is useful for pragmatic relations, for example, for the tool-type ones (linking the template with the tools for its processing or mechanical processing).

Thus, the stage of ontological modeling of precedents of templates is added to the existing life cycle of designing and manufacturing the tooling. At the same time, it should be noted that such modernization was possible due to the formalization of the ontology model and the separation of the main dictionaries of terms:

- template accessories;
- manufacture;
- control;
- regulatory and technological support.

These terms groups reflect all the production elements involved in the development and creation of a template snap. At the same time, to ensure the flexibility of the production system to changing conditions, the designed ontology is open and extensible in order to fully meet the requirements of aviation production.

Objects of template classes in the interactive classification are defined and investigated with reference to the technological production processes. Each of these objects has its own constructive components that allow the most rational determination of the ratio of the template to the manufactured part. These features of the template structure are realized as an ontology section and as a formal representation of the classifier tree.

The method of ontological support of the design of the template tooling takes into account the controlled accumulation of the experience of template development in the form of use case models prepared for reuse. The algorithms for designing the template equipment are integrated into the technological preparation of production and are characterized by an increased degree of automation of the design process achieved by programming part of the designer's typical operations in terms of the design of the geometry of the electronic template model.

Positive effects from the inclusion in the instrumental and technological support of the design of aircraft templates means of ontological support:

Controlled accumulation of experience in the development of templates in the form of use case models prepared for reuse.

Systematization of template models, based on interactive classification and binding using the mechanisms of systematization in ontologies.

Controlled use of vocabulary, including concepts, in documents developed in the course of work.

Creation of precedent models is carried out within the framework of the WIQA question-and-answer environment, which allows using a set of tools by all participants in the life cycle of designing and manufacturing the template rigging. Also, an extensive toolkit of WIQA capabilities allowed implementing an interactive classifier that helps to increase the rationality of the «detail-rigging» relationship.

An important feature of the proposed tools is that they rely on the results of experiments that are conducted with semantic models of templates and models of numerical control programs that are used in their production.

4 Conceptual experimentation based on the display of the template on the question-answer memory

The selection and implementation of that part of the template structure that complements the form (and dimensions) of the specific section of the part included in the template are derived for some design tasks that allow for alternative solutions. Even in cases where the library of their «best designs» was used to design the templates, which confirmed their «rationality» in the production of already known types of aircraft, in the transition to the production of a new aircraft, the adaptive design to create a new, of course, using the mastered templates, and developing new templates. That is, the creative nature of designing templates and their interconnection will remain. That is why the state of affairs and determining the appropriateness of including experiments in the life cycle of design (especially complex, configurable) templates.

At the initial stages of the design of templates, the use of conceptual experimentation is used to develop selected options for engineering solutions and obtaining new ones by conducting conceptual experimentation [5].

A conceptual experiment is a mental experiment, content, and process that quickly maps to the semantic memory, and also analyzes the results during the experimentation with useful goals.

The mapping of the design process to the question-answer memory allows for conceptual experimentation, which implies the possibility of stopping at any stage of the design, evaluating the decision taken, multiple visualizations and the possibility of correction. For the above algorithmic scheme, it is important to experiment with the template developer for the task chosen by him (not only related to laser cutting procedures). In the article for conducting such experiments a number of tools are presented, including tools for registering template models in the library and tools for dynamic visualization of laser cutting processes in time.

The basic possibilities of conceptual experimentation are based on the «cartographic» representation of aviation patterns. A cartographic model providing layer-by-layer structuring for the specialization of work for each layer (see Figure 4) represents configurable templates, carrying information about their geometry and a set of technological information, broadcasted to the program for their production. In general, the template, not only is the carrier of the geometry and shape of the part but also includes the structural and technological bases, contours and axes of internal parts that fall into this section, various structural and technological holes. In addition, various information is put on the templates: the name of the template, the cipher and the drawing number of the product, the brand and thickness of the material, the directions of the fold lines and the bead mills, the outline of the relief holes, the marking of the holes and other information.

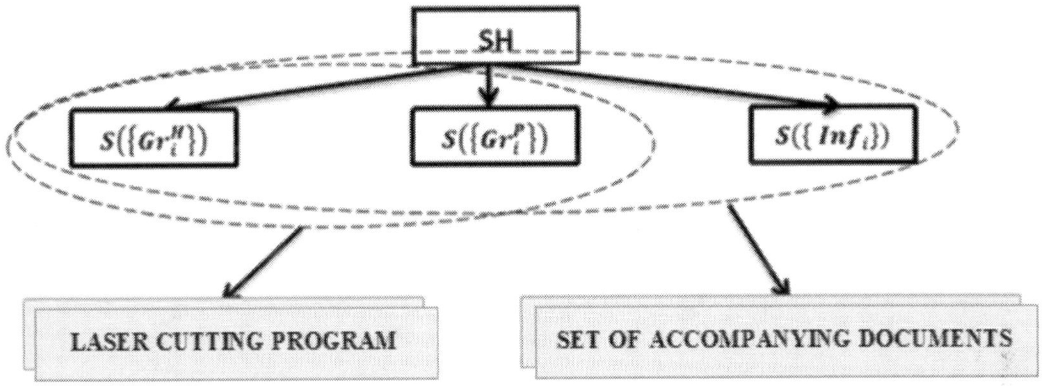

Figure 4. Application of cartographic layers in the design of templates

In addition to layering to form documentation and form elements of a laser processing program, this method allows you to divide the elements into the states that are accessed. Reducing the number of structural elements, allows you to solve the task without losing attention [8].

The set of techniques for conceptual experimentation with dynamics (design procedures) and the statics (state) of the life cycle of configurable templates, the study of which is carried out in the semantic question-answer memory and leads to an increase in the degree of automation of the design process and the quality of its result.

The conceptual experimentation technique presented in the form of pseudo-code programs, the execution of which can be performed in a step-by-step mode, with interrupts on any of the operators for performing experimental actions and procedures, if the template maker needs this.

In working with pseudo-code analogs of numerical control commands for laser cutting, two interpretations are supported, one of which is used to simulate cutting in the field of a specialized graphical editor of the WIQA tool. And the second interpretation allows switching from a pseudo-code program simulating a certain cutting to its version in working G-codes with instructions for preparation for cutting and operational interaction with laser equipment.

The second view is used not only to adapt the programs for manufacturing a template for equipment of different manufacturers but also can serve as the basis for optimizing the laser cutting process.

To minimize idling, it is necessary to maximize the unification of disparate elements in associated closed or open contours. The second aspect to minimize idling is the optimal determination of the insertion points in the contours so that the sum of the distances between the insertion points is minimal. Providing the designer of the templates with the possibility of experimenting with laser beam entry points in the processed contours, as well as dynamically visualizing the cutting in time, with the possibility of correcting these points, makes it possible to rationalize the movements of the head of the machine at idle. Thus, the purpose of experimentation is to reduce the time of making a template from a sheet by reducing idling (see Figure 5).

The progress of the experiment:

The template designer loads the prepared template model to form the control program.

Specify the starting point for processing, for example, the initial position of the carriage of the laser emitter in the machine.

A rational way of traversing the component contours is calculated programmatically, with the output of the basic metrics, such as the useful path and idling.

By a visual presentation, the designer decides whether to correct the calculation results and, if necessary, moves the start point or the cut-in points along the contours and causes a recalculation of the processing metrics until it receives the results satisfying it.

Based on the results of the experimental finding of contour traversing points, a machining program is developed for the CNC machine.

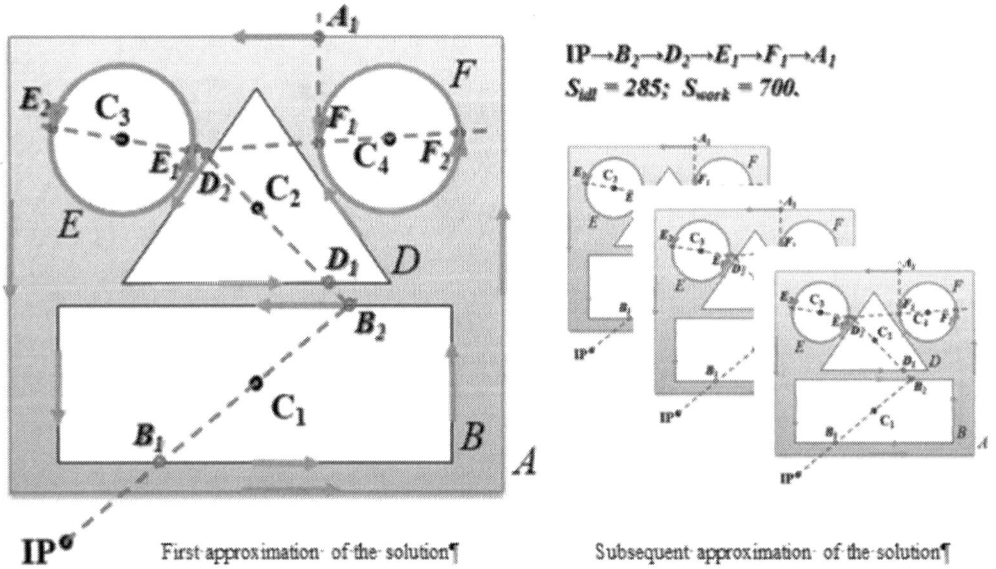

Figure 5. Step-by-step optimization of the path of the laser beam

Lastly, the external contour of the template is cut out, to determine the insertion point in which the nearest point of the outer contour from the insertion point to the last internal element is located. The point A1 on which the previous template was cut on the sheet plays the role of the starting point when calculating the traversal path for the next template.

To reduce the negative impact, it is necessary to purposefully reduce the reasons for deviating from the right decisions. The main purpose of experimentation is to identify random errors and defects caused by the actions of designers. The presence of errors is detected as a deviation of the result from the requirements in the process of verifying the materialization of the solution. All such anomalies are discovered during the life cycle of the template and significantly affect the result of the design process. Introduction of the mechanism of layers makes it easier to work on designing templates since each layer consists of fewer elements and carries logical completeness, which allows you to operate with fewer components. In addition, the design process from parallel to serial is converted, which also reduces its complexity. Reducing the number of deviations is also facilitated by a formalized approach to defining tasks and responsibilities within the organization.

The final step in the use of the WIQA system in the life cycle of the template is the design of the accompanying documentation. Documentation is one of the basic functionality of the WIQA toolkit, for the implementation of which any, also the document forms are mapped to semantic memory, with display units available for use in pseudo-code programming. So any result of experimentation, if it turned out to be working and embedded in the template project, can be quickly transferred to the appropriate document forms.

Operational documentation in the WIQA environment is transferred to QA-memory, and for each sample document, a question-answer presentation of its structure and a template for visual presentation of the document on the sample of its hard copy are formed.

The diagram on the previous figure reflects the fact that during the design of the template, it is required to generate certain normative documentation, which includes a map showing both the geometry of the designed template, the application variant (the sketch of the template application for the part), and the accompanying information to be entered into certain cells of a table document.

5 Conclusion

The practical value of the described work is a set of developed software on the basis of CAD NX 7.5 and the question-answer environment WIQA.NET providing an increase in the degree of automation of the process of designing leaf patterns due to:

- experimentation capabilities that allow you to adapt the models stored in the library and choose the best solutions already in existing solutions;
- libraries of typical configurable templates, including their semantic representations, and information about the production of templates and their use in forms that provide prompt access to semantic requests, as well as facilitating their reuse.
- formalization of the accumulated design experience, simplifying both the work of specialists with little experience, and relieving repetition of monotonous operations for more experienced ones.

Analysis of the results of experiments allows us to speak about the effectiveness of the proposed solution, since the complexity of designing the template equipment has decreased by an average of 20%, and the metal capacity of manufacturing - by 25%, in some cases - by 100%. The search relevance factor for the template snap varies from 0.6 to 0.85, which is a good indicator.

References

1. O.Chorakaev, P. Sosnin. Approach to a formal description of the structure of aircraft parts templates. Materials of the Russian Conference «Informatics, Modelling, Automation, Programming». -PP. 211-215(Ulyanovsk: UISTU, 2014)
2. Pseudo-Code Simulation of Designer Activity in Conceptual Designing Of Software Intensive Systems Proceedings - 27th European Conference on Modelling and Simulation, -PP. 85–92 (ECMS 2013)
3. P.Sosnin, Role «Intellectual Processor» in Conceptual Designing Of Software Intensive Systems, ICCSA'2013 – the 11-th International conference on Computational Science and Applications, Part III, LNCS 7973 Springer, Heidelberg, PP. 1–16 (2013)
4. P. Sosnin, Question-Answer Approach to Human-Computer Integration in Collaborative Designing/ P. I. Sosnin // Chapter in the book «Cognitively Informed Intelligent Interfaces: Systems Design and Development» Published IGI Global, pp. 157–176 (2012)
5. O.E.Chorakaev, P.I. Sosnin Conceptual experimenting in designing the configured patterns of aviation parts. Open semantic technologies for the design of intelligent systems (OSTIS-2015) Materials of the V international scientific and technical conference: Electronic resource. Belarusian State University of Informatics and Radioelectronics, State Institution «High Technologies Park Administration». P. 373-380, (2015)
6. M.Grishin, S. Larin,V. Kochergin Ontology as a design template tooling in a preparation of high-tech manufacturing / Automation of control processes. - 2015. - 1(39): 89–98. (In Russian).
7. M.Grishin, S.Larin, P. Sosnin Ontology of Designing Industrial Equipment within the Aviation Production / Ontology and Designing, **vol. 6**, no. 1(19), pp. 7–28, (2016)
8. R.M.Colombo, G.Guerra, F.Marcellini, M.HertyAhyperbolic model for the laser cutting process – Applied Mathematical Modelling, **37**, 14–15, 7810–7821(2013)

MATEC Web of Conferences **237**, 03014 (2018) https://doi.org/10.1051/matecconf/201823703014
D2ME 2018

The Design of Large Arc-shape High-precision Walking Device in TV (thermal-vacuum) Conditions

Xue Lian[a], Nan Hua, Liu Jiaqi, Liu Xin and Meng Gang

National Key Laboratory of Science and Technology on Test Physics & Numerical Mathematics

Abstract: The large-sized space environmental simulation test facility is mainly used for providing thermal radiation environment in high-vacuum, cold and dark space for satellites, spacecraft, lunar spacecraft to carry out whole satellite thermal vacuum test and for large equipment like antenna to carry out tests in TV conditions. Monitoring the spacecraft's surface optical image or temperature is a main task in space environment simulation test. In previous tests, optical image test mainly took place in a fixed position, so the measuring location and angle are limited. This paper focuses on large spherical space environmental simulation test facility, and designs a large arc-shape high-precision walking device in a space environment simulation test device. It introduces key technology in detail like structure design, thermal design, processes and manufacturing, etc. The test results show that this device can work steadily and reliably under simulation space environments, and the precision exceeds 0.5° .

1 Introduction

During the development of the spacecraft, in order to ensure correctness of spacecraft design and enhance reliability of on-orbit operation, it's a must to simulate cold and dark environments on the ground with solar radiation after the spacecraft is on the orbit, and carry out enough verification tests under such environments. The large-sized space environmental simulation test facility provides on-orbit environment simulation test conditions for aircrafts. Monitoring the spacecraft's surface optical image or temperature is a main task in space environment simulation test. In previous tests, optical image test mainly took place in a fixed position——observing through optical window outside the space environmental simulation test facility or in a fixed position inside space environmental simulation test facility —— so the measuring location and angle are limited. Therefore, we need to design a walking device that can work in space environmental simulation facility, allow large-range movements of the measuring instrument and adjust the vision direction of the measuring instrument. However, the emphasis and difficulty in the research are put on the structure design, thermal design, process design and other key elements of the large movement equipment that can work reliably in TV conditions [1].

2 Structure and Working Principles

[a] Corresponding author : casc_xuelian@163.com

The device consists of large arc-shape guide rail, track car, two-dimensional turntable, cable lead-out equipment and monitor system (including monitor software).The device's main body works in a large spherical space environmental simulation test facility (Φ 18 m), and the structure is shown in Figure 1 below.

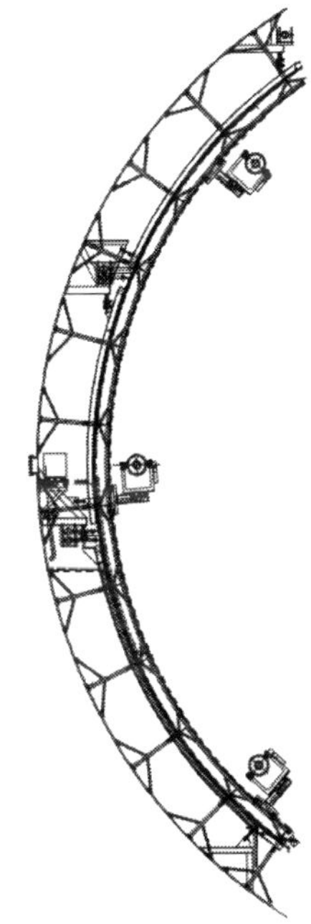

Figure 1. System Structure

The arc-shape guide rail permits the measuring instrument to rotate around the measuring aircraft, so as to do optical image measurements in different locations and at different angles. Two-dimensional turntable can enable test instruments to aim accurately. It's equipped with electric walking and posture adjustment to position rapidly and accurately. Considering that this device works in TV conditions, the design option should meet requirements of low temperature (≤100K) and high vacuum (≤10^{-3}Pa).

The Arc-shape track should be equipped in the longitude direction of the spherical space environmental simulation test facility, with a span of 500 mm between two tracks, an angle of 110 ° between both ends of rail and the center of the circle, and single arc-shape track length of 15 m. The arc-shape track is equipped with cable drag chain, and it can drive required power cables and signal cables to move with measuring instruments.

3 Device Option

3.1 Arc-shape Guide Rail

Arc-shape guide rail provides the track for the track car. It consists of arc-shape track, track support, track limiter, and rack, etc. The track is fixed in twelve supports (angle: 10 °) preset on the tank. Major parts of the guide rail are weldments designed with guide rails and arc-shape racks. All are made of stainless steel for ultra-low temperature.

Under normal temperature and normal pressure, all parts of the track can meet precision requirement by machining processing. However, because of working in TV conditions, materials may contract when cold and cause the whole track to change shape, which influences precision greatly. In order to decrease the influence of track's thermal contraction, the track design needs to consider necessary compensation process.

3.1.1 Compensation Process Design for the Track's Thermal Contraction

The Arch-shape track's installation precision on the environmental simulation test facility guarantees precision of measuring equipment. The main track compensation processes are as below:

1) The track is designed to be stretchable in the longitudinal direction. Fixing the position of the track on the longitude near the equator can let it stretch out and draw back freely along the longitude.

2) The track is divided into several sections during processing, and they are connected with bolts on site with certain stretching gaps for each section, so as to offset thermal contraction and prevent tracks' coming off.

3.1.2 Compensation Process Design for Track Supports' Thermal Contraction

The Track Support connects the arc-shape tracks and the wall of the test facility, and is the structural support of whole arc-shape track. In the facility, there is liquid nitrogen heat sink, which can cause a severe drop in the environment temperature. By means of simulation, under low-temperature working conditions, the minimum limit will be about 175 k, and the temperature of areas connecting with the tank wall is close to normal temperature. There is a big temperature difference, and it leads to huge local deformation.

For a loaded track, shear stress on the support is the main load and is the most demanding for the joints, so the load capacity of joints should be enhanced. deformation in Figure 2. The biggest force bearing point is the joint with the pipe wall, where reinforcement is necessary.

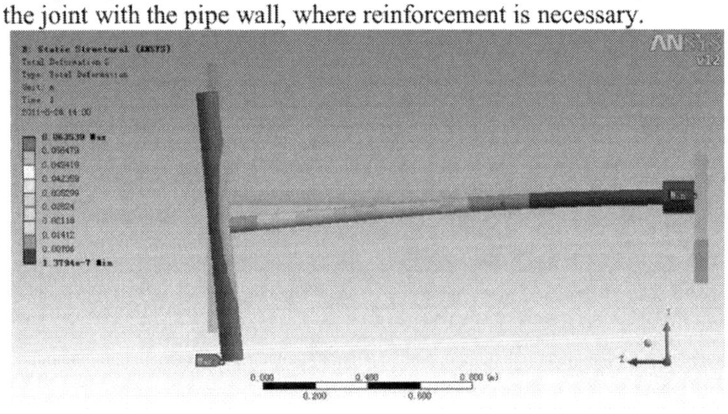

Figure 2. Deformation of the Whole Support

Regarding the value of deformation caused by temperature reduction and load, when there is a load, the track surface has a very big value of deformation up to 63 mm. So side supports should be added with the structure support design in Figure 3 below. It can decrease some deformation effectively. The track supports are arranged symmetrically and evenly, so they can balance the stress and reduce local deformation. According to calculation results of stress analysis, there is more stress on the joints, so they should be made of stainless steel, reinforced with flanges, and equipped with shorter support bars.

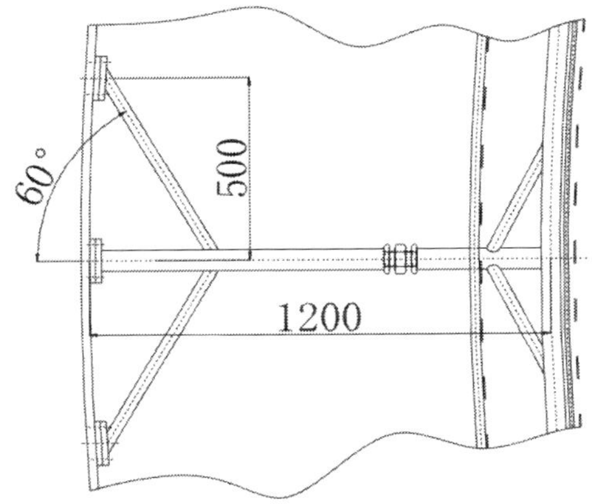

Figure 3. Arch-shape Track Support Structure

Main compensation processes for track supports include:

1) Use flanges to connect guide rails and the wall of the environmental simulation test facility. The arch-shape surface of the interior wall of the facility track is the installation base surface. However, there is dimension error in the arc-shape surface during production. When flanges are used for connection, gaskets can be added to adjust installation surfaces of the track to ensure the position tolerance between the track and installation base surface;

2) The connection rod of the track support is designed according to sections. In the middle there is adjustment equipment. The track's position tolerance is regulated by adjusting the screw nuts in the middle of the support.

3) The track support will not touch heat sink as there is a PTFE gasket in the middle. This can avoid heat leakage caused by heat conduction and too low local temperature of the track.

3.1.3 Compensation Process Design for Racks' Thermal Contraction

Rack is the transmission equipment for the whole device. Out-of-chain phenomenon should be avoided when the rack deforms due to thermal contraction. To avoid it, following compensation measures should be taken in the design:

1) The rack should be designed to be stretchable in the longitudinal direction. Fixing the position of the rack on the longitude near the equator can let it stretch out and draw back freely along the longitude. The fixation method is same with that of the arch-shape track.

2) Install the rack on the track support. Adjust the radius of the installation base of the rack through drilling slotted hole on the track. See Figure4.

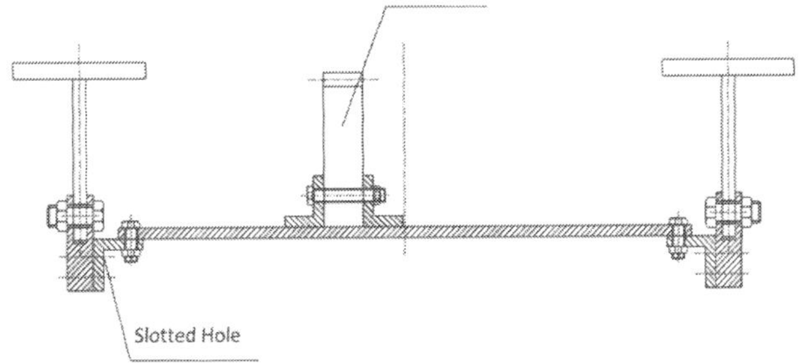

Figure 4. Installation Methods of Arc-shape Rack

3) The rack is divided into several sections during processing, and they are connected with bolts on site with certain stretching gaps for each section and in the way of slanted plane intersection, so as to offset thermal contraction and prevent racks' coming off.

3.2 Track Car and Two-dimensional Turntable

The track car carries the two-dimensional turntable to move on the arc-shape track. The car is equipped with guiding wheels, driving motor, reducer and driving gear, etc. Use an imported right-angle precision gear reducer and carry out dry lubrication for the reducer [2], in order to meet requirements of working at high vacuums. Cable of the measuring instrument is fixed on the arc-shape track car by hanging, and going into the drag chain with equipment cables. See the structure in Figure 5.

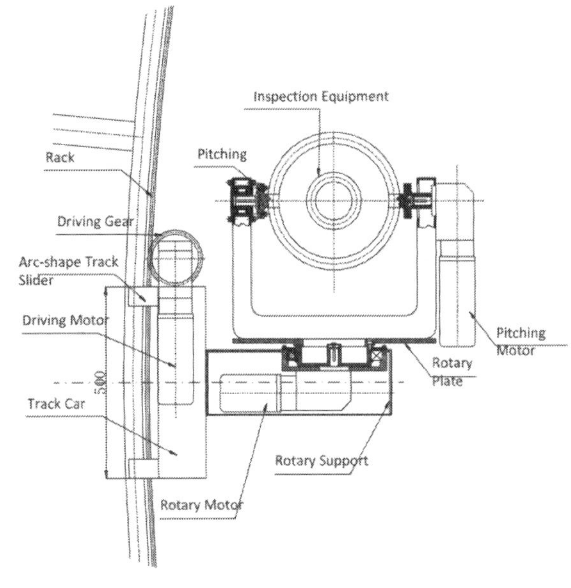

Figure5. Track Car and Two-Dimensional Turntable

Because of thermal contraction at low temperatures, when the track car and two-dimensional turntable are designed, the main concern is how to avoid the fact deformation of materials at low temperatures blocks the bearing or between driving sliders and the track.

When designing process options, we take following measures:

1) Adopt single-row roller bearings and thin wall enclosure cross roller bearings. These kinds of bearings can adjust the gap between the bearing roller with internal and external bushings in order to achieve operational effects at low temperatures.

2) The little car has four wheels on the track: two are fixed, and two are floating with 3 ~ 5 mm expansion gap. Meanwhile, there is also some allowance for distance of car tracks, to avoid blocking the car in the track.

3) Add a friction surface in a different material on the contact area between the track and the car. Once they touch when running, this can reduce the friction coefficient as well as prevent scratches on the contact area.

When the slider part's temperature decrease from normal temperature to minimum temperature, a huge shrinkage deformation will take place. See Figure6.

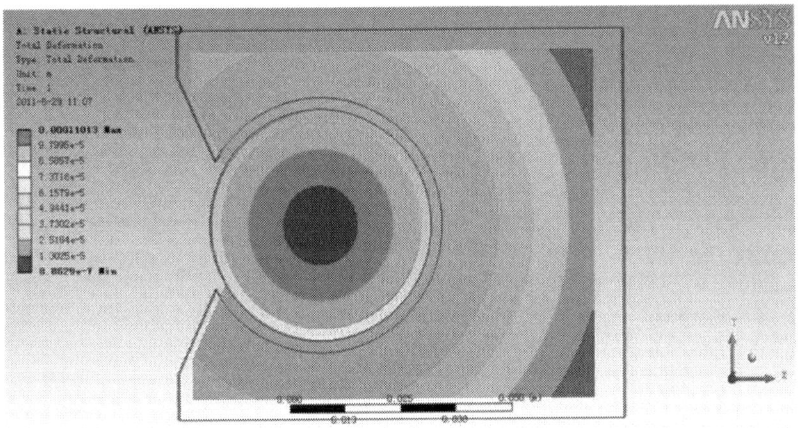

Figure 6. Deformation of Guiding Rail's sliders in Case of a Drop in temperature

In the figure, the black frame shows the original state before the drop in temperature, and colored areas indicate state after deformation. In terms of the drop in temperature and deformation, because the same stainless steel material is used, there is little deformation and the drop in temperature will not cause a block, and there is still a big moving space. The expansion reserved space can meet the requirements.

3.3 Cable Lead-out Equipment

Cables include power cord, electrical control lines, measuring equipment's signal lines, cables, etc. The working temperature of cables cannot be too low, so the cables require insulation and heating. Heating control is automatic in order to ensure the device's normal operation. Add fixed cables in cable bundles to bear tension and bending deformation, and to protect cables. There is also a shield layer and protection covers. Cable lead-out equipment is fixed in the middle of the guide rail, and extends horizontally (vertical to the track) to lead cables of the little car and test equipment. Near the lead-out equipment there is extension and retraction equipment which can meet cable's bending radius requirements, in order to avoid intertwining because of too long cables.

The extension and retraction of cables are controlled by drag chains made of stainless steel, in order to ensure normal operation and little air release at low temperatures.

3.4 Monitor System

The monitor system needs to control three axles for movements: walking (z), pitching (f), yaw (p). It needs to monitor technical parameters like the speed, location, etc. during loaded movement. To ensure the normal operation of the system, PMAC is employed to realize comprehensively control and monitor three motors and the whole movement system, to get real-time status of the whole movement system[3][4].

The monitor system has following main hardware:

1) 1 industrial control computer (including LCD)

2) PMAC and its accessories: control three servo motors, receive feedback signals to realize closed loop control, and have auxiliary I/O points for external control, etc;

3) Servo drive unit's electrical components (air switch, contactor, relay, related control's switch power, isolating transformer, and auxiliary I/O wiring board, etc.);

4) Connector plugs of external motors, encoders and I/O signals on the device side;

5) Cabinet and related control buttons;

6) The electric cabinet's design, assembly, wiring, debugging, etc.

Monitor software system is designed to control 3D motion simulator's walking, pitching, yaw and

 https://doi.org/10.1051/matecconf/201823703014

other equipment's movement, to control the speed and location of walking, pitching and yaw control, and to realize positioning movement of yaw, pitching and walking and reciprocating movement. Besides, amplitude and frequency of movement are adjustable.

4 Heat Control Option

To ensure stability, reliability and accuracy of the device in TV conditions, it's necessary to control the device's key heat sensitive components' stability. In this walking device, the motor, reducer and control cables belong to heat sensitive components. Heat control is required for these components, and the working temperature should be above 273 k, to ensure the operation reliability of the system.

To ensure that motors and reducers can be stable at 273 k in TV conditions of 100K, key parts of motors and reducers are pasted with thin heating films, and covered with insulating materials. According to the motor structure, it is simplified into following simulation calculation method [5]. In Figure 7, it designs size and power of each part's thin heating film.

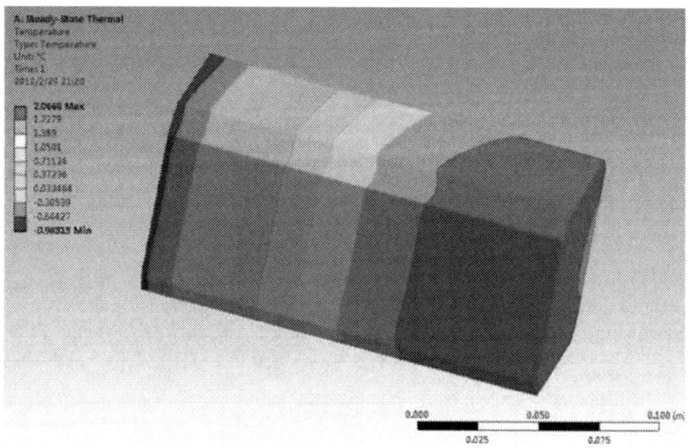

Figure7. Simulation Calculation of Motor's Heat Density

A T thermocouple is installed in the key area of every motor and reducer. The thermocouple is connected with thin heating films in series, PID intelligent temperature control device, and programmable DC power supply, to realize the closed loop control. Because of different heating boundary in each position and different power of heating films, 8 PID intelligent temperature control devices are used to control heat separately in 8 heating areas. Heating power cords and thermocouple connect exterior PID intelligent temperature control device by means of special connecting flanges to go through the tube wall by means. See principles as Figure8 below.

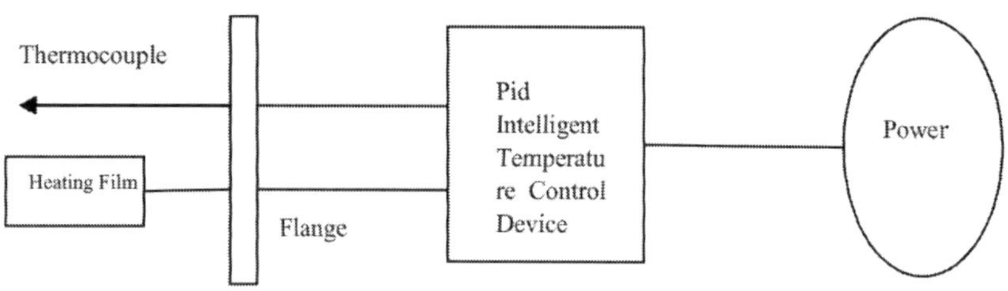

Figure8. Control Principle Chart

5 Test Application

This device, after the successful development, has been used in spacecraft environment simulation tests for many times. The system's movement is steady and reliable. See real object in Figure 9. By installing laser measurement equipment in the two-dimensional turntable center, it adjusts the laser beam to go through the center of the ball of the facility and measure the direction accuracy of the device. The test results show that the accuracy exceeds 0.5 °.

Figure 9. Photo of Large Arc-shape High-precision Walking Device

6 Conclusion

The successful development of Large Arc-shape High-precision Walking Device effectively solves problems in multi-angle and multi-direction measurement of optical images in large aircraft space environment simulation tests. Many test applications show that this device can move steadily and reliably without any blocking in TV conditions and with extremely accurate angle pointing.

ACKNOWLEDGMENTS

This work is supported by the National Natural Science Foundation of China (61671045).

References

1. XuGuohua， Tan Min. Status and trends of the mobile robot [J].Robot Technique and Application, 2001(3):6-14. (In Chinese)
2. Tanaka Y, ObaraS,SasakiA,etal.The effects of vibration in air and in ultrahigh vacuum on solid lubricated bearings[C]//IN: Proceedings of the 10th ESMATS.2003,524:49~56.
3. BorensteinJ.Control and kinematic design of multi-degree-of freedom mobile
4. Robots with compliant linkage [J].IEEE Transactions on Robotics and Automation,1995, 11(1):21-35.
5. ZhuangZhuo,YouXiaochuan,LiaoJianhui,et al. Finite element analysis and its application based on ABAQUS[M].Beijing:QinghuaUniversitiy Press,2009.(in Chinese)

AUTHOR INDEX

AUTHOR INDEX

EDP Sciences – Web of Conferences
17, Avenue du Hoggar
Parc d'Activité de Courtabouf
BP 112
F-91944 Les Ulis Cedex A
France

ISBN 978-1-5108-7562-3

9 781510 875623

CURRAN ASSOCIATES INC.
proceedings
.com